Computational Atomic Physics

Springer
Berlin
Heidelberg
New York
Barcelona
Budapest
Hong Kong
London
Milan
Paris
Santa Clara
Singapore
Tokyo

Klaus Bartschat (Ed.)

Computational Atomic Physics

Electron and Positron Collisions with Atoms and Ions

With 13 Figures,
78 Suggested Problems,
and a $3\frac{1}{2}''$ MS-DOS Diskette

Springer

Professor Dr. Klaus Bartschat

Department of Physics and Astronomy
Drake University
Des Moines
Iowa 50311, USA

ISBN 3-540-60179-1 Springer-Verlag Berlin Heidelberg New York

Library of Congress Cataloging-in-Publication Data. Computational atomic physics: electron and positron collisions with atoms and ions/ Klaus Bartschat, ed. p.cm. Includes bibliographical references. ISBN 3-540-60179-1 (hard: alk. paper) 1. Electron-atom scattering–Data processing. I. Bartschat, Klaus,
1956- .QC794.6.C6C655 1996 539.7'58–dc20 95-49512

Typesetting: Data conversion by Springer
SPIN 10014637 56/3144-5 4 3 2 1 0 – Printed on acid-free paper

Preface

Electron–atom collisions have been of central importance since the beginning of modern physics. A close interaction between experimental and theoretical efforts in this field has allowed for detailed investigations of the collision dynamics. In many cases, these collisions can be described to sufficient accuracy by non-relativistic quantum mechanics, and modifications to include relativistic effects do not represent a significant problem. It may thus seem surprising that the most fundamental of these collision processes, electron scattering from atomic hydrogen, has withstood a general solution until the beginning of the present decade – two thirds of a century after the presentation of the Schrödinger equation!

The reason for the problems in the numerical simulation of the collision dynamics lies in the quantum-mechanical nature of this so-called "three-body Coulomb problem", in which three particles, two electrons and a proton, interact with each other through the long-range Coulomb force. A reliable calculation of the ionization process, in particular, turns out to be very difficult. The same statement applies to positron scattering, where positronium formation as a re-arrangement channel causes additional difficulties for the theorist.

A special feature of electron (and positron) scattering from atomic hydrogen is the fact that this is the only collision system with a neutral target for which the non-relativistic target wavefunctions are known exactly. Consequently, it provides the rare opportunity to test only the collision part of various theoretical approaches to simulate the process. For all other neutral targets, there is always some uncertainty about the possible effects of an imperfect target description. Even an exact model for the collision itself may yield results that disagree with experimental findings, due to remaining inaccuracies in the target wavefunctions.

Apart from their fundamental role as a test ground for various computational approaches, electron–atom collisions are of critical importance in many fields, including the physics of stars, the upper atmosphere, lasers and plasmas. Reliable collision data are needed, for example, as input for programs to simulate energy transport in plasmas or the calculation of stellar opacities.

Only a small amount of the data needed has been or will ever be determined experimentally, due to the difficulties and costs associated with setting up and performing such experiments. A major goal of theoretical atomic physics, therefore, has been to develop reliable and efficient computational methods to calculate the outcome of the various collision processes of interest.

This book gives an overview of some of these methods, with example programs for a "hands on" instruction of the reader. The book is intended as a supplement to advanced undergraduate and general graduate courses on quantum mechanics, atomic physics, and scattering theory. However, the detailed explanation of the various methods, in connection with the computer programs, should also be useful to researchers in both theoretical and experimental collision physics. The computer codes can be and have been (by the authors) used for state-of-the-art research. It is assumed that the reader has a basic knowledge of quantum mechanics and collision theory. Some "tool routines" for Clebsch–Gordan coefficients, Racah symbols, (associated) Legendre and Laguerre polynomials, and basic linear algebra operations are used without detailed description.

Although the chapters are self-contained and thus may be read in any order, the reader is recommended to follow the sequence in the book. After an overview of the general problem in Chap. 1 with a program for potential scattering from spherically symmetric potentials, Chaps. 2 and 3 are devoted to atomic structure calculations. For all but truly one-electron targets such as H, He^{+}, Li^{++}, etc., these are an integral part of electron–atom collision calculations, since first of all the target has to be represented as well as possible. The two following chapters concentrate on perturbation methods based on the Born-series expansion, with Chap. 4 covering elastic scattering and excitation, and Chap. 5 dealing with electron impact ionization. An alternative method is based on the "close-coupling" expansion. The standard form of this approach is introduced for positron–atom scattering (without positronium formation) in Chap. 6, followed by the R-matrix method in Chap. 7, and the "convergent-close-coupling" (CCC) approach in Chap. 8. Whereas the R-matrix method is a very efficient way to calculate data for a large number of collision energies, as needed in many applications, the CCC approach has been used successfully to provide some benchmark results for electron scattering from quasi one-electron systems as well as helium.

Most of the program packages presented in this book use Bessel and Coulomb functions, usually for matching the numerical solution to its known asymptotic form in the region where the projectile is far away from the target. The numerical aspects of calculating these functions are discussed in Chap. 9. Furthermore, recent developments in experimental techniques have allowed for very detailed angle-differential measurements, such as differential cross sections, spin polarization parameters, and various observables from electron–photon coincidence experiments. In some cases even, the "perfect" or "complete" scattering experiment has been performed, where all relevant quantum-mechanical scattering amplitudes have been determined. Data from such measurements present the ultimate test of atomic collision theory – a much more detailed test than is possible, for example, by comparing total cross section results alone. Chapter 10 discusses the calculation of scattering amplitudes from numerical transition or reactance matrix elements. Finally, Chap. 11 presents an introduction to the density matrix theory that allows for a systematic description of a large variety of scattering experiments.

As mentioned above, the detailed description of the individual computational models in each chapter is accompanied by a package of programs distributed on disk. The codes and example input data are organized in subdirectories (one for each chapter) and can be installed on any IBM compatible computer that can handle DOS (5.0 or higher) files. The programs are compressed but will be expanded on the user's hard drive during the installation procedure. A `README` and an `INSTALL.BAT` file are provided in the main directory of the disk. The codes have been been tested on several machines, including IBM risc processors, DEC Vax and Alpha, the HP-735 series, Suns, CRAYs, and IBM compatible PCs (with 16 Mb of RAM using Microsoft Fortran Powerstation). They are written in standard `FORTRAN 77`, except for some double precision complex arithmetic (in most programs) and lower-case characters as well as extended variable names (in the `CCC` program). Except for a few insignificant compiler warnings on some computers, none of these extensions has caused execution problems on any of the above-mentioned machines. A separate `README` file is also presented in each chapter, containing basic information on how to run the program. An example is printed on the next page. Problems with the codes should be reported directly to the authors.

A list of suggested problems is given at the end of each chapter; generally the problem set starts with a few suggestions for running the programs for some example cases, before modifications of the source codes become necessary. Since only these source codes and some input data are provided on the disk, users will need to create their own executable files on the computers they want to use for their studies.

I would like to thank my colleagues who contributed to this book. They have not only taken up the challenge to explain in detail how the standard equations seen in many textbooks on this subject are actually put on a computer and solved to a certain degree of accuracy, but have also released their computer codes. Finally, I would like to thank Dr. V. Wicks for careful reading of the manuscript and making the final changes to the TEX files.

Des Moines, October 1995 Klaus Bartschat

Example: The README file for Chap. 4

```
                How to run the DWBA program in this subdirectory.
                =================================================
                          K. Bartschat:  July 19, 1995

   1) Compilation:  Create an executable file by compiling the
                    source code dwba.f

      Typical commands to do this would be:
      a) Unix-based compilers:  f77 -o DWBA dwba.f
      b) Vax under VMS:         for dwba.f
                                link dwba
   2) Execution:

       e-H:   Typical command:  DWBA < HDWBA.IN > HDWBA.OUT

              Note:
              HDWBA.IN  is the standard input file
                          (read from IRD=5).
              HDWBA.OUT is the standard output file
                          (written to IWRI=6).
              DBW1.AMP will contain the scattering amplitudes.

       e-Na:  Preparatory commands: Copy NA3S.WF WF.INI
                                    Copy NA3P.WF WF.EXI
                                    Copy NA3S.POT WF.INI
                                    Copy NA3P.POT POT.EXI
                                    Copy NAMESH.DWB MESH.DAT
              Now the files for the wavefunctions (WF) and
              potentials (POT) in the initial (INI) and exit
              (EXI) channels are defined on a numerical mesh
              (MESH.DAT).

              Execution command: DWBA < NADWBA.IN > NADWBA.OUT

              Note:
              NADWBA.IN  is the standard input file.
              NADWBA.OUT is the standard output file.
              DBW1.AMP will contain the scattering amplitudes.
```

Contents

List of Contributors

A.R. Barnett
Department of Physics
and Astronomy,
University of Manchester,
Manchester, M13 9PL,
England
`ar@mags.ph.man.ac.uk`

Klaus Bartschat
Department of Physics
and Astronomy,
Drake University,
Des Moines
Iowa 50311, USA
`kb0001r@acad.drake.edu`

Igor Bray
Electronic Structure
of Materials Centre,
The Flinders University
of South Australia,
G.P.O. Box 2100,
Adelaide 5001, Australia
`igor@esm.ph.flinders.edu.au`

P.G. Burke
Department of Applied Mathematics
and Theoretical Physics,
The Queen's University of Belfast,
Belfast BT7 1NN,
Northern Ireland

Alan Hibbert
Department of Applied Mathematics
and Theoretical Physics,
The Queen's University of Belfast,
Belfast, BT7 1NN,
Northern Ireland
`A.Hibbert@Queens-Belfast.AC.UK`

Don H. Madison
Department of Physics and LAMR,
University of Missouri-Rolla,
Rolla, Missouri 65401, USA
`madison@apollo.physics.umr.edu`

Ian E. McCarthy
Institute for Atomic Studies,
The Flinders University of
South Australia,
Bedford Park, SA 5042,
Australia
`ian@esm.ph.flinders.edu.au`

R.P. McEachran
Department of Physics
and Astronomy,
York University, Toronto,
M3J 1P3, Canada
`bob@theory1.phys.yorku.ca`

M.P. Scott
Department of Applied Mathematics
and Theoretical Physics,
The Queen's University of Belfast,
Belfast BT7 1NN,
Northern Ireland
amg0263@Queens-Belfast.AC.UK

Andris Stelbovics
Centre for Atomic, Molecular
and Surface Physics,
School of Mathematical
and Physical Sciences,
Murdoch University,
Perth 6150, Australia
Stelbovi@atsibm.csu.murdoch.edu.au

Zhang Xixiang
Department of Modern Physics,
The University of Science
and Technology of China,
Hefei, Anhui, China
Xixiang.Zhang@Queens-Belfast.AC.UK

1. Electron–Atom Scattering Theory: An Overview

Klaus Bartschat

Department of Physics and Astronomy, Drake University,
Des Moines, Iowa 50311, USA

Abstract

A summary of the basic equations that are used in the electron–atom collision calculations discussed in this book is presented. We begin with the simplest case of potential scattering, before we introduce integral equation approaches for the transition operator and, finally, the close-coupling expansion.

1.1 Introduction

"The most important experimental technique in quantum physics is the scattering experiment". This statement is the opening sentence of a well-known textbook [1.1] and highlights the central role that scattering experiments continue to play in the development of modern physics. A schematic diagram of a generic scattering experiment is shown in Fig. 1.1. The particles A and B are allowed to interact. When emerging from the interaction region, the collision partners generally have changed directions and may be in quantum states different from the initial ones. For electron and positron scattering from atoms and ions, the situation is somewhat simplified since the light mass of the projectile suggests a coordinate system that is centered at the target, which itself is considered to be a point. In standard gas beam experiments, this coordinate system is essentially fixed in space and coincides with the laboratory frame. Hence, one may regard the particle B as a heavy target at rest and the particle A as the projectile that is scattered by an angle θ, which is the angle between the initial and final linear momenta of the projectile far away from the target. Except for forward and backward scattering, these two momenta are used to define the "scattering plane".

The aim of theoretical quantum physics is to model as accurately as possible the development of the system in the interaction region, for confrontation of the predictions with actual observables. The experimentalist aims to precisely define the incident channel, as well as to characterize the collision products as closely as possible. For many important processes in nature, typical observables are averages over key parameters, such as the incident directions,

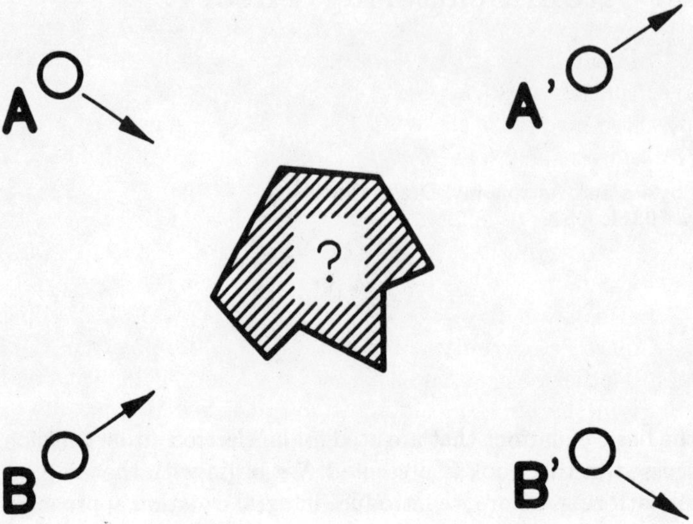

Fig. 1.1. Schematic diagram of a collision process.

scattering angle, velocity, temperature, etc. However, the ultimate goal is to establish *uniquely* the relationship between the ket vectors $|\psi\rangle_{\text{in}}$ and $|\psi\rangle_{\text{out}}$, which determine the initial and final states of the system. A complete description in the quantum-mechanical sense is succinctly expressed in terms of a corresponding set of complex-valued scattering amplitudes.

The accurate calculation of scattering amplitudes has, therefore, been of great importance in atomic physics since the first experiments were performed. For electron (and positron) scattering, most of the efforts have been concentrated on quantal methods, although in some cases classical or semiclassical methods have been used as well. This book deals with quantum-mechanical attempts to simulate these collisions, based on the non-relativistic Schrödinger equation. To illustrate the methods, the target is usually atomic hydrogen, since the structure part of this problem is known exactly in non-relativistic quantum mechanics. The generalization to more complex systems is, in principle, straightforward, and the structure codes presented in Chaps. 2 and 3 provide the basis for such an extension.

The numerical methods discussed in this book generally fall into one of two categories, namely the perturbation-series expansions based on variations of the Born series or the non-perturbative close-coupling approach that is based on the expansion of the trial wavefunction into a set of basis functions. Traditionally, Born-series expansions were used successfully for higher collision energies where the projectile–target interaction is a relatively small perturbation of the free-particle motion with large kinetic energy. On the other hand, close-coupling expansions were applied to simulate low-energy collisions where the incident energy is such that only elastic scattering or at most excitation of a few low-lying target states is possible. As shown in Chap. 8 on the "convergent close-coupling approach", however, this method may be extended to deal with higher collision energies as well, provided the effect of the target continuum states is accounted for properly.

In this chapter, we give an overview of the basic equations used in the electron–atom collision calculations discussed in this book. The chapter is intended for readers with a basic knowledge of quantum mechanical scattering theory and should not be regarded as a replacement for standard textbooks on atomic collision theory. Instead, it is setting the stage for the following chapters in this book. Since it seems impossible to even give a nearly complete list of general references, we only mention the recent book by McCarthy and Weigold [1.2] for a more detailed discussion of the general formulation.

If not mentioned otherwise, atomic units are used throughout this and most other chapters in the book, i.e., charges are measured in units of the elementary charge $|q_{\rm el}| \approx 1.602 \times 10^{-19}$ C, lengths in units of the Bohr radius $a_0 \approx 0.529 \times 10^{-10}$ m, masses in units of the electron mass $m_{\rm el} \approx 9.108 \times 10^{-31}$ kg, and energies in Hartrees, where 1 Hartree = 2 Rydberg ≈ 27.21 eV ≈ 4.359×10^{-18} J.

1.2 Potential Scattering

We consider electron or positron scattering from an atom or ion. In the simplest case, the effect of the target atom on the projectile motion is represented by a potential $V(\boldsymbol{r})$ in the Schrödinger equation for the projectile with a total energy E,

$$H\Psi_E^+(\boldsymbol{k},\boldsymbol{r}) = [T + V(\boldsymbol{r})]\,\Psi_E^+(\boldsymbol{k},\boldsymbol{r}) = E\,\Psi_E^+(\boldsymbol{k},\boldsymbol{r})\,, \tag{1.1}$$

where T is the kinetic energy operator in the hamiltonian H.

We label a potential as "short range", if it either vanishes beyond a certain point ($V \equiv 0$ for $r > a$ where a is a constant) or at least falls off faster than $1/r^2$ for large r. To simplify the situation further, we also specialize to the case of a spherically symmetric potential $V(\boldsymbol{r}) = V(r)$. For such potentials, (1.1) needs to be solved subject to the boundary condition

$$\Psi_E^+(\boldsymbol{k},\boldsymbol{r}) \longrightarrow \mathrm{e}^{\mathrm{i}\boldsymbol{k}\cdot\boldsymbol{r}} + f(\theta)\,\frac{\mathrm{e}^{\mathrm{i}kr}}{r} \quad \text{for } r \longrightarrow \infty\,, \tag{1.2}$$

where $k^2 = E/2$ and $\boldsymbol{k}$ is the linear momentum of the incident projectile. This boundary condition expresses the physical situation of an incident projectile beam described by a plane wave that is scattered by a potential located at $r = 0$. The scattered wave, as seen by a detector far away from the scattering region, is an outgoing spherical wave (superscript +) with amplitude $f(\theta)$. This "scattering amplitude" depends on the scattering angle θ; recall that this is the angle between the incident beam direction $\boldsymbol{k}$ and the direction

$\boldsymbol{k}'$ where the detector is located and the scattered projectiles are detected. There is no dependence on the azimuthal angle ϕ in the scattering amplitude, due to the cylindrical symmetry of the problem around the incident beam (z) axis that we use as the quantization axis.

The scattering amplitude $f(\theta)$ is related to the differential cross section $\sigma(\theta)$ through

$$\frac{\mathrm{d}\sigma(\theta)}{\mathrm{d}\Omega} = |f(\theta)|^2 \,, \tag{1.3}$$

and the total collision cross section is given by

$$\sigma_{\mathrm{tot}} = \int_\Omega \mathrm{d}\Omega |f(\theta)|^2 = 2\pi \int_0^\pi \mathrm{d}\theta \sin\theta f(\theta)|^2 = \frac{4\pi}{k} \mathrm{Im}\{f(0^\circ)\} \,, \tag{1.4}$$

where the second equality holds for spherically symmetric potentials and the third represents the "optical theorem". The latter relates the total cross section to the imaginary part (Im) of the forward scattering amplitude.

The standard way to calculate the scattering amplitude is based on the "partial-wave expansion" of

$$\Psi_E^+(\boldsymbol{k},\boldsymbol{r}) = \sqrt{\frac{2}{\pi}} \frac{1}{kr} \sum_{\ell m} \mathrm{i}^\ell F_\ell(k,r) Y_{\ell m}(\hat{\boldsymbol{r}}) Y^*_{\ell m}(\hat{\boldsymbol{k}}) \,, \tag{1.5}$$

where $Y_{\ell m}(\hat{\boldsymbol{r}})$ is a spherical harmonic for the angles associated with the direction of the position vector $\boldsymbol{r}$, and the function $F_\ell(k,r)$ describes the radial motion of the projectile.

Inserting the expansion (1.5) into the Schrödinger equation (1.1) yields the radial Schrödinger equation

$$\left(\frac{\mathrm{d}^2}{\mathrm{d}r^2} - \frac{\ell(\ell+1)}{r^2} - 2V(r) + k^2\right) F_\ell(k,r) = 0 \,. \tag{1.6}$$

In the asymptotic region, the potential is supposed to be negligible. Consequently, $F_\ell(k,r)$ in this region must be a linear combination of the regular and the irregular solutions of (1.6). Transforming to standing wave boundary conditions allows $F_\ell(k,r)$ to be regarded as a real instead of a complex function. Its asymptotic behaviour is then given by

$$\lim_{r\to\infty} F_\ell(\rho) = A_\ell j_\ell(\rho) - B_\ell \eta_\ell(\rho) \,, \tag{1.7}$$

where $j_\ell(\rho)$ and $\eta_\ell(\rho)$ are regular and irregular Riccati–Bessel functions with argument $\rho = kr$ (see Chap. 8 for more details).

Using the incident beam axis as the quantization axis together with the asymptotic form of the functions $j_\ell(\rho)$ and $\eta_\ell(\rho)$, we find the scattering amplitude can be expressed as

$$f(\theta) = \frac{1}{2ik} \sum_{\ell=0}^{\infty} (2\ell + 1)\, T_\ell\, P_\ell(\cos\theta)\,, \tag{1.8}$$

where

$$T_\ell = \mathrm{e}^{2\mathrm{i}\delta_\ell} - 1 = S_\ell - 1 \tag{1.9}$$

is the transition matrix element, $S_\ell = \mathrm{e}^{2\mathrm{i}\delta_\ell}$ is the scattering matrix element, and δ_ℓ is the phase shift due to the potential $V(r)$. This phase shift is related to the asymptotic form (1.7) through

$$\tan\delta_\ell = K_\ell = B_\ell / A_\ell\,. \tag{1.10}$$

Due to the centrifugal barrier in the radial equation (1.6), the potential phase shift must converge to zero for high angular momenta and short-range potentials, thereby guaranteeing the convergence of the sum (1.8).

Elementary complex algebra reveals the connection

$$S_\ell = 1 + T_\ell = (1 + \mathrm{i}K_\ell)/(1 - \mathrm{i}K_\ell)\,. \tag{1.11}$$

This can be extended to give a more general relationship between the **S**, **T** and **K** matrices in many-channel problems (see Sect. 1.4 below).

The above treatment can be generalized in a straightforward way to the case of Coulomb scattering where the potential $V(r)$ contains both a short-range term and a long-range Coulomb part of the form Z_{asym}/r. In this case, a Coulomb phase shift σ_ℓ needs to be introduced, and the Riccati–Bessel functions in (1.7) are replaced by regular and irregular Coulomb functions (see Chap. 8) multiplied by ρ.

In practice, the phase shift for potential scattering problems can be obtained quite easily. A simple method consists of integrating the radial Schrödinger equation (1.6) outward from $r = 0$ to $r = a$ where a is chosen large enough for the asymptotic form (1.7) to be valid. The requirement of a finite physical solution at the origin leads to

$$F_\ell(k, r)|_{r=0} = 0 \tag{1.12a}$$

as the first starting condition. Secondly, an arbitrary (but non-zero) value may be chosen for the first derivative, i.e.,

$$\frac{\mathrm{d}}{\mathrm{d}r} F_\ell(k, r)|_{r=0} = C \neq 0\,. \tag{1.12b}$$

Different choices of C only affect the magnitude of the numerical solution as an overall scaling factor. The phase shift, however, can be determined by comparing the *logarithmic derivative* of the analytical form (1.7) and the numerical solution at $r = a$, and this derivative is independent of such scaling. The result is

$$\tan \delta_l = \frac{kf' - Df}{kg' - Dg} \,. \tag{1.13}$$

Here f' and g' are the derivatives of the spherical Bessel functions or the Coulomb functions multiplied by their argument ρ with respect to ρ, and D is the logarithmic derivative of the numerically integrated solution with respect to r.

The computer program provided for this chapter, which is described in more detail in Sect. 1.5, uses the above recipe. A simple outward integration, however, may exhibit numerical problems, especially for high partial-wave angular momenta. For any non-zero angular momentum, the numerical solution will increase exponentially until the classical turning point $r_{\rm cl}$ is reached. This point is determined by the condition

$$k^2 = \frac{\ell(\ell+1)}{r_{\rm cl}^2} + 2V(r_{\rm cl}) \,. \tag{1.14}$$

For high angular momenta, $r_{\rm cl}$ may be so far away from $r = 0$ that the solution has "blown up" in this radial range and prohibits an accurate determination of the phase shift by the above procedure. Other problems may be related to the choice of the radial mesh and the differential equation algorithm for the integration. Whereas enough points are needed to ensure an accurate integration, too many points and too large a choice for the matching point $r = a$ may unnecessarily accumulate errors. In order to increase the accuracy, therefore, a new version of the program `BASFUN` used in the R-matrix packages (see, for example, Berrington *et al* [1.3] and Chap. 7) is employed in the radial integration. The essential parts are the use of de Vogelaere's method for the radial integration and a matching of the outward integration with two independent inward integrations to obtain an improved numerical function [1.3]. The program prints warning messages if the mesh is obviously too coarse or the potential has not fallen off enough at the matching radius.

1.3 Perturbation Approaches

Whereas the determination of phase shifts for potential scattering is, in principle, straightforward, the situation becomes more complicated when the target is represented by an N-electron wavefunction and excitation or even ionization is possible in addition to elastic scattering of the projectile. In such cases, perturbative approaches are often used to provide approximate solutions of the collision problem. These methods will be discussed in detail in Chap. 4 for elastic scattering and excitation, and in Chap. 5 for ionization.

To illustrate the basis ideas within the framework of potential scattering, we follow McCarthy and Weigold [1.2] and rewrite (1.1) in the form

$$(E^{(+)} - T)\,\Psi_E^+(\boldsymbol{k},\boldsymbol{r}) = V(r)\,\Psi_E^+(\boldsymbol{k},\boldsymbol{r}) \,. \tag{1.15}$$

Equation (1.16) involves the free-particle operator $(E^{(+)} - T)$ and an inhomogeneous term on the right-hand side. It can formally be solved by using standard Green's function approaches. The solution with the appropriate boundary condition (1.2) is given by

$$\Psi_E^+(\boldsymbol{k},\boldsymbol{r}) = \mathrm{e}^{\mathrm{i}\boldsymbol{k}\cdot\boldsymbol{r}} + \int \mathrm{d}^3\boldsymbol{r}' G_0(E^{(+)};\boldsymbol{r},\boldsymbol{r}') V(r')\, \Psi_E^+(\boldsymbol{k},\boldsymbol{r}')\,, \tag{1.16}$$

where the free-particle Green's function is given by

$$G_0(E^{(+)};\boldsymbol{r},\boldsymbol{r}') = -\frac{1}{2\pi}\frac{\mathrm{e}^{\mathrm{i}k|\boldsymbol{r}-\boldsymbol{r}'|}}{|\boldsymbol{r}-\boldsymbol{r}'|}\,. \tag{1.17}$$

Equation (1.16) is the Lippmann–Schwinger equation for the exact scattering wavefunction $\Psi_E^+(\boldsymbol{k},\boldsymbol{r})$. Note that this is an integral equation, since the unknown wavefunction appears not only on the left-hand side, but also in the integral on the right-hand side. Before discussing approximate solutions, however, it is useful to investigate the asymptotic form of (1.17) for large r and its consequences for (1.16). The final result is the expression

$$f(\theta) = -\frac{1}{2\pi}\int \mathrm{d}^3\boldsymbol{r}'\mathrm{e}^{-\mathrm{i}\boldsymbol{k}\cdot\boldsymbol{r}'} V(r')\Psi_E^+(\boldsymbol{k},\boldsymbol{r}) \tag{1.18}$$

for the scattering amplitude [1.2]. Defining

$$|\boldsymbol{k}\rangle \equiv \frac{1}{(2\pi)^{3/2}}\,\mathrm{e}^{\mathrm{i}\boldsymbol{k}\cdot\boldsymbol{r}} \tag{1.19a}$$

and

$$|\boldsymbol{k}^+\rangle \equiv \frac{1}{(2\pi)^{3/2}}\,\Psi_E^+(\boldsymbol{k},\boldsymbol{r})\,, \tag{1.19b}$$

we can abbreviate (1.18) as

$$f(\theta) = -4\pi^2\langle\boldsymbol{k}'|V|\boldsymbol{k}^+\rangle\,. \tag{1.20}$$

This equation may be used to *define* the T operator for potential scattering according to

$$\langle\boldsymbol{k}'|V|\boldsymbol{k}^+\rangle \equiv \langle\boldsymbol{k}'|T|\boldsymbol{k}\rangle\,, \tag{1.21}$$

which is the momentum-space Lippmann–Schwinger equation for the transition operator.

The accurate numerical solution of the Lippmann–Schwinger equation through basis-function expansions is discussed in Chap. 8. Here we want to concentrate on another important aspect of the above formalism, namely the possibility of using perturbative methods. The simplest approach consists of replacing the exact scattering wavefunction $|\boldsymbol{k}^+\rangle$ by the (very) approximate plane wave $|\boldsymbol{k}\rangle$ in (1.20). This yields the "first Born approximation"

$$\langle \boldsymbol{k}'|T|\boldsymbol{k}\rangle \approx \langle \boldsymbol{k}'|V|\boldsymbol{k}\rangle\,. \tag{1.22}$$

Equation (1.22) is expected to be a good approximation if the potential $V(r)$ is a small perturbation compared to the total energy – in other words, if the kinetic energy is much larger than the potential energy everywhere in space. (For more detailed discussions, we refer again to textbooks on collision theory.) To obtain a better approximation, one might want to split up the potential as $V = V_1 + V_2$ and treat V_1 more accurately than V_2 by including it on the left-hand side of (1.15); in fact, this *must* be done for any long-range Coulombic term in the potential [1.2]. This procedure corresponds to the "first-order distorted-wave approximation", since the plane waves are replaced by distorted waves that are calculated with the potential V_1. Another possibility is the next iteration in the solution of the Lippmann–Schwinger equation with the full V included, a procedure called the "second Born approximation". Again, this can be improved by splitting up the potential, thereby defining the "second-order distorted-wave approximation". Partial extensions to account for even higher orders have been considered in the literature, but the computational effort becomes very extensive beyond first-order approaches.

Finally, we note that the partial-wave T-matrix element in this formalism is usually defined as [1.2]

$$T_\ell = -\frac{1}{2\mathrm{i}\pi k}(S_\ell - 1)\,, \tag{1.23}$$

and thus differs from the definition in (1.9) by a complex, energy-dependent factor. The reason for the different choices is the convenience in the individual approaches, here in particular the simple form of the first Born matrix element (1.22). The example demonstrates, however, that special care must be taken when using formulas from different authors for T-matrix elements, scattering amplitudes, and cross sections.

1.4 The Close-Coupling Expansion

The close-coupling approximation has been a standard method of treating low-energy scattering, both elastic and inelastic, for many years. This non-perturbative method is based on an expansion of the total wavefunction for a collision system in terms of a sum of products which are constructed from target states Φ_i that diagonalize the N-electron target Hamiltonian

$$H_T^N = \sum_{i=1}^{N}\left[-\frac{1}{2}\nabla_i^2 - \frac{Z}{r_i} + \frac{1}{2}\sum_{j\neq i}^{N}\frac{1}{|\boldsymbol{r}_i - \boldsymbol{r}_j|}\right] \tag{1.24}$$

according to

$$\langle \Phi_{i'} \mid H_T^N \mid \Phi_i\rangle = E_i\,\delta_{i'i}\,, \tag{1.25}$$

and unknown functions F_i describing the motion of the projectile. If relativistic effects are neglected, the wavefunction for each total orbital angular momentum L, total spin S, and parity π is expanded as

$$\Psi^{LS\pi}(\boldsymbol{r}_1,\ldots,\boldsymbol{r}_{N+1}) = \mathcal{A} \sum\!\!\!\!\!\!\int_i \Phi_i^{LS\pi}(\boldsymbol{r}_1,\ldots,\boldsymbol{r}_N,\hat{\boldsymbol{r}})\,\frac{1}{r}\,F_i(r)\,. \tag{1.26}$$

Here $\sum\!\!\!\!\!\!\int$ denotes a sum over all discrete and an integral over all continuum states of the target, and $\mathcal{A}$ is the antisymmetrization operator that accounts for the indistinguishability of the projectile and the target electrons. Furthermore, the angular and spin coordinates of the projectile electron (collectively denoted by $\hat{\boldsymbol{r}}$) have been coupled with the target states to produce the "channel functions" $\Phi_i^{LS\pi}(\boldsymbol{r}_1,\ldots,\boldsymbol{r}_N,\hat{\boldsymbol{r}})$.

After some algebraic manipulations, the unknown radial wavefunctions F_i are determined from the solution of a system of coupled integro-differential equations given by

$$\left[\frac{\mathrm{d}^2}{\mathrm{d}r^2} - \frac{\ell_i(\ell_i+1)}{r^2} + k^2\right] F_i(r) = 2\sum\!\!\!\!\!\!\int_j V_{ij}(r)\,F_j(r) + 2\sum\!\!\!\!\!\!\int_j W_{ij}\,F_j(r)\,, \tag{1.27}$$

with the direct coupling potentials (q_{proj} is the charge of the projectile)

$$V_{ij}(r) = q_{\mathrm{proj}}\left(\frac{Z}{r}\,\delta_{ij} + \sum_{k=1}^{N}\langle\Phi_i\,|\,\frac{1}{|\boldsymbol{r}_k-\boldsymbol{r}|}\,|\,\Phi_j\rangle\right)\,, \tag{1.28}$$

and the exchange terms (for electrons only)

$$W_{ij}F_j(r) = \sum_{k=1}^{N}\langle\Phi_i\,|\,\frac{1}{|\boldsymbol{r}_k-\boldsymbol{r}|}\,|\,(\mathcal{A}-1)\,\Phi_j F_j\rangle\,. \tag{1.29}$$

For each "i", several sets of independent solutions (labeled by a second subscript "j") must be found, subject to the appropriate boundary conditions. For scattering from and excitation of neutral targets, these are given by [1.4]

$$F_{ij}|_{r=0} = 0\,; \tag{1.30a}$$

$$\lim_{r\to\infty} F_{ij} = \delta_{ij}\,\sin\left(k_i r - \tfrac{1}{2}\ell_i\pi\right) + K_{ij}\,\cos\left(k_i r - \tfrac{1}{2}\ell_i\pi\right);\; i = 1, n_{\mathrm{open}}\,; \tag{1.30b}$$

$$\lim_{r\to\infty} F_{ij} = C_{ij}\,\exp(-|k_i|r);\; i > n_{\mathrm{open}}\,. \tag{1.30c}$$

In (1.30), $k_i = \sqrt{2(E-E_i)}$ is the linear momentum in channel i for the total energy E and the channel energy E_i, and n_{open} is the number of "open" channels for which k_i is a real number. For the "closed" channels, k_i is purely imaginary.

The coupled integro-differential equations (1.27) can be generalized to a relativistic framework, and the collision problem essentially consists of finding the solution to this system for each total energy. This can be achieved by various iterative, non-iterative or algebraic methods. An introductory overview is given, for example, by Burke and Seaton [1.4], and Chaps. 6–8 of this book are devoted to this problem.

Without going into details, we note that the nature of the physical problem makes some simplifications possible. For example, no exchange effects need to be considered for positron scattering discussed in Chap. 6. (On the other hand, positronium formation as a re-arrangement channel makes a more accurate treatment of positron scattering extremely difficult.) Even for electron scattering, such effects can be neglected outside a sphere of radius a, approximately the "size" of the target. This is an important aspect of the R-matrix method discussed in Chap. 7. Furthermore, the "centrifugal barrier" associated with partial waves of large angular momenta ensures that these waves only "see" the long-range part of the interaction potential. Consequently, the analytic "effective range formula" [1.5] and Born-type approximations [1.6] can be used to speed up the calculation.

Such approximate methods for high partial-wave angular momenta are often necessary in the calculation of scattering amplitudes (see Chap. 10) to give converged angle-differential results for the various observables discussed in Chap. 11. The construction of these scattering amplitudes and the observable parameters of interest is the final step of the calculation. It involves the scattering (**S**), transition (**T**), or reactance (**K**) matrices related through

$$\mathbf{S} = \mathbf{1} + \mathbf{T} = [\mathbf{1} + \mathrm{i}\,\mathbf{K}]\,[\mathbf{1} - \mathrm{i}\,\mathbf{K}]^{-1}\,, \tag{1.31}$$

which is the multi-channel generalization of (1.11).

1.5 Computer Program for Potential Scattering

The computer program is written in FORTRAN77. Given a radial mesh, a potential defined on this mesh, the charge of the projectile, an input energy, and a maximum angular momentum, the program uses the matching algorithm outlined in Sect. 1.2 to determine the potential scattering phase shifts δ_ℓ for the partial-wave angular momenta of interest. The present version is restricted to static potentials that can be given in the analytic form

$$V(r) = \sum_{i=1}^{M} A_i r^{N_i} \exp\{C_i r\}\,. \tag{1.32}$$

A Coulomb part may be included to obtain the phase shifts, but differential cross sections are only calculated for scattering from neutral targets. It is also possible to include a polarization potential of the form

$$V_{\text{pol}}(r) = -\frac{\alpha_{\text{d}}}{2r^4}\left[1 - \exp\left\{-\left(\frac{r}{r_{\text{c}}}\right)^6\right\}\right] . \tag{1.33}$$

where α_{d} is the dipole polarizability and r_{c} is a "cut-off radius". Such potentials will be further discussed in Chap. 2. Extensions, for example, to read in numerical potentials are straightforward and, if desired, should be implemented by the user.

1.5.1 Description of the Input Data

Except for the title (`FORMAT A80`), the input data are read in free format from unit 5 (`IREAD`). The individual records are:

1. `TITLE` Title of the run.

2. `NBUG1, NBUG2, NBUG3, NBUG4`

`NBUG1`	Normally set to 0. For details, see description in `BASFUN`.
`NBUG2`	Normally set to 0. For details, see description in `BASFUN`.
`NBUG3`	If > 0, the radial mesh is printed. If > 1, the potential array is printed.
`NBUG4`	If > 0, the phase shifts are printed. If > 1, details of the matching procedure are printed. If > 2, the numerical wavefunction is printed.

3. `LMAX, NQPROJ`

`LMAX`	Phase shifts for angular momenta $0 \le \ell \le$ `LMAX` are calculated.
`NQPROJ`	Charge of the projectile.

4. `HHALF, NIX1`

`HHALF`	Basic step size on which the potential is defined.
`NIX1`	Number of integrals with different stepsizes.

5. `(IHX(I),I=1,NIX)`

   In interval #`I`, the basic stepsize is `HHALF` · `IHX(I)`.

6. `(IRDX(I),I=1,NIX)`

   The last meshpoint number in interval #`I` is `IRDX(I)`+1. This is also the matching point. Since the radial function is only calculated on every other meshpoint, another mesh with stepsize `HINT = 2*HHALF` is defined in the program. The last interval meshpoints on that mesh are defined by `IRX(I) + 1 = IRDX(I)/2 + 1`.

7. `ZASYM, ENERGY`

`ZASYM`	Asymptotic charge seen by the projectile.
`ENERGY`	Collision energy in eV.

8. `MTERMS` Number of terms in the expansion (1.32) of the potential.

9. `COEFF(I), IPOWER(I), EXPO(I)`

`COEFF(I)`	A_i in (1.32).

IPOWER(I) N_i in (1.32).
EXPO(I) C_i in (1.32).
Altogether, MTERMS of these records must be read.

10. ALPHAD, RC
ALPHAD α_d in (1.33).
RC r_c in (1.33).

1.5.2 Test-Run Data

The input file listed below is used for electron scattering from atomic hydrogen at an incident energy of 10 eV. The run includes both the static potential

$$V(r) = q_\mathrm{proj}\left(1+\frac{1}{r}\right)\mathrm{e}^{-2r} \tag{1.34}$$

and a polarization term with $\alpha_\mathrm{d} = 4.5$ and $r_\mathrm{c} = 2.0$. The static potential is obtained by restricting (1.27) to only include the 1s orbital of hydrogen, neglect the exchange term (1.29), and perform the angular and radial integrations in (1.28).

```
POTENTIAL SCATTERING CALCULATION FOR ELECTRONS ON H-1S
 0    0    0    1
15   -1
  0.0005    7
 1    2    4    8   16   32   64
80  160  240  320  400  600 3000
  0.0  10.0
 2
  1.0  -1    -2.0
  1.0   0    -2.0
  4.5  2.0
```

Results for selected phase shifts are shown in Table 1.1.

Table 1.1. Potential phase shifts δ_ℓ (in radians) for electron–hydrogen scattering at 10 eV.

ℓ	0	1	2	3	5	8	15
δ_ℓ	1.0887879	0.3216620	0.1006706	0.0348725	0.0081298	0.0021440	0.0003485

1.6 Summary

We have presented an introduction to the more sophisticated methods discussed in the following chapters of this book. A computer program is provided to solve electron or positron scattering from spherically symmetric potentials.

1.7 Suggested Problems

1. Run the program for several collision energies between 0.1 and 100 eV, including only the static potential (1.34).
2. Repeat problem 1, but change the mesh parameters and the matching radius. Discuss the mesh-dependence of your results.
3. Repeat the above problems but include a polarization potential with $\alpha = 4.5$ and $r_c = 2.0$. By comparing with the results from problem 1, show that phase shifts for high angular momenta become more important in the calculation of the cross section, and that these phase shifts can be approximated by the effective-range formula [1.5]

$$\tan \delta_\ell = \frac{\pi \alpha k^2}{(2\ell+3)(2\ell+1)(2\ell-1)} . \tag{1.35}$$

4. Repeat the above problems but also include an exchange potential of the Furness–McCarthy form [1.7]

$$V_{\text{exch}}(r,E) = \pm\frac{1}{2}\left\{\left[\left(E - V(r)\right)^2 + \rho(r)\right]^{\frac{1}{2}} - \left[E - V(r)\right]\right\}, \tag{1.36}$$

where $\rho(r) = P_{1s}^2(r)/r^2$ is the radial electron density. This approach simulates scattering in the singlet (+) and triplet (–) total spin channels. Compare your results with highly accurate phase shifts (see, for example, Scholz *et al* [1.8] and references therein).
5. Replace the **BASFUN** routine by a Runge–Kutta integration package [1.9] (see, for example, Chap. 2). Check the accuracy of the approaches by calculating phase shifts for pure Coulomb potentials and for no potential at all. (In both cases, the potential phase shifts should be 0.)
6. Calculate the static potential of He^+ and run the program for this problem.
7. Produce differential cross sections by using the phase shift results from problem 6. (The total cross section diverges.) To do this, the result (1.8) for the scattering amplitude needs to be modified by including a Coulomb phase shift and a Coulomb amplitude [1.2]. Compare the results with the Rutherford formula for charges $Z = 1$ and $Z = 2$.

Acknowledgments

This chapter was written during a visit to The Queen's University of Belfast. I would like to thank the Department of Applied Mathematics for their hospitality and Queen's University for financial assistance. This work has also been supported, in part, by the United States National Science Foundation under grant PHY-9318377.

References

1.1 J.R. Taylor, *Scattering Theory*, (Malabar: Krieger Publishing, 1987)
1.2 I.E. McCarthy and E. Weigold, *Electron–Atom Collisions*, (Cambridge: University Press, 1995)
1.3 K.A. Berrington, P.G. Burke, M. LeDourneuf, W.D. Robb, K.T. Taylor and VoKyLan, Comp. Phys. Commun. **14** (1978) 367
1.4 P.G. Burke and M.J. Seaton, Meth. Comput. Phys. **10** (1971) 1
1.5 T.F. O'Malley, L. Spruch and L. Rosenberg, J. Math. Phys. **2** (1961) 491
1.6 M.J. Seaton, Proc. Phys. Soc. **77** (1961) 174
1.7 J.B. Furness and I.E. McCarthy, J. Phys. B **6** (1973) 2280
1.8 T. Scholz, P. Scott and P.G. Burke, J. Phys. B **21** (1988) L139
1.9 W.H. Press, B.P. Flannery, S.A. Teukolsky and W.T. Vetterling *Numerical Recipes: The Art of Scientific Computing*, 2nd edition (New York: Cambridge University Press, 1992)

2. Core Potentials for Quasi One-Electron Systems

Klaus Bartschat

Department of Physics and Astronomy, Drake University,
Des Moines, Iowa 50311, USA

Abstract

A method is presented to calculate core potentials for quasi one-electron systems and the corresponding single-electron orbitals. It is shown that the approximate inclusion of exchange effects between the valence electrons and the core removes the unphysical structure in the potential function that is characteristic for potentials calculated by only including the effect of core polarization due to the valence electrons.

2.1 Introduction

Many numerical calculations for electron scattering, photoionization and particle impact ionization can be simplified significantly by using a semi-empirical core potential, i.e., by effectively treating the target as a quasi one-electron system. The method has been applied very successfully, for example, to electron scattering from sodium [2.1] or cesium atoms [2.2–4], as well as to photoionization of alkaline earth systems such as barium [2.5,6], where the electron scattering problem must be solved for Ba^+ as the target.

In most of the work concerning such core potentials, the only effect included explicitly is the polarization of the atomic charge cloud by the scattered projectile. In an *ab initio* way, this effect can be accounted for by including target states in a close-coupling expansion where the core is opened up. From a computational point of view, however, this is a very expensive, slowly converging, and in many cases not even accurate, approach [2.1]. Alternatively, the effect can be represented in a phenomenological way by introducing an additional term in the scattering potential that behaves asymptotically as $-\alpha_d/2r^4$ where α_d is the static dipole polarizability of the core. However, this term must be "cut off" at small radii $r < r_c$, where r_c is usually determined by optimizing the one-electron ionization energies of the valence orbitals.

The core potential presented in this chapter includes an additional and equally important effect, namely the exchange between a valence electron and the core electrons [2.7]. This is achieved through a variant of the local exchange potential introduced by Furness and McCarthy [2.8]. While it seems possible to obtain approximately the same accuracy in the one-electron ionization potentials from optimizing the "cut-off" parameter r_c in the core potential alone, the main advantage of the present approach lies in the fact that unphysical "kinks" in the potential due to small values of r_c are almost entirely removed (for details, see Sect. 2.5). Hence, the potential looks much more physical, and it seems advantageous to use this potential and the corresponding wavefunctions in scattering calculations from quasi one-electron systems.

We begin with a brief outline of the theory in Sect. 2.2 before the algorithm is described in Sect. 2.3. Details of the computer program are provided in Sect. 2.4, followed by examples in Sect. 2.5 and some conclusions in Sect. 2.6.

2.2 Theory

We calculate the valence orbitals as eigenfunctions of the modified core potential

$$V^{\ell}_{\text{core}}(r) = V_{\text{static}}(r) + V_{\text{exch}}(r) + V^{\ell}_{\text{pol}}(r)\,, \tag{2.1}$$

where the static potential is the standard Hartree potential [2.9]

$$V_{\text{static}}(r) = -\frac{Z}{r} + \sum_{n_c,\ell_c} N_{n_c\ell_c} \int_0^\infty \mathrm{d}r' \, \frac{P^2_{n_c\ell_c}(r')}{\max(r,r')}\,. \tag{2.2}$$

In (2.2), Z is the nuclear charge and $N_{n_c\ell_c}$ is the number of electrons in the core orbital $P_{n_c\ell_c}$.

Furthermore, we include the *local* exchange potential of Furness and McCarthy [2.8]:

$$V_{\text{exch}}(r,E) = -\tfrac{1}{2}\left\{ [(E - V_{\text{static}}(r))^2 + 4\pi\rho(r)]^{\frac{1}{2}} - (E - V_{\text{static}}(r)) \right\}, \tag{2.3}$$

where

$$\rho(r) = \sum_{n_c,\ell_c} N_{n_c\ell_c} \frac{P^2_{n_c\ell_c}(r)}{4\pi r^2} \tag{2.4}$$

is the electron density in the core. We then choose the energy $E \equiv 0.0$ (ionization threshold) to ensure orthogonal orbitals.

Following Norcross [2.1], we finally include the polarization effect of the valence electron on the core through a potential of the form

$$V^{\ell}_{\text{pol}}(r) = -\frac{\alpha_{\text{d}}(\ell)}{2r^4}\left[1 - \exp\left\{-\left(\frac{r}{r_{\text{c}}(\ell)}\right)^6\right\}\right]. \tag{2.5}$$

Both the dipole polarizability $\alpha_d(\ell)$ and the cut-off radius $r_c(\ell)$ are then treated as adjustable parameters and are varied to achieve optimum agreement with experimental ionization potentials of the various Rydberg series. Since we use a non-relativistic hamiltonian, we average the experimental values obtained from Moore's tables [2.10] with a weighting factor of $2J+1$ over the fine-structure splitting for a given orbital angular momentum ℓ in a state described as $(n\ell)^2L_J$. Note that the use of different potentials for each orbital angular momentum does not cause any difficulties for the orthogonality of the valence orbitals but gives some additional flexibility in the optimization process. This flexibility is also the main reason for allowing even different dipole polarizabilities for different ℓ values; the error introduced in a subsequent scattering calculation can be expected to be very small, since the quasi one-electron system will generally have a significantly larger dipole polarizability than does the closed core. An alternative approach would be to obtain a fairly reliable value for the dipole polarizability from fits to Rydberg series with high angular momentum ℓ (for which the short-range part of the potential is of little importance), then fix this value and introduce another variation parameter in the exchange potential. This will be further discussed below.

2.3 The Algorithm

The program described in this chapter calculates wave functions and eigenenergies for atoms based on a frozen-core approximation. The eigenenergies and wave functions of a loosely bound electron are calculated with respect to a core consisting of the nucleus and the other electrons. The program uses the core potential given in (2.1), except that the exchange term (2.3) is multiplied by a weighting factor $a_{\text{exch}}(\ell)$. This further increases the flexibility of the method, and the program could be modified to vary this parameter as well. This becomes particularly important if one would like to use a fixed dipole polarizability for the core, i.e., optimization of $a_{\text{exch}}(\ell)$ could replace the variation procedure for the core polarizability.

The program finds solutions of the non-relativistic radial Schrödinger "equation"

$$\left[\frac{\mathrm{d}^2}{\mathrm{d}r^2} - \frac{\ell(\ell+1)}{r^2} - 2V_{\text{core}}^{\ell}(r)\right] P_{n\ell}(r) = -2\,E_{n\ell}P_{n\ell}(r) \tag{2.6}$$

in an iterative process. Given a lower limit for the bound-state energy $E_{n\ell}$ (the upper limit, by default, is 0.0), the fourth-order Runge-Kutta method [2.11] is used to propagate the solution of the differential equation with the initial conditions $P_{n\ell}(r{=}0) = 0$ and $P'_{n\ell}(r{=}0) = 1$ from the origin beyond the classical turning point r_{class} for zero angular momentum. (This point is defined through $V_{\text{core}}(r) > E_{n\ell}$ for $r > r_{\text{class}}$.) Since the correct bound-state solution must decrease exponentially, the propagation is halted as soon as the absolute value of the solution starts to increase in this classically forbidden region. The number of nodes of the resulting function is counted and compared to the desired number. This allows for the definition of revised upper and lower bounds, and a new trial energy is taken as the mean value of the current upper and lower limits. Once these two limits agree within a specified error, the bound-state energy and wavefunction have been found to the desired accuracy.

When the bound-state energies for a series of n values with a given orbital angular momentum ℓ have been found, these energies $E_{n\ell}$ are compared to the experimental values and the total relative error, including the sign, is calculated. If, for example, the result is negative, this indicates that the phenomenological polarization potential (or, though unlikely, the exchange potential) is too attractive. Consequently, the cut-off radius r_{c} is increased to reduce the strength of this potential. Similarly, r_{c} is decreased if the error is positive. A few variations of the r_{c} value are usually sufficient to bring the total error to less than $10^{-6}\%$ in this second iteration process.

In the outermost (third) iteration, the results for the *absolute* error are determined as a function of the dipole polarizability $\alpha_{\mathrm{d}}(\ell)$ which is then varied to find the minimum. Note that this procedure implies that the minimum absolute error occurs at a point where the total error including the sign is basically zero. While this is an approximation, it was found empirically that the assumption is very well fulfilled, thereby simplifying the minimization process. As an option for the future, the iterations could be modified so that the program finds the minimum of the absolute error with respect to three variables, $r_{\mathrm{c}}(\ell)$, $\alpha_{\mathrm{d}}(\ell)$, and $a_{\mathrm{exch}}(\ell)$, in a simultaneous, three-dimensional search.

2.4 Computer Program

The computer program is written in standard FORTRAN77. As mentioned before, it calculates wave functions and eigenenergies for one valence electron based on a potential that is changed through an iteration process. Through comparison of these energies with experimental values, the program tries to optimize the parameters $r_{\mathrm{c}}(\ell)$ and $\alpha_{\mathrm{d}}(\ell)$. The final result provides a local, energy independent approximation for the core potential of the atom. It can be used in further applications, such as electron scattering.

2.4.1 Program Structure

The program begins with the reading and printing of some basic input data. Next, atomic wave functions in the form given by Clementi and Roetti [2.12] are read in to provide the core orbitals, together with experimental ionization energies.

The program uses several radial meshes for internal calculations and final output. The array RHALF contains the finest mesh and RFULL contains every other meshpoint in RHALF. These meshes are necessary, for example, for standard Simpson's rule integration with an integrand given on the RHALF mesh, where the result at every other meshpoint (i.e., on the RFULL mesh) can be stored. The RFULL mesh is obtained as follows. A basic initial stepsize is read in as HINT. Then NIX intervals are defined with the last meshpoint number given by IRX(I)+1, I=1,NIX. In each interval, the stepsize is given by IHX(I)·HINT.

Following the convention of many other computer codes (see also Chaps. 4 and 7), we set RFULL(1) = RHALF(1) = 0.0. Note that the number of meshpoints on the RFULL mesh is NSTEP = IRX(NIX)+1, while the number of points on the RHALF mesh is 2*IRX(NIX)+1. In the same way, the program also defines a radial output mesh (stored in ROUT) onto which the final potentials and wavefunctions can be interpolated and printed out for further use, for example in the DWBA program described in Chap. 4.

After all the basic information is calculated, the subroutine CALEXT is called to read the wavefunction parameters for the core orbitals from the tables of Clementi and Roetti [2.12]. The coefficients are then transformed to Slater form [2.13], and the wavefunctions, calculated according to the formula

$$P_{n_c,\ell_c}(r) = \sum_{i=1}^{M} A_i r^{N_i} \exp\{-C_i r\} \, , \tag{2.7}$$

with r defined on the RHALF mesh, are stored in the array WF(N,L,I). Using the occupation numbers of the core orbitals, the program then calls subroutine VINNER to obtain the contribution of each orbital to the Hartree potential given in (2.2), as well as to the electron density (2.4) used in the exchange potential (2.3). This is an example where the output of VINNER is given on the RFULL mesh, i.e., it is subsequently interpolated onto the RHALF mesh. It should also be noted that the static potential is a rapidly varying function of r near the origin. Consequently, the much smoother function $r\,V_{\text{static}}$ is actually used in the interpolation.

At this point, all the input information has been processed and the iterations described above can be performed for a number of Rydberg series with fixed ℓ values. The final wave functions are normalized to unity, stored in the array WAVES and, if desired, the results (defined on the output mesh) are written to output files. To achieve maximum compatibility with the DWBA program described in Chap. 4, the code also allows for the calculation of Hartree potentials *including* one electron in the valence orbitals, and the output potentials are all multiplied by $2r$. Hence, they are effectively given in Rydberg units with limiting values of $-2\,$ZNUC near the origin and $-2\,$(ZNUC − NELEC) far away from the nucleus, where ZNUC is the nuclear charge and NELEC is the number of electrons in the system. This format should be changed depending on the individual user's needs.

2.4.2 Description of the Input Parameters

The following input parameters need to be provided by the user on unit 5 (standard input). Furthermore, one file with input parameters for the core wavefunctions and another file with experimental energies relative to the ionization threshold need to be set up (see below). All input is read in free format.

1. `ITIN1, ITIN2, ITOUT1, ITOUT2, ITOUT3, ITOUT4`

`ITIN1` Input file for the experimental bound state energies of the $P_{n\ell}$ orbitals. It should contain the following columns:
`N, L, ENERGY`.
The first two lines of this file are not used; they may, for example, contain titles and other related information.

`ITIN2` Input file for the Clementi–Roetti wavefunctions. For each core orbital, this file should contain (cf. Eq. 2.7):
`M`
`A(I), N(I), C(I)` for `I=1,M`.

`ITOUT1` Output file for the final wave functions $P_{n\ell}$; only used if `ITOUT1` $\neq 0$.

`ITOUT2` Output file for the Hartree potentials including one electron in $P_{n\ell}$; only used if `ITOUT2` $\neq 0$.

`ITOUT3` Output file for $V^{\ell}_{\rm core}$; only used if `ITOUT3` $\neq 0$.

`ITOUT4` Output file with the radial mesh.

2. `ZNUC, ESTART, ERROR`

`ZNUC` Charge of the nucleus.

`ESTART` Lower limit for the bound orbital energies; this should be chosen safely below the expected result for the lowest energy.

`ERROR` If the difference between the upper (`EHIGH`) and lower (`ELOW`) limit for the orbital energy is less than `ERROR`, the $E_{n\ell}$ iteration is assumed to have converged.

3. `NITM1, NITM2, NITM3, IBUG1, IBUG2, IBUG3`

`NITM1` Maximum number of iterations to find a bound orbital.

`NITM2` Maximum number of iterations to bring the total percentage error (including the sign) to less than 10^{-6}.

`NITM3` Maximum number of iterations to minimize the absolute percentage error.

`IBUG1` Debug parameter for determination of a bound orbital.
$= 0$: No output.
> 0 : Print every $E_{n\ell}$ that is found, and overlap integrals of orbitals for different n values in the Rydberg series. (An overlap is also printed if its absolute value exceeds 10^{-4}.)
> 1 : Print values of upper and lower limits in the bound state iteration.
> 2 : Print normalized wave function.

> 3 : Print temporary output of propagation. (Note that this output may become very large !)

IBUG2 Debug parameter for $r_c(\ell)$ and $\alpha_d(\ell)$ iteration.
= 0 : No output.
> 0 : Print result for absolute percentage errors as a function of $\alpha_d(\ell)$.
> 1 : Print intermediate results for bound-state energies in comparison with experimental results.

IBUG3 Debug parameter for subroutine CALEXT.
= 0 : No output.
= 1 : Print original overlap integrals of core orbitals. (An overlap is also printed if its absolute value exceeds 10^{-4}.)

4. LMIN, LMAX, LRANG1, IEXCH

LMIN Minimum ℓ value for which bound orbitals are calculated.

LMAX Maximum ℓ value for which bound orbitals are calculated.

LRANG1 Maximum ℓ value for which core orbitals are occupied.

IEXCH1 If set to 0, the exchange potential is not included.

5. NMINL(L), NMAXL(L), NMIOUT(L), NMAOUT(L)

(LMAX-LMIN+1) of these records are needed for L = LMIN,LMAX.

NMINL(L) Minimum n value for this ℓ for which orbitals are calculated.

NMAXL(L) Maximum n value for this ℓ for which orbitals are calculated.

NMIOUT(L) Minimum n value for this ℓ for which orbitals and Hartree potentials are written to units ITOUT1 and ITOUT2.

NMAOUT(L) Maximum n value for this ℓ for which orbitals and Hartree potentials are written to units ITOUT1 and ITOUT2.

6. RCMIN, RCMAX

RCMIN Minimum value allowed for $r_c(\ell)$ in the iteration; if $r_c(\ell)$ needs to be smaller than this value, the program stops.

RCMAX Maximum value allowed for $r_c(\ell)$ in the iteration; if $r_c(\ell)$ needs to be bigger than this value, the program stops.

7. (ALSTRT(L),L=0,LMAX)

Starting values $\alpha_d(\ell)$ for each ℓ.

8. (RCSTRT(L),L=0,LMAX)

Starting values for $r_c(\ell)$ for each ℓ.

9. (RCSTEP(L),L=0,LMAX)

Starting stepsizes for $r_c(\ell)$ for each ℓ.
If RCSTEP(ℓ) = 0.0, no $r_c(\ell)$ iteration is performed for this ℓ. This feature can be used, for example, to recalculate core orbitals with the parameter set obtained for the valence orbitals. Such a calculation could become necessary, if a complete (core plus valence) set of orthogonal orbitals is needed in a particular application.

10. (EXWEIG(L),L=0,LMAX)

Exchange potential weighting factor $a_{\text{exch}}(\ell)$ for each ℓ.

11. (MAXNHF(L),L=0,LRANG1)
    Maximum n value of the occupied core orbitals for each ℓ.

12. (NOCC(N,L),N=L,MAXNHF(L))
    LRANG+1 of these records are needed for L = 0,LRANG1.
    They give the number of electrons in each core orbital $P_{n_c\ell_c}$.

13. NIX,HINT

NIX Number of intervals in the RFULL mesh.

HINT Basic stepsize of the RFULL mesh.

14. (IHX(I),I=1,NIX)
    In interval #I, the basic stepsize is HINT · IHX(I).

15. (IRX(I),I=1,NIX)
    The last full meshpoint number in interval #I is IRX(I) + 1.

16. MIX,HOUT

MIX Number of intervals in the ROUT mesh.

HOUT Basic stepsize of the ROUT mesh.

17. (JHX(I),I=1,MIX)
    In interval #I, the basic stepsize is HOUT · JHX(I).

18. (JRX(I),I=1,MIX)
    The last full meshpoint number in interval #I is JRX(I) + 1.

2.5 Test Run

The input file listed below is used to apply the method to the Rydberg series

$$\begin{aligned} &[\mathrm{Ne}]n\mathrm{s}(^2\mathrm{S}); \quad n = 3, 4 \ldots, 9 \\ &[\mathrm{Ne}]n\mathrm{p}(^2\mathrm{P}); \quad n = 3, 4, \ldots, 9 \\ &[\mathrm{Ne}]n\mathrm{d}(^2\mathrm{D}); \quad n = 3, 4, \ldots, 9 \end{aligned} \tag{2.8}$$

in atomic sodium where [Ne] denotes the neon-like configuration of the core, i.e. [$1s^22s^22p^6$]. After the core potentials have been optimized based on the ionization potentials of all the above orbitals, the 3s, 4s, 3p, and 4p wavefunctions as well as the corresponding Hartree potentials (i.e., core potential plus one electron in these orbitals) are tabulated and printed.

```
-0.5D0      20.0D0      0.0D0   50   0
0     2    11     12    13    14
11.0D0   -100.0D0   1.0D-9
100   30   20      0     1     0
0      2     1     1
3      9     3     4
3      9     3     4
```

```
  3    9     0     0
  0.3D0      25.0D0
  1.000D0    1.300D0    1.200D0    1.000D0    1.000D0
  1.500D0    1.500D0    2.000D0    2.000D0    2.000D0
  0.300D0    0.400D0    0.400D0    0.500D0    0.500D0
  1.000D0    1.000D0    1.000D0    1.000D0    1.000D0
  2    2
  2    2
  6
  7      0.0012D0
  1      2     4     8    16    32    64
 40     80   120   160   200   300  4996
 13      0.00075D0
  1      2     4     8    16    32    64   128   256   512  1024  2048  4096
 40     80   120   160   200   240   280   320   360   400   440   480   520
```

The results are in Table 2.1. As expected, the ionization potentials obtained with the static potential alone are far too small in all cases, whereas very good agreement with experimental results is achieved by including the semi-empirical polarization potential. Interestingly, this is true with or without the exchange contribution to the core potential. In several cases, the agreement is better than six significant figures. An indication about the remaining discrepancies between experiment and our semi-empirical predictions can be found in the "error" column where the percentage error is given.

While the optimum values of $\alpha_d(\ell)$ are very similar in the two calculations with the polarization potential, the final values of $r_c(\ell)$ are always larger in the "exchange calculation" due to the additional, short-range attractive exchange potential. Hence, a smaller value of r_c can be used, to some extent, to simulate such an exchange effect. This may explain the apparent success of previous approaches without the exchange term.

The potential curves obtained *without* the exchange term, however, show an irregularity in the region of $r \approx r_c$, hence indicating possible problems with the method. This feature is demonstrated in Fig. 2.1 which shows the core potentials (multiplied by $2\,r$) in the region near the cut-off parameter r_c, as obtained without the exchange term in the core potential. Note the significant "kinks" in the region near r_c, particularly for the D series with $\ell = 2$. As can be seen from Fig. 2.2, this irregularity is completely removed in the potential curves obtained with the inclusion of the exchange effect. Hence, the separate treatment of the exchange and the polarization effects clearly improves the physical appearance of the core potential.

The core potentials obtained with the method described in this chapter have been used very successfully in calculations of bound-state properties such as oscillator strengths as well as for scattering processes [2.4].

Table 2.1. Bound orbital energies (in atomic units with respect to the ionization threshold) for atomic sodium.

$n\ell$	experiment	static	% error	polarization	% error	pol.+exch.	% error
				$\alpha_d = 1.281$	$r_c = 0.664$	$\alpha_d = 1.412$	$r_c = 1.717$
3s	−0.188859	−0.158935	15.84	−0.188594	−0.0002	−0.188859	−0.0000
4s	−0.071579	−0.064543	9.83	−0.071586	−0.0101	−0.071586	−0.0091
5s	−0.037585	−0.034878	7.20	−0.037589	−0.0103	−0.037589	−0.0102
6s	−0.023133	−0.021814	5.70	−0.023134	−0.0061	−0.023134	−0.0063
7s	−0.015663	−0.014924	4.72	−0.015664	−0.0040	−0.015664	−0.0043
8s	−0.011306	−0.010849	4.04	−0.011305	0.0081	−0.011305	0.0077
9s	−0.008544	−0.008242	3.54	−0.008542	0.0226	−0.008542	0.0222
				$\alpha_d = 1.105$	$r_c = 0.651$	$\alpha_d = 1.634$	$r_c = 1.621$
3p	−0.111548	−0.090538	18.84	−0.111548	−0.0000	−0.111555	0.0002
4p	−0.050935	−0.044350	12.93	−0.050935	0.0009	−0.050932	0.0062
5p	−0.029196	−0.026299	9.92	−0.029195	0.0037	−0.029195	0.0047
6p	−0.018919	−0.017393	8.06	−0.018919	−0.0027	−0.018920	−0.0036
7p	−0.013253	−0.012352	6.80	−0.013254	−0.0053	−0.013254	−0.0069
8p	−0.009800	−0.009224	5.88	−0.009800	0.0006	−0.009800	−0.0013
9p	−0.007540	−0.007149	5.18	−0.007540	0.0027	−0.007540	0.0008
				$\alpha_d = 1.048$	$r_c = 0.383$	$\alpha_d = 1.038$	$r_c = 2.278$
3d	−0.055937	−0.055586	0.63	−0.055937	−0.0000	−0.055939	0.0001
4d	−0.031443	−0.031268	0.56	−0.031442	0.0029	−0.031442	0.0024
5d	−0.020106	−0.020011	0.47	−0.020106	0.0004	−0.020106	0.0002
6d	−0.013953	−0.013895	0.41	−0.013953	0.0029	−0.013952	0.0030
7d	−0.010246	−0.010208	0.37	−0.010245	0.0087	−0.010245	0.0088
8d	−0.007840	−0.007815	0.31	−0.007840	−0.0049	−0.007840	−0.0047
9d	−0.006192	−0.006175	0.28	−0.006192	−0.0099	−0.006193	−0.0097

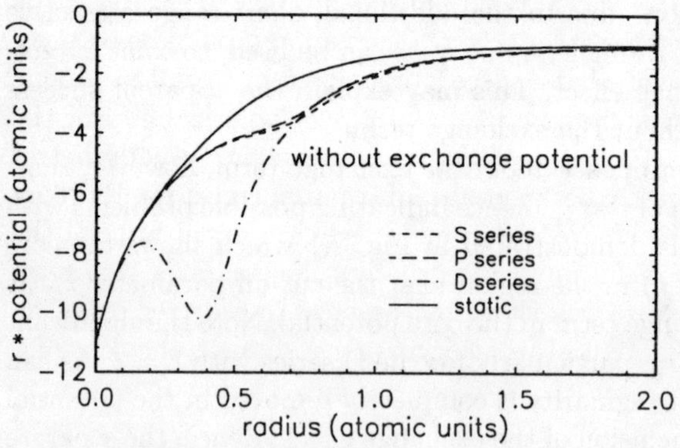

Fig. 2.1. Core potentials (multiplied by r) in atomic units for the Rydberg series with $\ell = 0$ (S), $\ell = 1$ (P), and $\ell = 2$ (D) of Na^+ without accounting for exchange between the valence and the core electrons.

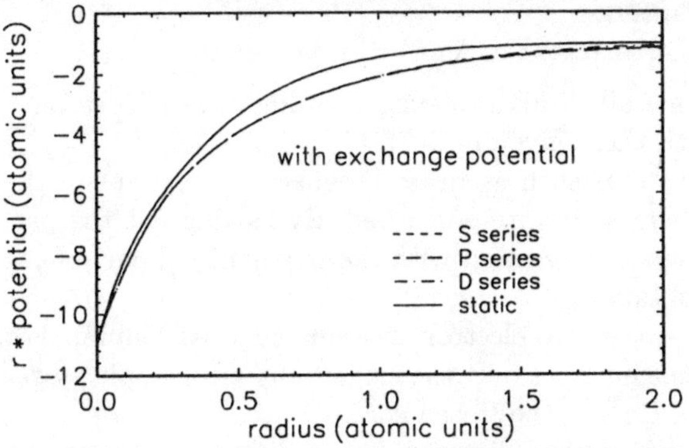

Fig. 2.2. Same as Fig.2.1, except for the inclusion of an exchange potential.

Generally, significant improvements over previous calculations without the exchange term can be expected in energy regions where the form of the potential is still critical but where the energy is high enough for the electron to "see" the unphysical structure in the potentials without the exchange terms.

2.6 Summary

In this chapter, we have presented an approach to obtain core potentials for quasi one-electron systems that includes the effects of charge cloud polarization of the core due to the valence electrons and, in addition, an approximate representation of exchange between core and valence electrons. The computer program developed for this approach uses an approximate core potential and core wavefunctions that can be obtained from standard atomic structure codes to find optimum values for the dipole polarizability of the core and a "cut-off" parameter $r_c(\ell)$. Very good agreement (usually better than 0.01 % deviations) with experimental ionization energies for Rydberg states up to $n \approx 10$ can be achieved by using a semi-empirical form of the polarization potential. The inclusion of the usually neglected exchange term in the core potential, however, was shown to be essential for removing an unphysical structure in the potentials that occurs for radii near the "cut-off" parameter $r_c(\ell)$.

2.7 Suggested Problems

1. Apply the method to other alkali-like systems, including ions such as Ba^+. Compare the results with those given in [2.7].
2. Apply the method to systems such as In or Tl where some terms of the quasi one-electron Rydberg series are perturbed. By leaving out the perturbers, show that very good agreement with the unperturbed part of the Rydberg series can be obtained.
3. Extend the method to quasi two-electron systems such as helium. For excited states, average the energies over the singlet and triplet spin states and use the 1s orbital of He^+ as the frozen core.
4. Modify the program to use a fixed dipole polarizability α_d and optimize the exchange factor $a_{exch}(\ell)$ instead.
5. With this modification, return to the helium problem and perform the calculation separately for the excited singlet and triplet spin states.
6. Extend the procedure to allow for optimization of all three parameters, α_d, r_c, and a_{exch}. You may want to use a program like Powell's [2.14] to minimize the total absolute error directly.

Acknowledgments

I would like to thank my former students, B.J. Albright, K.M. Bloom, P.R. Flicek, and R. Schwienhorst who assisted in the development of earlier versions of this program. This work has been supported, in part, by the National Science Foundation under grants PHY-9014103 and PHY-9318377.

References

2.1 D.W. Norcross, Phys. Rev. Lett. **32** (1974) 192
2.2 N.S. Scott, K. Bartschat, P.G. Burke, O. Nagy and W.B. Eissner, J. Phys. B **17** (1984) 3775
2.3 U. Thumm U and D.W. Norcross, Phys. Rev. Lett. **67** (1991) 3495
2.4 K. Bartschat, J. Phys. B **26** (1993) 3595
2.5 K. Bartschat, M.R.H. Rudge and P. Scott, J. Phys. B **19** (1986) 2469
2.6 C.H. Greene and M. Aymar, Phys. Rev. A **44** (1991) 1773
2.7 B.J. Albright, K. Bartschat and P.R. Flicek, J. Phys. B **26** (1993) 337
2.8 J.B. Furness and I.E. McCarthy, J. Phys. B **6** (1973) 2280
2.9 D.R. Hartree, Proc. Camb. Phil. Soc. **24** (1928) 111
2.10 C.E. Moore, *Atomic Energy Levels; NBS Circular Vol. 3, No. 467* (1970)
2.11 W.H. Press, B.P. Flannery, S.A. Teukolsky and W.T. Vetterling *Numerical Recipes: The Art of Scientific Computing*, 2nd edition (Cambridge U. Press, New York, 1992)
2.12 E. Clementi and C. Roetti, Atomic Data and Nuclear Data Tables **14** (1974) 177
2.13 J.C. Slater, Phys. Rev. **8** (1951) 385
2.14 M.J.D. Powell, Computational Journal **7** (1964) 155

3. Energies and Oscillator Strengths Using Configuration Interaction Wave Functions

Alan Hibbert

Department of Applied Mathematics and Theoretical Physics,
The Queen's University of Belfast, Belfast, BT7 1NN, Northern Ireland

Abstract

The method of configuration interaction (CI) is developed from its basis of the Hartree–Fock approximation, which makes use of one-electron orbitals akin to hydrogen-like wave functions. The method is applied to the calculation of oscillator strengths of allowed electric dipole transitions. The general configuration interaction code `CIV3` is presented, together with an interactive program `SETCIV3`, which generates the input data for `CIV3`, and the program `GENCFG`, which determines the configuration data. Simple examples of the use of these codes are given and exercises are suggested for the construction and use of CI wave functions for atoms and ions with a few electrons. The codes can be used for a wide range of atoms and ions, within the LS coupling scheme.

3.1 Introduction

The calculation of any atomic property requires knowledge of the wave functions of the relevant atomic states. If we neglect the effects of relativity, these wave functions Ψ_n are solutions of Schrödinger's equation

$$H\Psi_n = \mathrm{i}\hbar\frac{\partial\Psi_n}{\partial t} . \tag{3.1}$$

The Hamiltonian H consists of the total kinetic energy of the N electrons plus the total potential energy of the electrons due to their electrostatic interactions with the nucleus and with each other:

$$H = \sum_{i=1}^{N}\left(-\frac{1}{2}\nabla_i^2 - \frac{Z}{r_i}\right) + \sum_{i<j}\frac{1}{r_{ij}} , \tag{3.2}$$

where we have used atomic units (or Hartrees), for which $\hbar$, the electronic charge (e), and the reduced mass (μ) are all set equal to unity. In (3.2), r_{ij}

$= |\boldsymbol{r}_i - \boldsymbol{r}_j|$ and the nucleus is taken as the origin of coordinates; Z is the nuclear charge.

Stationary-state solutions of (3.1) take the form

$$\Psi_n(\{\boldsymbol{r}_i\}, t) = \Psi_n(\{\boldsymbol{r}_i\}) \exp(-\mathrm{i}E_n t), \tag{3.3}$$

so that

$$|\Psi_n(\{\boldsymbol{r}_i\}, t)|^2 = |\Psi_n(\{\boldsymbol{r}_i\})|^2,$$

where

$$H\Psi_n(\{\boldsymbol{r}_i\}) = E_n \Psi_n(\{\boldsymbol{r}_i\}) \tag{3.4}$$

is the stationary-state (time-independent) form of Schrödinger's equation and E_n is the energy eigenvalue of the Hamiltonian H corresponding to the wave function Ψ_n.

It is possible to obtain *exact* solutions of (3.4) only for one-electron systems, i.e., for hydrogen-like ions. Even for two-electron systems, the presence of the interelectronic distance r_{ij} in the Hamiltonian prevents this operator from being amenable to the separation of variables technique. It is therefore necessary either to simplify (approximate) the Hamiltonian operator so that its eigenfunctions and eigenvalues can be found or to simplify (approximate) the form of the wave functions so that (3.4) is satisfied only approximately.

In practice, either perturbation theory or (more frequently) the variational principle has been used to systematize these approximations. Two-electron ions have formed one of the major systems of interest (because they are the simplest for which the solution of (3.4) is non-trivial) in the development of methods that will yield accurate wave functions. Even as early as 1929, Hylleraas [3.1] showed that a fairly simple form of the wave function, explicitly involving the interelectronic distance r_{12}, could result in quite an accurate value for the ground-state energy of helium. More elaborate forms of wave functions subsequently appeared in the literature, culminating in the researches of Pekeris and coworkers [3.2], who obtained the non-relativistic ground-state energy of helium correct to one part in 10^9. For excited states of two-electron ions, this method was not quite as successful as for the ground state. The introduction of r_{12} as a coordinate can be important when, as in the ground state, the mean radii of the two electrons are similar. In terms of terminology we will define later, it is an efficient means of incorporating short-range electron correlation. But for excited states, the mean radii of the two electrons are quite different from each other and short-range electron correlation is less significant.

The Hylleraas method has proved difficult to extend to many-electron systems, mainly for technical reasons. There have been some attempts [3.3] to expand a many-electron wave function in terms of groupings of two-electron functions (geminals) and to express each geminal in terms of coordinates involving r_{ij}. However, the vast majority of calculations for many-electron systems have involved the expansion of the wave functions in terms of one-electron functions (orbitals – more correctly, spin orbitals). In particular, nearly all the recent, very accurate calculations have employed this method. It has as its base the conceptually simple Hartree–Fock method, and can be extended – in principle to any accuracy – by the method of configuration interaction (or some form of it).

We discuss these methods below in a way that allows them to be applied to any atomic system. First of all, though, it is useful to recall the essential characteristics of hydrogen-like wave functions and the one-electron Hamiltonian, since these form the "building blocks" of wave functions for larger systems.

3.2 Hydrogen-Like Ions

For these one-electron systems, the Hamiltonian in (3.1) consists of just one term from the first summation in (3.2). It is convenient to express Schrödinger's equation in spherical polar coordinates r, θ, and ϕ:

$$H\Psi = -\frac{1}{2}\left[\frac{1}{r^2}\frac{\partial}{\partial r}\left(r^2\frac{\partial\Psi}{\partial r}\right) + \frac{1}{r^2\sin\theta}\frac{\partial}{\partial\theta}\left(\sin\theta\frac{\partial\Psi}{\partial\theta}\right) + \frac{1}{r^2\sin^2\theta}\frac{\partial^2\Psi}{\partial\phi^2}\right] + V(r)\Psi = E\Psi\,. \tag{3.5}$$

This equation may be solved exactly by the separation of variables method whereby the radial and angular parts of Ψ may be factorized as

$$\Psi(r,\theta,\phi) = R_{nl}(r)Y_l^m(\theta,\phi)\,. \tag{3.6}$$

Substitution of (3.6) into (3.5) leads to the following equations for Y_l^m and R_{nl}:

$$-\left[\frac{1}{\sin\theta}\frac{\partial}{\partial\theta}\left(\sin\theta\frac{\partial}{\partial\theta}\right) + \frac{1}{\sin^2\theta}\frac{\partial^2}{\partial\phi^2}\right]Y = l(l+1)Y \tag{3.7}$$

and

$$-\frac{1}{2}\left[\frac{1}{r^2}\frac{\mathrm{d}}{\mathrm{d}r}\left(r^2\frac{dR}{dr}\right) - \frac{l(l+1)R}{r^2}\right] + (V-E)R = 0 \tag{3.8}$$

with $V = -Z/r$. The requirement that Y_l^m is single valued leads to the condition that l and m are integers satisfying $-l \le m \le l$, while the square integrability of R_{nl} necessitates n also being an integer with $n > l$. The explicit form of R_{nl} is

$$R_{nl}(r) = \mathcal{N}\exp\left(-\frac{Zr}{n}\right)\left(\frac{2Z}{n}\right)^l L_{n+l}^{2l+1}(r)\left(\frac{2Z}{n}\right),$$

where $L_{n+l}^{2l+1}(r)$ is an associated Laguerre polynomial and $\mathcal{N}$ is a normalization constant.

The functions $Y_l^m(\theta,\phi)$ are known as *spherical harmonics* and take the form

$$Y_l^m(\theta,\phi) = (-1)^m \left[\frac{(2l+1)}{4\pi}\frac{(l-m)!}{(l+m)!}\right]^{\frac{1}{2}} P_l^m(\cos\theta)\mathrm{e}^{\mathrm{i}m\phi}\,; \qquad m \geq 0\,, \tag{3.9}$$

where the $P_l^m(\cos\theta)$ are associated Legendre polynomials. Furthermore,

$$Y_l^{-m}(\theta,\phi) = (-1)^m \left[Y_l^m(\theta,\phi)\right]^*\,.$$

The interesting feature about (3.8) is that for *any* potential energy V that is a function of r only (i.e., a central potential), and not of θ or ϕ, the angular dependence of the wave function would be a spherical harmonic. As we shall see later, one of the fundamental approximations used by Hartree in dealing with many-electron problems was to replace the potential energy by a set of central potentials, thereby retaining the spherical harmonics as crucial in the construction of many-electron wave functions.

3.2.1 Angular Momentum

In classical mechanics a quantity that is often constant (e.g., for central forces) is the angular momentum:

$$\boldsymbol{l} = \boldsymbol{r} \times \boldsymbol{p}\,.$$

The equivalent quantum mechanical operator is the *orbital* angular momentum

$$\boldsymbol{l} = \boldsymbol{r} \times (-\mathrm{i}\nabla)\,.$$

In particular, in spherical polar components

$$l_z = -\mathrm{i}\frac{\partial}{\partial\phi} \tag{3.10}$$

and

$$\boldsymbol{l}^2 = l_x^2 + l_y^2 + l_z^2 = -\left[\frac{1}{\sin\theta}\frac{\partial}{\partial\theta}\left(\sin\theta\frac{\partial}{\partial\theta}\right) + \frac{1}{\sin^2\theta}\frac{\partial^2}{\partial\phi^2}\right]\,. \tag{3.11}$$

Comparing with the expression for ∇^2 or with (3.7), we see that $-\boldsymbol{l}^2$ is just the angular part of ∇^2, so Y_l^m is an eigenfunction of $\boldsymbol{l}^2$ with

$$\boldsymbol{l}^2 Y_l^m(\theta,\phi) = l(l+1)Y_l^m(\theta,\phi)\,. \tag{3.12}$$

Moreover, by simple differentiation,

$$l_z Y_l^m = mY_l^m\,. \tag{3.13}$$

Hence we see that l and m are related to the eigenvalues of the orbital angular momentum and its z component. The integer n is closely related to the energy E:

$$E_n = -\frac{Z^2}{2n^2} \,. \tag{3.14}$$

It is unusual that the energy depends only on n and not on l. This is known as *accidental degeneracy*. For larger atoms, the energy *does* depend on the total angular momentum eigenvalues, but not on the eigenvalues of their z components.

3.2.2 Commuting Operators

Since the angular operators in ∇^2 are identical with $-\boldsymbol{l}^2$ and independent of r, then

$$[H, \boldsymbol{l}^2] = 0 \,. \tag{3.15}$$

We may also show that

$$[H, l_x] = [H, l_y] = [H, l_z] = 0$$

and

$$[\boldsymbol{l}^2, l_x] = [\boldsymbol{l}^2, l_y] = [\boldsymbol{l}^2, l_z] = 0 \,,$$

but

$$[l_x, l_y] = \mathrm{i} l_z \,, \tag{3.16}$$

so that the components of $\boldsymbol{l}$ do not commute with each other.

An important property of commuting operators is that they have a common set of eigenfunctions. Hence, by (3.5) and (3.12–16), we see that H, $\boldsymbol{l}^2$, and l_z form a set of commuting operators with Ψ as a common eigenfunction, and so l, m, and n are "good" quantum numbers.

The *parity* operator Π is also conserved:

$$\Pi : \ \boldsymbol{r} \to -\boldsymbol{r}$$

gives

$$(r, \theta, \phi) \to (r, \pi - \theta, \pi + \phi)$$

and

$$\Pi Y_l^m(\theta, \phi) = Y_l^m(\pi - \theta, \pi + \phi) = (-1)^l Y_l^m(\theta, \phi) \,,$$

so Y_l^m, and therefore Ψ, is also an eigenfunction of Π. If l is even, $\Pi\Psi = \Psi$ and the parity is said to be *even*; if l is odd, $\Pi\Psi = -\Psi$ and the parity is said to be *odd*. Moreover, Π may be added to the set of commuting operators.

3.2.3 Fine Structure

In practice, most electronic energy levels of hydrogenic ions are not single isolated values, but rather pairs of values, each pair of levels being very close together. This small separation between the members of a pair is an example of what is known as *fine structure*. But it is not explained by the model of the atom or ion given by Schrödinger's equation. It can be explained semi-empirically, but the correct way to introduce a model to account for this effect is to set up a relativistic treatment of the electron using Dirac's equation rather than Schrödinger's equation.

It is beyond the scope of this chapter to do this, but a compromise can be achieved by expanding Dirac's equation in powers of v^2/c^2, where v is the speed of the electron and c is the speed of light. If only the leading terms in this expansion are retained then Dirac's equation reduces to a Schrödinger-type equation with a Hamiltonian of the form

$$H_{\text{fs}} = H_{\text{Sch}} + H_{\text{so}} = \left(-\frac{1}{2}\nabla^2 - \frac{Z}{r}\right) + \frac{1}{2}\alpha^2\frac{Z}{r^3}\boldsymbol{l}\cdot\boldsymbol{s} \tag{3.17}$$

(plus some other terms which do not contribute to the splitting of the energy level). In (3.17), α is the (dimensionless) fine-structure constant, with value approximately 1/137 and $\boldsymbol{s}$ is the *spin* of the electron. Hence H_{so} is known as the *spin-orbit operator*. The spin operator $\boldsymbol{s}$ has the properties of the angular momentum, (3.16), and is closely related to the Pauli spin matrices:

$$\boldsymbol{s} = \tfrac{1}{2}\boldsymbol{\sigma},$$

where

$$\sigma_x = \begin{pmatrix} 0 & 1 \\ 1 & 0 \end{pmatrix},$$

$$\sigma_y = \begin{pmatrix} 0 & -\mathrm{i} \\ \mathrm{i} & 0 \end{pmatrix},$$

$$\sigma_z = \begin{pmatrix} 1 & 0 \\ 0 & -1 \end{pmatrix}.$$

The eigenvectors of σ_z are $\binom{1}{0}$ and $\binom{0}{1}$, respectively.

We now find that the orbital angular momentum components l_x, l_y, l_z do not commute with H_{fs}. For example,

$$[l_z, H_{\text{so}}] = \frac{1}{2}\alpha^2\frac{Z}{r^3}[l_z, l_x s_x + l_y s_y + l_z s_z] \neq 0.$$

However, the full Hamiltonian H_{fs} *does* commute with the vector sum of $\boldsymbol{l}$ and $\boldsymbol{s}$:

$$\boldsymbol{j} = \boldsymbol{l} + \boldsymbol{s}. \tag{3.18}$$

This vector sum $\boldsymbol{j}$ also satisfies the angular momentum commutation relations:

$$[j_x, j_y] = \mathrm{i} j_z \tag{3.19}$$

and

$$[\boldsymbol{j}^2, j_x] = [\boldsymbol{j}^2, j_y] = [\boldsymbol{j}^2, j_z] = 0\,.$$

Then, for example,

$$[j_z, H_{fs}] = 0\,.$$

The set of commuting operators is then H_{fs}, j_z, $\boldsymbol{j}^2$, $\boldsymbol{l}^2$, $\boldsymbol{s}^2$, and Π, but *not* l_z or s_z.

The eigenvalues of $\boldsymbol{j}^2$ and j_z are $j(j+1)$ and m_j, where

$$|l - s| \le j \le (l + s)\,, \tag{3.20}$$

with $s = \frac{1}{2}$; j is in unit steps, and $m_j = m_l + m_s$. So in practice $j = l \pm \frac{1}{2}$ with $j = \frac{1}{2}$ if $l = 0$.

The changes in the energy arising from the inclusion of H_{so} can be calculated by perturbation theory and, since α is small, first-order perturbation theory suffices. The energy correction is then

$$\langle \Psi | H_{\mathrm{so}} | \Psi \rangle\,,$$

where Ψ is the "zero-order" wave function, which is the eigenfunction of the Schrödinger Hamiltonian. Then

$$\begin{aligned} \Delta E_n &= \frac{\alpha^2}{2} Z \left\langle \frac{1}{r^3} \right\rangle \langle \boldsymbol{l} \cdot \boldsymbol{s} \rangle \\ &= \frac{1}{2}\alpha^2 Z \left\langle \frac{1}{r^3} \right\rangle \left[\frac{1}{2} \langle \boldsymbol{j}^2 - \boldsymbol{l}^2 - \boldsymbol{s}^2 \rangle \right] \\ &= \frac{1}{2}\alpha^2 Z \left\langle \frac{1}{r^3} \right\rangle \left[\frac{1}{2} \{ j(j+1) - l(l+1) - s(s+1) \} \right]\,, \end{aligned} \tag{3.21}$$

where

$$\left\langle \frac{1}{r^3} \right\rangle = \int_0^\infty [R_{nl}(r)]^2 \frac{1}{r^3} r^2 \mathrm{d}r\,.$$

Hence the separation in energy between the two fine-structure levels corresponding to $j = l \pm \frac{1}{2}$ is

$$\begin{aligned} \delta E_n &= \Delta E_n \left(j = l + \frac{1}{2} \right) - \Delta E_n \left(j = l - \frac{1}{2} \right) \\ &= \frac{1}{2}\alpha^2 Z \left\langle \frac{1}{r^3} \right\rangle \left(l + \frac{1}{2} \right)\,. \end{aligned}$$

The separation is proportional to the larger value of j, which is a specific example of the Landé interval rule.

3.3 Two-Electron Atoms and Ions

Equation (3.18) is an example of the *coupling* of two angular momenta. We meet a similar situation when we come to consider two electrons. If we go back to Schrödinger's equation, then the Hamiltonian is

$$H = -\frac{1}{2}\nabla_1^2 - \frac{1}{2}\nabla_2^2 - \frac{Z}{r_1} - \frac{Z}{r_2} + \frac{1}{r_{12}}. \tag{3.22}$$

Schrödinger's equation can no longer be solved exactly. The difficulty is the final term, since it links the coordinate systems of the two electrons.

There are difficulties from the point of view of the angular momenta also. We find, for example, that

$$[H, l_{1x}] \neq 0.$$

To see this, consider

$$[l_{1x}, f(r_{12}^2)]\psi = l_{1x}(f\psi) - f(l_{1x}\psi) = \psi \cdot (l_{1x}f), \tag{3.23}$$

since l_{1x} is a differential operator. Now

$$\begin{aligned} & l_{1x}f(r_{12}^2) \\ &= -\mathrm{i}\left(y_1\frac{\partial}{\partial z_1} - z_1\frac{\partial}{\partial y_1}\right)[f\{(x_1-x_2)^2 + (y_1-y_2)^2 + (z_1-z_2)^2\}] \\ &= -2\mathrm{i}f'(r_{12}^2)\{y_2z_1 - y_1z_2\}. \end{aligned}$$

Then, by symmetry,

$$l_{2x}f(r_{12}^2) = -2\mathrm{i}f'(r_{12}^2)\{y_1z_2 - y_2z_1\}.$$

Therefore, we find that the sum of l_{1x} and l_{2x} *does* commute with H even though individually the two operators do not. From (3.23),

$$[l_{1x} + l_{2x}, H] = [l_{1x} + l_{2x}, f(r_{12}^2)] = (l_{1x} + l_{2x})f = 0,$$

with $f(r^2) = \dfrac{1}{\sqrt{r^2}} = \dfrac{1}{r}$. A similar argument applies for the other components. Hence the components of

$$\boldsymbol{L} = \boldsymbol{l}_1 + \boldsymbol{l}_2 \tag{3.24}$$

commute with H, and so do the components of the total spin,

$$\boldsymbol{S} = \boldsymbol{s}_1 + \boldsymbol{s}_2. \tag{3.25}$$

The coupling of angular momenta associated with (3.24) and (3.25) is known as *LS* or *Russell–Saunders coupling*.

The wave functions, even though we cannot calculate them exactly, should at least be eigenfunctions of the appropriate angular momentum operators.

Let us suppose that $\psi_{j_1m_1}$ is an eigenfunction of $\boldsymbol{j}_1^2$ and j_{1z}, while $\psi_{j_2m_2}$ is an eigenfunction of $\boldsymbol{j}_2^2$ and j_{2z}. (In this context, we use j to denote a general angular momentum.) Then we can construct linear combinations of their products,

$$\psi_{jm} \equiv \sum_{m_1,m_2} C(j_1j_2j;m_1m_2m)\psi_{j_1m_1}\psi_{j_2m_2}\,, \tag{3.26}$$

so that ψ_{jm} is an eigenfunction of $\boldsymbol{j}^2$ and j_z, where

$$\boldsymbol{j} = \boldsymbol{j}_1 + \boldsymbol{j}_2\,.$$

The coefficients C are known as Clebsch–Gordan coefficients. In this way, we can construct eigenfunctions of $\boldsymbol{L}^2$, L_z and of $\boldsymbol{S}^2$, S_z from the eigenfunctions of $\boldsymbol{l}_i^2$, l_{iz}: [i.e., the spherical harmonics $Y_{l_i}^{m_i}(\theta_i,\phi_i)$] and of $\boldsymbol{s}_i^2$, s_{iz}: [i.e., $\binom{1}{0}$ and $\binom{0}{1}$, which it is customary to denote by α and β]. If, in addition, we wish to describe fine structure, we would need to use *intermediate (or LSJ) coupling* in which the total electronic angular momentum is given by

$$\boldsymbol{J} = \boldsymbol{L} + \boldsymbol{S}\,.$$

There is another requirement of many-electron wave functions: they should be *antisymmetric* with respect to the interchange of any two sets of coordinates.

3.3.1 Wave Functions for Two-Electron Ions

Let us now look at a few specific situations. It is convenient to label the electrons in a way similar to that used for hydrogen: i.e., by (nl). We shall see later that, unlike those for hydrogen, these values of n and l do not correspond to "good" quantum numbers, but they do represent a useful approximation to this situation: in particular n gives a general idea of the region of space in which the electron has the highest probability of being found.

Consider two-electron states in which one electron is described as 1s, while the other is 2s. (We use the notation "s, p, d, f, g, ..." to denote $l = 0,1,2,3,4,\ldots$). Then $l_1 = l_2 = 0$, therefore $L = 0$. But $s_1 = s_2 = \frac{1}{2}$ so we may have $S = 0$ or $S = 1$.

If we use (3.26) and further require that the wave function be antisymmetric, we find that for $S = 0$:

$$\Psi(1,2) = \frac{1}{2}\left[u(1)v(2) + v(1)u(2)\right]\frac{1}{\sqrt{2}}\left[\alpha(1)\beta(2) - \alpha(2)\beta(1)\right]. \tag{3.27}$$

(Notice the change of sign in the replacement $1 \leftrightarrow 2$.) For $S = 1$:

$$\Psi(1,2) = \frac{1}{2}\left[u(1)v(2) - v(1)u(2)\right]\begin{cases}\alpha(1)\alpha(2)\,,\\ \dfrac{1}{\sqrt{2}}\left[\alpha(1)\beta(2) + \alpha(2)\beta(1)\right],\\ \beta(1)\beta(2)\,.\end{cases} \tag{3.28}$$

So for $S = 1$ there are three possible wave functions, which correspond to $M_S = 1, 0, -1$, respectively. Then $S = 0$ is a "singlet" state, and $S = 1$ is a "triplet" state. It is customary to label a state in a way that includes the angular momenta. In LS coupling, the label includes ${}^{2S+1}L$, where $(2S+1)$ takes its numerical value and L is denoted by a letter according to S, P, D, F, G, ... for $L = 0, 1, 2, 3, 4, \ldots$, respectively. In addition, for odd parity states, it is customary to add the superscript "o" following the letter denoting L. If LSJ coupling is used, the J value appears as a subscript to that letter. Hence the singlet and triplet states above are written as ^{1}S, ^{3}S. The 1s2p state with L=1, S=1, J=2 would be written as ${}^3\mathrm{P}^{\mathrm{o}}_2$.

The functions u and v in (3.27) and (3.28) describe the 1s- and 2s-labeled electrons. But suppose now that *both* electrons are labeled as 1s: so $v = u$. Then Ψ for $S = 1$ vanishes identically (so that this state has a zero probability of occurring), whereas that for $S = 0$ does not. In this case, the electrons are said to be *equivalent* (same n and l). Then

$$\Psi({}^1\mathrm{S}) = \mathrm{u}(1)\mathrm{u}(2)\frac{1}{\sqrt{2}}\{\alpha(1)\beta(2) - \alpha(2)\beta(1)\}\,. \tag{3.29}$$

3.3.2 Variational Principle

We now need to consider how the functions u and v might be determined. Suppose $\psi_0, \psi_1, \ldots$ are the complete set of eigenfunctions of the Schrödinger equation based on (3.22) with eigenvalues $E_0, E_1, \ldots$ ordered so that E_0 is the lowest (for a particular L and S). Let $\tilde{\psi}$ be any function. Then formally

$$\tilde{\psi} = \sum_n c_n \psi_n$$

is a complete set expansion, and

$$\begin{aligned}\langle\tilde{\psi}|H|\tilde{\psi}\rangle &= \sum_n \sum_m c_n^* c_m \langle\psi_n|H|\psi_m\rangle \\ &= \sum_n |c_n|^2 E_n \geq E_0 \sum_n |c_n|^2 = E_0 \langle\tilde{\psi}|\tilde{\psi}\rangle\,.\end{aligned}$$

Therefore,

$$\frac{\langle\tilde{\psi}|H|\tilde{\psi}\rangle}{\langle\tilde{\psi}|\tilde{\psi}\rangle} \geq E_0 \tag{3.30}$$

for any $\tilde{\psi}$. That is, the expression on the left of (3.30) will always be greater than the lowest-energy eigenvalue.

Hence we may allow $\tilde{\psi}$ to depend on various parameters, and then vary these parameters, so that the ratio on the left of (3.30) is as low as possible. This may then be a criterion for what should be the "best" choice of $\tilde{\psi}$, since if $\tilde{\psi}$=ψ_0, the equality in (3.30) will hold. The expression (3.30) therefore constitutes a variational principle for the lowest-energy eigenvalue of a given L and S. In particular, for $\tilde{\psi}$ expressed in the form (3.27) or (3.28), we may use this principle to determine the optimal u and v.

3.3.3 Application to the Two-Electron Ground-State Energy

For the ground-state wave function (3.29), there is just one spatial function u. The integrals in (3.30) can be written as

$$\langle\tilde{\psi}|H|\tilde{\psi}\rangle = 2\left\langle u\left|-\frac{1}{2}\nabla^2 - \frac{Z}{r}\right|u\right\rangle \langle u|u\rangle + \left\langle u(1)u(2)\left|\frac{1}{r_{12}}\right|u(1)u(2)\right\rangle$$

and

$$\langle\tilde{\psi}|\tilde{\psi}\rangle = \langle u|u\rangle^2 .$$

Perhaps the simplest approximation for u is

$$u = (4\alpha^3)^{\frac{1}{2}} \mathrm{e}^{-\alpha r} Y_0^0(\theta, \phi) , \tag{3.31}$$

which has the same general form as the one-electron function (3.6). Then $\langle u|u\rangle = 1$, and

$$2\left\langle u\left|-\frac{1}{2}\nabla^2 - \frac{Z}{r}\right|u\right\rangle = \alpha^2 - 2Z\,\alpha ,$$

$$\left\langle u(1)u(2)\left|\frac{1}{r_{12}}\right|u(1)u(2)\right\rangle = \frac{5}{8}\alpha ,$$

so that

$$\frac{\langle\tilde{\psi}|H|\tilde{\psi}\rangle}{\langle\tilde{\psi}|\tilde{\psi}\rangle} = \alpha^2 - 2\alpha Z + \frac{5}{8}\alpha \equiv f(\alpha) .$$

Minimizing $f(\alpha)$ leads to

$$\alpha_{\mathrm{opt}} = Z - \frac{5}{16} ,$$

and then

$$f(\alpha_{\mathrm{opt}}) = -\left(Z - \frac{5}{16}\right)^2$$

is the optimal value of E_0 for a trial function given by (3.29) and (3.31). Hence, for neutral helium, the lowest-energy state (ground state), usually written as $(1\mathrm{s}^2)^1\mathrm{S}$, has energy

$$-\left(2 - \frac{5}{16}\right)^2 = -2.8476 \text{ a.u.}$$

The optimal exponent is $\alpha = \frac{27}{16} = 1.6875$. Had we chosen the hydrogenic value of $\alpha = Z = 2$, then the energy would have been –2.75 a.u.

Hence the use of the variational principle (3.30) has resulted in a much lower (and therefore much improved) energy. The difference, $\frac{5}{16}$, between Z and the optimal exponent is known as the *screening* constant. In effect, instead of each electron seeing the full nuclear charge Z, the nuclear charge is screened in part by the other electron, so that effectively each electron sees only a charge of $(Z - \frac{5}{16})$.

We have taken in (3.31) a very simple approximation for u. In the next section we take more terms of the same form:

$$u = \left(\sum_j c_j r^{I_j} \mathrm{e}^{-\zeta_j r} \right) \times Y_0^0(\theta, \phi) .$$

The best $1s^2$ energy is then –2.86168 a.u.

3.4 Many-Electron Atoms and Ions

Our discussion in the previous section of two-electron systems was limited to very simple forms of the wave functions. Other, still quite simple forms (see [3.1]) can lead to much improved energy values. But the forms (3.27) and (3.28) can be considered as special cases of a type of wave function which can be readily determined for atoms and ions with many electrons, and which provides the starting point for many state-of-the-art calculations currently being undertaken.

3.4.1 Hartree's Method

The spin-dependent factors in (3.27) and (3.28) contribute a factor of unity to both integrals appearing in the variational principle (3.30). This factorization of the space and spin parts of the wave function is not a property that carries over from two- to many-electron systems. Hartree [3.4] in effect proposed that a wave function form be constructed which retains this factorization. He suggested writing the total N-electron wave function $\tilde{\psi}$ as a product of one-electron functions u_i, each depending on the coordinates of just one electron:

$$\tilde{\psi}(1, 2, .., N) = u_1(1)u_2(2)...u_N(N) , \tag{3.32}$$

so that

$$|\tilde{\psi}|^2 = |u_1|^2|u_2|^2...|u_N^2| . \tag{3.33}$$

Recall that the probability of a sequence of independent events is the product of their separate probabilities. Then (3.33) is implying that Hartree's model supposes each electron to be moving independently in the field of the other electrons; that field for electron i being given by the potential

$$V_i(\boldsymbol{r}_i) = \sum_{j \neq i} \int \frac{|u_j(\boldsymbol{r}_j)|^2}{r_{ij}} \mathrm{d}\tau_j \,. \tag{3.34}$$

In order to solve the corresponding Schrödinger-type equation, Hartree made the further approximation of replacing $V_i(\boldsymbol{r}_i)$ by its spherical average,

$$V_i(r_i) = \int V_i(\boldsymbol{r}_i) \mathrm{d}\Omega_i \,, \tag{3.35}$$

with the resulting set of equations

$$\left(-\frac{1}{2}\nabla_i^2 - \frac{Z}{r_i} \right) u_i + V_i(r_i) u_i = \varepsilon_i u_i \tag{3.36}$$

for the one-electron functions u_i. Based on the discussion of (3.8), the spherically symmetric nature of the potential leads immediately to the requirement that the angular dependence of u_i is a spherical harmonic.

3.4.2 The Hartree–Fock Method

The basic flaw in Hartree's method is that the wave function (3.32) is not antisymmetric with respect to interchange of two coordinate sets (i) and (j). This problem was overcome by Fock [3.5] by replacing the simple product of one-electron functions (3.32) by an antisymmetrized product, which can be written as the determinant. For example, the wave function (3.29) can be written as a determinant:

$$\Psi(1\mathrm{s}^2\ {}^1\mathrm{S}) = \frac{1}{\sqrt{2}} \begin{vmatrix} u(1)\alpha(1) & u(1)\beta(1) \\ u(2)\alpha(2) & u(2)\beta(2) \end{vmatrix} \,. \tag{3.37}$$

The antisymmetric character follows from the properties of determinants. Moreover (3.27) and (3.28) can also be written as linear combinations of determinants:

$$\Psi(1\mathrm{s}2\mathrm{s}\ {}^{3,1}\mathrm{S}) = \frac{1}{2\sqrt{2}} \left[\begin{vmatrix} u(1)\alpha(1) & v(1)\beta(1) \\ u(2)\alpha(2) & v(2)\beta(2) \end{vmatrix} \pm \begin{vmatrix} u(1)\beta(1) & v(1)\alpha(1) \\ u(2)\beta(2) & v(2)\alpha(2) \end{vmatrix} \right] \tag{3.38}$$

for $M_s = 0$.

In general, we can extend this to any number of electrons. If we write the wave functions in terms of one-electron functions

$$\phi_i(\boldsymbol{r}_i, m_{si}) = u(i)[\alpha(i) \text{ or } \beta(i)] \,, \tag{3.39}$$

then, when a single determinant is sufficient to give eigenfunctions of the angular momentum operators, the wave function will take the form

$$\Psi(LS) = \frac{1}{\sqrt{N!}} \begin{vmatrix} \phi_1(1) & \cdots & \phi_N(1) \\ \vdots & \ddots & \vdots \\ \phi_1(N) & \cdots & \phi_N(N) \end{vmatrix} . \tag{3.40}$$

In general a sum of determinants, as in (3.38), will be required to ensure an eigenfunction of the angular momentum operators.

By the properties of determinants, we may add any multiple of one column to another, so we may choose $\langle \phi_i | \phi_j \rangle = 0$ if $i \neq j$, and we can normalize for the ϕ_i so that $\langle \phi_i | \phi_i \rangle = 1$. Therefore we may require

$$\langle \phi_i | \phi_j \rangle = \delta_{ij} . \tag{3.41}$$

The normalization factor $1/\sqrt{N!}$ in (3.40) then ensures that $\langle \Psi | \Psi \rangle = 1$.

The functions $\{\phi_i\}$ are known as *spin orbitals* or sometimes simply *orbitals*. It is customary to write the $\{u_i\}$ as a product of a radial function and a spherical harmonic:

$$u_i(\boldsymbol{r}) = R_{n_i l_i}(r) Y_{l_i}^{m_{l_i}}(\theta, \phi) \equiv \frac{1}{r} P_{n_i l_i}(r) Y_{l_i}^{m_{l_i}}(\theta, \phi) . \tag{3.42}$$

This simple product expression (3.42) is a restriction that was not present in Fock's formalism, but it maintains the form used by Hartree. In addition, since the radial functions are independent of m_l and m_s, the number of unknown functions – now only the $P_{n_i l_i}(r)$ – is substantially reduced.

The orbitals $\{u_i\}$ (or more specifically the radial functions $P_{n_i l_i}$) are determined by requiring that $\langle \Psi | H | \Psi \rangle$ be minimized subject to the condition $\langle \Psi | \Psi \rangle = 1$, or, equivalently (3.41), where the Hamiltonian is given by (3.2).

This is a calculus of variations problem in which the first-order change

$$\delta \left[\langle \Psi | H | \Psi \rangle - \sum_{i,j} \varepsilon_{ij} \langle u_i | u_j \rangle \right] ,$$

with respect to small changes in $\{u_i\}$, is set to zero. Formally, it leads to a set of coupled integro-differential equations [3.6]:

$$\begin{aligned} &\left(-\frac{1}{2} \nabla_i^2 - \frac{Z}{r_i} \right) u_i(\boldsymbol{r}_i) \\ &\quad + \sum_{j \neq i} \left[u_i(\boldsymbol{r}_i) \int \frac{|u_j(\boldsymbol{r}_j)|^2}{r_{ij}} \mathrm{d}\tau_j - \delta(m_{s_i}, m_{s_j}) u_j(\boldsymbol{r}_i) \int \frac{u_j^*(\boldsymbol{r}_j) u_i(\boldsymbol{r}_j)}{r_{ij}} \mathrm{d}\tau_j \right] \\ &= \varepsilon_i u_i(\boldsymbol{r}_i) , \end{aligned} \tag{3.43}$$

where $\varepsilon_i = \varepsilon_{ii}$. This process is known as the Hartree–Fock (HF) method, and (3.43) are the Hartree–Fock equations for the wave function (3.40). Because of (3.42), we could calculate the integrals over the angles, and obtain corresponding equations for $\{P_{n_i l_i}(r)\}$.

In (3.43), the first term corresponds to the sum of the kinetic energy and the nucleus-electron potential energy. The remaining terms on the left-hand-side correspond to the electron–electron interaction. Of these, the first is of the form $u_i(\boldsymbol{r}_i)V(\boldsymbol{r}_i)$ and so is a *local* potential energy term. It is known as the *direct term* and physically corresponds to the potential associated with the electron charge density of all the other electrons. But in the second term, we have a multiple of $u_j(\boldsymbol{r}_i)$, not $u_i(\boldsymbol{r}_i)$, and so we do not now have a multiplicative potential. Formally we can still write (3.43) as

$$\left[-\frac{1}{2}\nabla_i^2 - \frac{Z}{r_i} + V'\right] u_i(\boldsymbol{r}_i) = \varepsilon_i u_i(\boldsymbol{r}_i) \, ,$$

i.e., as a Schrödinger-type equation, but because of the last term on the left-hand-side in (3.43), the potential V' is *non-local*. This last term is known as the *exchange term*. It corresponds physically to the possibility of interchanging the indistinguishable electrons. It is a characteristic feature of the antisymmetry of the wave function. Hartree's equations (3.36), for which the wave function is the simple (rather than antisymmetrized) product of the orbitals (3.32), does not contain this term. The parameter ε_i is known as the orbital energy.

In practice, the equations (3.43) – or their equivalents for $P_{n_i l_i}$ – must be solved iteratively. The simplest form of this process consists of the following tasks.

1. Choose an initial guess for $\{u_i\}$.
2. Construct the integrals in (3.43) and solve the resulting (eigenvalue) differential equations for a new set of $\{u_i\}$.
3. Recompute the integrals and repeat until a satisfactory degree of convergence – or *self-consistency* – is reached.

This process is known as the *self-consistent field* (SCF) *method*. It can be approached in two ways:

(a) numerically; in which the equations are solved numerically, yielding numerical tabulations of $P_{nl}(r)$, typically at equal intervals of $\log_e r$ to ensure a larger number of tabulation points where the functions are varying most rapidly. The most widely used computer program to generate such functions is that of Froese Fischer [3.7].

(b) analytically; for example, by representing $P_{nl}(r)$ as a linear combination of analytic basis functions:

$$P_{nl}(r) = \sum_{j=1}^{k} c_{jnl} r^{I_{jnl}} \exp(-\zeta_{jnl} r) \tag{3.44}$$

or

$$P_{nl}(r) = \sum_{j=1}^{k} c'_{jnl} \chi_{jnl}(r) \, ,$$

where χ_j is a normalized Slater-type orbital (STO) of the form

$$\chi_{jnl}(r) = \left[\frac{(2\zeta_{jnl})^{2I_{jnl}+1}}{(2I_{jnl})!}\right]^{\frac{1}{2}} r^{I_{jnl}} \exp(-\zeta_{jnl} r) \,. \tag{3.45}$$

The $\{\zeta_{jnl}\}$ are treated as variational parameters, and the equations (3.43) are transformed into algebraic equations for the $\{c_i\}$ which are solved self-consistently. A number of tabulations of the parameters in (3.45) for HF orbitals exist in the literature. The largest collection is that of Clementi and Roetti [3.8], mostly for the lowest-energy states of atoms and ions with $Z \leq 54$. These have been extended to higher Z by McLean and McLean [3.9], and by Snijders *et al* [3.10]. A new evaluation of these parameters has been reported by Koga *et al* [3.11]. More specific sets of orbital parameters exist for the ground and excited states of Li [3.12] and for ions of C,N,O [3.13]. For the exercises given at the end of the chapter, we include the parameters provided by Tatewaki *et al* [3.21] for Be-like ions.

3.5 Configuration Interaction Methods

Although we can use (3.44) instead of a very simple, single STO form such as (3.31) and thereby lower the energy, we are still constrained by the form of the HF wave function, which consists of a single determinant (or a linear combination in order to ensure an eigenfunction of the angular momentum operators) with the elements as one-electron orbitals, to each of which is associated a *single* spherical harmonic. We now wish to consider how to lift such restrictions.

Before we consider the general problem, let us explore the two-electron case and treat the HF method as a first approximation. The ground state of helium is usually denoted by $(1s^2)^1S$. This is really a representation of the HF approximation. The individual l_i of the electrons are coupled to give L, where

$$|l_1 - l_2| \leq L \leq (l_1 + l_2) \,.$$

We can therefore obtain $L = 0$ by having $l_1 = l_2 = 0$. But we can also obtain $L = 0$ for any values of l_1 and l_2 for which $l_1 = l_2$. So, $2p^2$ or 3d4d are equally possible from the point of view of the angular momenta. The labels "2", "3", and "4" may simply serve to distinguish between radial functions of the same l, rather than indicate the size of their mean radii (as in the case of the HF orbital functions). But still the angular momenta of these orbitals must couple to form ^1S.

So a more flexible form of wave function for the state labeled $(1s^2)^1S$ in HF would be

$$\begin{aligned}\Psi(^1\mathrm{S}) = & a_1\Phi_1(1\mathrm{s}^2\ {}^1\mathrm{S}) + a_2\Phi_2(1\mathrm{s}2\mathrm{s}\ {}^1\mathrm{S}) + a_3\Phi_3(2\mathrm{s}^2\ {}^1\mathrm{S}) + \ldots \\ & + a_4\Phi_4(2\mathrm{p}^2\ {}^1\mathrm{S}) + a_5\Phi_5(2\mathrm{p}3\mathrm{p}\ {}^1\mathrm{S}) + \ldots \\ & + a_6\Phi_6(3\mathrm{d}^2\ {}^1\mathrm{S}) + \ldots\end{aligned}$$

Similarly for the $(1\mathrm{s}2\mathrm{p})^1\mathrm{P}^{\mathrm{o}}$ state:

$$\begin{aligned}\Psi(^1\mathrm{P}^{\mathrm{o}}) = & b_1\Phi_1(1\mathrm{s}2\mathrm{p}\ {}^1\mathrm{P}^{\mathrm{o}}) + b_2\Phi_2(1\mathrm{s}3\mathrm{p}\ {}^1\mathrm{P}^{\mathrm{o}}) + \ldots + b_3\Phi_3(2\mathrm{s}2\mathrm{p}\ {}^1\mathrm{P}^{\mathrm{o}}) \\ & + b_4\Phi_3(2\mathrm{s}3\mathrm{p}\ {}^1\mathrm{P}^{\mathrm{o}}) + \ldots + b_5\Phi_5(2\mathrm{p}3\mathrm{d}\ {}^1\mathrm{P}^{\mathrm{o}}) + \ldots ,\end{aligned}$$

where we now have, in all cases, $|l_1 - l_2| = 1 = L$.

The assignments of the electrons to orbitals (1s, 2p, ...), together with the angular momentum coupling, are known as *configurations*, and the process for finding the coefficients $\{a_i\}$, $\{b_i\}$ as well as the radial functions involved in these one-electron orbitals is known as the method of *configuration interaction* (CI). The $\{\Phi_i\}$ are often called *configuration state functions* (CSF).

These wave functions can be written in the general form

$$\Psi(LS) = \sum_{i=1}^{M} a_i\Phi_i(\alpha_i LS) , \qquad (3.46)$$

where $\{\alpha_i\}$ denote all the distinguishing features of Φ_i other than L and S. The summation in (3.46) should in theory be infinite, but in practice a finite value has to be used.

3.5.1 Optimization of the Expansion Coefficients

We shall first of all see how to determine the expansion coefficients $\{a_i\}$. If we use (3.46) in a variational principle and assume the $\{\Phi_i\}$ (including their radial functions) are fixed, and hence that the only variational parameters are the $\{a_i\}$, then

$$\delta[\langle\Psi|H|\Psi\rangle - E\{\langle\Psi|\Psi\rangle - 1\}] = 0$$

becomes

$$0 = \delta\left[\sum_i\sum_j a_i a_j \langle\Phi_i|H|\Phi_j\rangle - E\left\{\sum_i\sum_j a_i a_j\langle\Phi_i|\Phi_j\rangle - 1\right\}\right] .$$

Therefore,

$$0 = \sum_j a_j\left(H_{ij} - E\delta_{ij}\right) , \qquad (3.47)$$

assuming $\langle\Phi_i|\Phi_j\rangle = \delta_{ij}$, and we have written

$$H_{ij} = \langle\Phi_i|H|\Phi_j\rangle . \qquad (3.48)$$

Then the possible values for the Lagrange multipliers E are the eigenvalues of the *Hamiltonian matrix* with general element H_{ij}, and the $\{a_i\}$ are the components of the corresponding eigenvector.

3.5.2 Optimization of the CSF

We can derive an alternative form of the variational principle, which allows us to optimize the CSF – specifically, the radial functions of the one-electron orbitals. To do this, we need to use the *Hylleraas–Undheim theorem* [3.14]. Let

$$E_1^{(M)} < E_2^{(M)} < \ldots < E_M^{(M)}$$

be the M-ordered eigenvalues of the Hamiltonian matrix arising from (3.46). Suppose we add one further configuration function Ψ_{M+1}. The new eigenvalues are then (again ordered)

$$E_1^{(M+1)} < E_2^{(M+1)} < \ldots < E_{M+1}^{(M+1)} .$$

The Hylleraas–Undheim theorem gives

$$\ldots E_{i-1}^{(M)} < E_i^{(M+1)} < E_i^{(M)} < E_{i+1}^{(M+1)} \ldots , \tag{3.49}$$

i.e., the eigenvalues for the $(M+1)$ case interleave those for M.

Now consider adding more configurations. Then (3.49) implies that

$$E_i^{(M)} > E_i^{(M+1)} > E_i^{(M+2)} > \ldots$$

If we were to extend the expansion (3.46) to a complete set, we would obtain

$$E_i^{(M)} > E_i^{\text{exact}} . \tag{3.50}$$

We have already seen that (3.30) provides a variational principle for the lowest-energy state for each LS symmetry. As well as giving a physical meaning to the eigenvalues of the matrix (as upper bounds to the corresponding energies), (3.50) provides a variational principle for any excited state and in particular for the orbitals on which it depends. It has been customary to assume that, as with the HF method, the orbitals form an orthonormal set, although some lifting of that restriction has recently been investigated and used in configuration interaction calculations [3.15]. The calculated eigenvalue $E_i^{(M)}$ will depend on the radial functions used in $\{\Phi_i\}$. If we treat the parameters of these functions as variational parameters, we can use any eigenvalue as the variational functional. So we can optimize these parameters to give as good a representation as possible for any atomic state.

This process is employed in the program `CIV3` [3.16] which is associated with this chapter. This code therefore uses an analytic representation of the radial functions as in (3.44). Alternatively, (3.50) could be used to set up integro-differential equations for the radial functions, in a manner similar to the derivation of the HF equations. This is the basis of the multi-configurational Hartree–Fock (MCHF) method. Again the `MCHF` code of Froese Fischer [3.17] has been very widely used. As with her HF code

[3.7], the radial functions are determined numerically, by solving the integro-differential equations self-consistently. One major difference between the HF and MCHF methods is the need in the latter to solve the secular equation (3.47) at each stage of the iteration.

3.5.3 Correlation Energy

If we include the Hartree–Fock configuration as Φ_1 in (3.46), then we will do at least as well as the HF approximation. In practice, we will do better. The improvement in energy is known as the *correlation energy*:

$$E^{\mathrm{corr}} = E^{\mathrm{exact}} - E^{\mathrm{HF}} . \tag{3.51}$$

For the ground state of helium, the energies (in a.u.) are

$$\begin{aligned} E^{\mathrm{HF}} &= -2.86168 , \\ E^{\mathrm{exact}} &= -2.90372 , \\ E^{\mathrm{corr}} &= -0.04204 . \end{aligned}$$

It is interesting to observe how this is achieved as more and more configurations are included. Let us write, in keeping with (3.46):

$$\begin{aligned} \Psi(1\mathrm{s}^2\,{}^1\mathrm{S}) = {} & \sum_{m,n} a_{mn}\Phi(msns) + \sum_{m,n} b_{mn}\Phi(mpnp) \\ & + \sum_{m,n} c_{mn}\Phi(mdnd) + \sum_{m,n} d_{mn}\Phi(mfnf) + \dots . \end{aligned} \tag{3.52}$$

Then the cumulative contribution to E^{corr} is given in Table 3.1.

Table 3.1. Energy of the He $(1\mathrm{s}^2)^1\mathrm{S}$ ground state in configuration interaction expansions.

	Bunge [3.18]	up to m, n=6
s limit	-2.87903	-2.87900
+ p limit	-2.90051	-2.90045
+ d limit	-2.90275	-2.90270
+ f limit	-2.90331	-2.90323
+ g limit	-2.90347	-2.90339
⋮	⋮	⋮
exact (non-rel.)	-2.90372	

Each row contains the cumulative effect of adding the summations in (3.52). The calculations of Bunge [3.18] are very extensive, including configurations with $m, n \leq 10$. A less extensive set of calculations is recommended as one of the exercises at the end of this chapter. If we take orbitals up to $m, n = 6$, then the final column of results is obtained. Two features emerge:

(a) convergence with respect to m, n is reasonably fast;
(b) convergence with respect to l is comparatively slow.

3.5.4 Choice of Configurations

In the case of the ground state of helium, the type of configurations possible is fairly straightforward to determine. As the number of electrons increases, the choice of configurations becomes more of an art. It is convenient to classify them, and we follow the scheme of Sinanoğlu and coworkers [3.19].

Let us consider the specific example of carbon: the HF configuration for the ground state is $(1s^2 2s^2 2p^2)^3P$. There are three types of *correlation effect.*

1. *Internal correlation*
   The configurations are constructed by using the set of orbitals (known as the *Hartree–Fock sea*) that are either occupied in the HF configuration or have the same n value as those which are. In the case of carbon, since the total parity must be preserved, we would include the configuration $(1s^2 2p^4)^3P$, obtained from the HF configuration by the replacement $2s^2 \rightarrow 2p^2$. In principle, the replacement $1s^2 \rightarrow 2p^2$ would also be possible, giving the configuration $(2s^2 2p^4)^3P$. However, since the shell model of the atom is fairly good, at least in the sense that the mean radii of the radial functions describing the shells are quite different, this configuration would have a small coefficient a_i in (3.46), and it would normally be omitted if only fairly simple wave functions were required.
2. *Semi-internal correlation*
   This type of correlation is included for an N-electron atom or ion by using configurations constructed so that $(N-1)$ orbitals are from the HF sea and one is from outside it. For example, for the ground state of carbon, configurations could include $(1s^2 2s3d2p^2)^3P$, corresponding to a straight replacement $2s \rightarrow 3d$ from the HF configuration, or $(1s^2 2p^3 4f)^3P$ which involves a different choice of orbitals from the HF sea from that of the HF configuration.
3. *All-external correlation*
   The configurations associated with this type of correlation contain two or more orbitals from outside the HF sea. For example, for the ground state of carbon, they could include $(1s^2 2s^2 3p^2)^3P$ or $(1s^2 2p^2 3p^2)^3P$.

Hence a fairly simple CI wave function for the ground state of carbon might be

$$\begin{aligned}\Psi(^3P) = {} & a_1\Phi_1(1s^2 2s^2 2p^2) + a_2\Phi_2(1s^2 2p^4) + a_3\Phi_3(1s^2 2s3d2p^2) \\ & + a_4\Phi_4(1s^2 2p^3 4f) + a_5\Phi_5(1s^2 2s^2 3p^2) + a_6\Phi_6(1s^2 2s^2 3d^2) \\ & + a_7\Phi_7(1s^2 2p^2 3s^2) + a_8\Phi_8(1s^2 2p^2 3p^2) + a_9\Phi_9(1s^2 2p^2 3d^2) \\ & + a_{10}\Phi_{10}(1s^2 2s2p3s3p)\,. \end{aligned} \tag{3.53}$$

In fact, this wave function is more extensive than it might seem. For example, in Φ_3, three possible couplings are allowed: $[1s^2(2s3d)^{1,3}D\ 2p^2(^3P)]^3P$ and $[1s^2(2s3d)^3D\ 2p^2(^1D)]^3P$. The expansion (3.53) includes only the HF sea together with the orbitals 3s, 3p, 3d, 4f. Comparison with the case of the ground state of helium discussed in the previous subsection indicates that such an expansion would give an energy that is far from converged with respect either to n or to l. If energies converged to several decimal places are desired, then it is necessary to include orbitals (nl) with high values of n and l. Bearing in mind that each assignment of electrons can be associated with several sets of angular momentum couplings (as in Φ_3 above), the size of the CI expansion (3.46) can grow rapidly. Current state-of-the-art atomic structure calculations routinely include several thousand or even several tens of thousands of CSF, even for light atoms and ions.

However, if one wishes to consider properties such as oscillator strengths (see the next section) which involve the *difference* between two energies rather than the two energies separately, an accurate value for this difference can frequently be obtained with fewer CSF than would be needed for the energies individually. Moreover, the dominant terms in (3.46), i.e., those with the largest coefficients $\{a_i\}$, come from internal and semi-internal correlation effects. The former necessarily give rise to only a few configurations. The latter can also be represented by limited CI. If the radial functions of the orbitals outside the HF sea are determined variationally, normally only one is needed for each l value (or perhaps one per shell for each l value), because for semi-internal correlation configurations these orbitals occur only in a linear way. It is the all-external correlation which gives rise to the large numbers of orbitals and configurations and the slow convergence of the energies. For many calculations, particularly of oscillator strengths, much of the physics can be obtained by including the internal and semi-internal correlation fully and only a limited amount of all-external correlation.

3.5.5 Calculation of Several States Simultaneously

Up to now we have been concerned with the calculation of a wave function for a single state. In many situations, wave functions for several states are required in a single calculation. An obvious example is the calculation of the oscillator strength of an atom undergoing a transition from one state to another. In other cases, there may be two or more states of the same LS symmetry required.

Although the CI expansion (3.46) for an isolated state may have one dominant term (Φ_1, the HF configuration) with $a_1 \simeq 0.9$, the CI expansion of states which have the same LS symmetry might involve two or more configurations with $|a_i| \geq 0.5$. For example, the two ^2D states ($1s^22s^22p^63s3p^2$ and $1s^22s^22p^63s^23d$) of Si II interact strongly [3.20], giving rise to a CI expansion (even with just these two configurations included) for the former state as

$$\Psi(3s3p^2) = 0.82\Phi_1(3s3p^2) - 0.57\Phi_2(3s^23d)\,.$$

In undertaking calculations that involve several states, it is necessary to include correlation effects to the same extent in each state. This can be difficult to achieve, but the optimization of different orbital radial functions on the energy eigenvalues of different states, by using (3.50), greatly facilitates this.

In our suggested exercises at the end of the chapter, we will not include calculations involving strongly interacting states, since rather extensive calculations are needed if the results are to be of useful accuracy. However, we will include the calculation of oscillator strengths, and the importance of this flexibility in optimizing the radial functions over the two states of the transitions will become apparent.

3.6 Transition Probabilities and Oscillator Strengths

So far we have considered only the calculation of wave functions and hence energies. We now turn to the calculation of another atomic property: the probability that an atom undergoes a transition from one state to another.

Atoms can change state either by losing energy (emission) or by gaining energy (absorption). We shall first consider the *spontaneous* emission of radiation.

If in an assembly of atoms, N_j is the number of atoms in state $|j\rangle$, then the rate of decay (reduction in N_j) will be proportional to N_j. The constant of proportionality A_{ji} for a transition to state $|i\rangle$ is known as the transition probability:

$$\frac{\mathrm{d}N_j}{\mathrm{d}t} = -A_{ji}N_j\,, \tag{3.54}$$

with the solution

$$N_j(t) = N_j(0)\exp(-A_{ji}t)\,. \tag{3.55}$$

From (3.54), A_{ji} has the dimension of 1/time; strictly, A_{ji} is the transition probability per unit time. The mean *lifetime* of the state $|j\rangle$ is then

$$\tau_j = \frac{1}{A_{ji}}\,. \tag{3.56}$$

Sometimes, an atom can decay into one of a number of states. Then (3.54–56) are replaced by

$$\frac{\mathrm{d}N_j}{\mathrm{d}t} = -\left(\sum_i A_{ji}\right)N_j\,, \tag{3.57}$$

$$N_j(t) = N_j(0) \exp\left(-\left\{\sum_i A_{ji}\right\} t\right), \tag{3.58}$$

and

$$\tau_j = \frac{1}{\sum_i A_{ji}}. \tag{3.59}$$

The sum over i is over all the accessible lower states.

An atom can also emit or absorb radiation because of outside influences. Einstein's theory of radiation then introduces two further coefficients associated with induced energy changes: B_{ji}, the induced emission probability; B_{ij}, the induced absorption probability. Then (3.54) is replaced by

$$\frac{\mathrm{d}N_j}{\mathrm{d}t} = -A_{ji}N_j - B_{ji}\varrho(\nu)N_j + B_{ij}\varrho(\nu)N_i, \tag{3.60}$$

where $\varrho(\nu)$ is the energy (frequency) density of the radiation field.

If ν_{ji} is the frequency of the emitted radiation, so that

$$h\nu_{ji} \equiv \hbar\omega_{ji} = E_j - E_i, \tag{3.61}$$

then according to Einstein's theory

$$A_{ji} = \frac{8\pi}{c^3}\nu_{ji}^2 h\nu_{ji} B_{ji}. \tag{3.62}$$

The individual emission and absorption coefficients are also related. The energy of a particular state depends on L and S but not on M_L and M_S, so there are $g = (2L+1) \times (2S+1)$ wave functions (the number of M_L, M_S values) all with the same energy; g is known as the statistical weight of the state. For LSJ coupling, $g = (2J+1)$. Then,

$$g_j B_{ji} = g_i B_{ij}. \tag{3.63}$$

A very convenient alternative measure of the transition probability is the *(absorption) oscillator strength*

$$f_{ij} = \frac{\mu}{\pi e^2} B_{ij} h\nu_{ji}, \tag{3.64}$$

so that $f_{ij} > 0$, $|i\rangle$ being the state with the lower energy. Then by (3.62) and (3.63)

$$g_i f_{ij} = -g_j f_{ji}. \tag{3.65}$$

Also, by using (3.62),

$$g_j A_{ji} = \frac{8\pi}{c^3}\nu_{ji}^2 h\nu_{ji} g_j B_{ji} = \frac{8\pi}{c^3}\nu_{ji}^2 h\nu_{ji} g_i B_{ij} = \frac{8\pi^2 e^2}{\mu c^3}\nu_{ji}^2 g_i f_{ij}. \tag{3.66}$$

We will concern ourselves solely with what are termed *electric dipole transitions*. We shall not derive the expressions for f_{ij} or the Einstein coefficients. We simply note that the radiation field can be expressed as a sum of Fourier components of the form

$$\boldsymbol{A}_0 \cos(\omega t - \boldsymbol{k} \cdot \boldsymbol{r}),$$

where ω is the angular frequency associated with the transition (ω_{ji}) and $\boldsymbol{k}$ is the associated wave vector with magnitude

$$k = \frac{2\pi}{\lambda_{ji}}. \tag{3.67}$$

The radiation field is treated as a small perturbation, so time-dependent perturbation theory is required, in which the total wave function is expanded in terms of the unperturbed (free-atom) wave functions:

$$\Psi = \sum_n c_n(t)\psi_n \exp\left[-\frac{i}{\hbar}E_n t\right].$$

Then for a one-electron atom, the transition probability per unit time for a transition from state $|j\rangle$ to state $|i\rangle$ is

$$\frac{1}{t}\left|c_i(t)\right|^2 = B_{ij}\varrho(\nu_{ji})$$

and

$$B_{ij} = \frac{2\pi}{3}\frac{c^2}{h^2\nu_{ji}^2}\left|\left\langle \psi_j \middle| \frac{e}{\mu c}\boldsymbol{p}\,\mathrm{e}^{\mathrm{i}\boldsymbol{k}\cdot\boldsymbol{r}} \middle| \psi_i \right\rangle\right|^2, \tag{3.68}$$

where $\boldsymbol{p}$ is the momentum of the electron. (For a many-electron atom, the operator consists of a sum of the operators in (3.68), one for each electron.)

3.6.1 Dipole Approximation

Let us expand

$$\mathrm{e}^{\mathrm{i}\boldsymbol{k}\cdot\boldsymbol{r}} = 1 + \mathrm{i}\boldsymbol{k}\cdot\boldsymbol{r} + \cdots$$

and evaluate the mean value of the second term. It is of order

$$k\langle r\rangle = \frac{2\pi}{\lambda_{ji}}\langle r\rangle,$$

where $\langle r\rangle$ is the mean radius of the electron. Typically, $\langle r\rangle$ is of the order of a few Å (1 a.u. of length $\simeq$ 0.529 Å), whereas λ_{ji} is a few hundred or even a few thousand Å. So $\langle \boldsymbol{k}\cdot\boldsymbol{r}\rangle \ll 1$, and to a good approximation we can replace $\mathrm{e}^{\mathrm{i}\boldsymbol{k}\cdot\boldsymbol{r}}$ by 1. This is known as the *dipole approximation*.

Combining (3.64) and (3.68) with the dipole approximation, we have

$$f_{ij} = \frac{2\mu c^2}{3e^2} \frac{1}{h\nu_{ji}} \left| \left\langle \psi_j \left| \frac{e}{\mu c} \boldsymbol{p} \right| \psi_i \right\rangle \right|^2 . \tag{3.69}$$

The oscillator strength f is a dimensionless quantity, so we use atomic units, giving

$$f_{ij}^{\mathrm{v}} = \frac{2}{3} \frac{1}{\Delta E} \left| \langle \psi_j | \boldsymbol{p} | \psi_i \rangle \right|^2 = \frac{2}{3} \frac{1}{\Delta E} \left| \langle \psi_j | \nabla | \psi_i \rangle \right|^2 , \tag{3.70}$$

where $\Delta E = E_j - E_i$ is the *transition energy*.

3.6.2 Length and Velocity Forms

If H is the free-atom (Schrödinger) Hamiltonian then in atomic units

$$[\boldsymbol{r}, H] = \mathrm{i}\, \dot{\boldsymbol{r}} = \mathrm{i}\, \boldsymbol{p} . \tag{3.71}$$

Therefore

$$\begin{aligned} \langle \psi_j | \boldsymbol{p} | \psi_i \rangle &= -\mathrm{i} \langle \psi_j | \boldsymbol{r} H - H \boldsymbol{r} | \psi_i \rangle \\ &= -\mathrm{i} [(E_i - E_j) \langle \psi_j | \boldsymbol{r} | \psi_i \rangle] = \mathrm{i} \Delta E \langle \psi_j | \boldsymbol{r} | \psi_i \rangle . \end{aligned} \tag{3.72}$$

Hence (3.70) may also be written as

$$f_{ij}^{\mathrm{l}} = \frac{2}{3} \Delta E \left| \langle \psi_j | \boldsymbol{r} | \psi_i \rangle \right|^2 . \tag{3.73}$$

Expressions (3.70) and (3.73) are referred to as the *velocity* and the *length* forms of the oscillator strength, respectively. Their equality is valid whenever (3.71) applies and the wave functions are eigenfunctions of the Hamiltonian H.

For many-electron systems, the exact wave functions are generally not known. But it is possible for both these conditions still to be satisfied, for example when the functions ψ_i and ψ_j are eigenfunctions of the same approximate and local Hamiltonian (as for Hartree's method). However, since this is an approximation, the agreement between f^{l} and f^{v} does not imply that their common value is correct.

In the HF approximation, (3.71) is not satisfied because of the non-local potential. In general, the two forms will not agree. But in a CI framework, they should converge to the correct result as the number of configurations in the expansions (3.46) increases.

3.6.3 The Effect of Electron Correlation

The choice of configurations in a CI calculation of oscillator strengths must be made carefully. A large calculation is in itself no guarantee of accuracy. In general, it is essential to include all possible forms of internal and semi-internal correlation. Since this can be done with a limited amount of CI, it does not present a major difficulty. For transitions between isolated states of few-times ionized systems, all-external effects are relatively small and a reasonably accurate value can be obtained with at most a very limited number of all-external configurations. However, close to the neutral end of an isoelectronic sequence or if there are two or more interacting states, it is important to introduce all-external effects in a systematic way and to examine the convergence of the oscillator strengths as the CI expansions are enlarged.

There is therefore no monotonic convergence to the correct value of the oscillator strength. One merely has indicators:

(a) the extent of the agreement between length and velocity forms;
(b) the changes in the values of these forms as the CI expansions are increased;
(c) the convergence of the energy levels of (more specifically, the energy differences between) the states involved.

We will explore the convergence in the examples and exercises at the end of the chapter.

3.6.4 Selection Rules

There are certain limitations (known as *selection rules*) for the angular momenta of the states, which can be connected through the dipole operator.

1. This operator is independent of spin. Hence for a non-zero oscillator strength, we require that ψ_i and ψ_j have the same total spin.
2. The operator is a vector (i.e., a tensor of unit rank). Hence the orbital angular momenta L_i and L_j can differ by no more than 1, and the result for $L_i = L_j = 0$ is also zero. (In LSJ coupling, all that is required in place of these first two rules is that J_i and J_j differ by no more than 1 and again $J_i = J_j = 0$ is not allowed.)
3. Since $\boldsymbol{r}$ or $\boldsymbol{p}$ is an operator of odd parity, ψ_i and ψ_j must be of opposite parity.

3.7 The Codes

There are three computer codes associated with this chapter.

1. `CIV3`

   This is a general configuration interaction package which calculates CI wave functions, the corresponding energies, and oscillator strengths of transitions between the states described by these wave functions. Details of the working of the code are given in [3.16]. We give here just the essentials. The present version of the code assumes LS coupling only. It can be used for any atom or ion for which this approximation is considered adequate. A version which incorporates relativistic effects via the Breit–Pauli approximation is available on request from the author.

The wave functions take the form of CI expansions (3.46), with coefficients $\{a_i\}$ being the components of the normalized eigenvectors arising from the solution of (3.47). The corresponding eigenvalues are the calculated state energies. The radial functions are expressed in the analytic form (3.44), i.e., as sums of Slater-type orbitals. The orbitals are assumed to form an orthonormal set. The orthogonality of orbitals with different l is guaranteed by the orthogonality of the spherical harmonics. However, for orbitals with the same l, the radial functions must themselves form an orthonormal set:

$$\int_0^\infty P_{nl}(r)P_{n'l}(r)\mathrm{d}r = \delta_{nn'} . \tag{3.74}$$

The exponents $\{\zeta_{jnl}\}$ can be treated as variational parameters. For a specific l, the $\{P_{nl}\}$ are input in order of increasing n, and the coefficients $\{c_{jnl}\}$ are calculated to ensure orthogonality to those already input (i.e., for $n' < n$). If $k = n - l$ in (3.44) the coefficients are uniquely determined by (3.74). However, if $k > n-l$, there are more coefficients than conditions of the form (3.74). In that case, the $\{c_{jnl}\}$ over and above those needed to satisfy (3.74) can also be treated as variational parameters.

The determination of the wave functions then proceeds in three stages:

(a) Optimize the radial functions.
(b) Decide on which configurations to include.
(c) Construct and diagonalize the Hamiltonian matrix.

The ensuing wave functions can then be used to calculate oscillator strengths. `CIV3` calculates both length and velocity forms.

The code `CIV3` is provided in "variable dimension" form. A global edit is needed to set just two parameters defining dimensions of arrays:

xx The upper limit to the number of configurations.
Choose a suitably large value, such as 100 or 300, ..., but not too large, since there are two square matrices (the Hamiltonian matrix and the matrix of eigenvectors) of size (*xx*,*xx*).

zz The upper limit of the number of two-electron (Slater) radial integrals.
This limit will depend upon the number of radial functions used. There are only four single arrays of this size; $zz = 30000$ should be sufficient for most cases you may wish to try. The actual number

calculated is given in the output to CIV3. If this limit is exceeded, the code flags this problem.

2. SETCIV3

   The input data for CIV3 is rather intricate, largely because it is a *general* program, applicable to a wide range of atomic systems. The input data has two main sections:

   (a) the radial functions;
   (b) the configurations data.

   These are surrounded by various parameters which define the calculation to be undertaken (e.g. whether the radial functions are to be optimized, whether oscillator strengths are to be calculated, etc.). The full details of the input data are given in [3.16]. They are complex and to avoid the reader having to sort out the data needed for any particular calculation, we have written the interactive program SETCIV3 which creates an input file according to the user's responses to a set of questions. Examples of the use of SETCIV3 are given below. The output from SETCIV3 consists of three files:

   (a) a "driver" file containing the data that define the type of calculation;
   (b) a file containing the radial functions;
   (c) a file that defines the configurations (really the electron occupancies, but not the angular momentum couplings). The configurations generated will be *all* those that can be obtained by means of single and double replacements from a basic configuration by the set of orbitals for which there are radial functions.

3. GENCFG

   This code picks up the last of the three output files from SETCIV3 and generates the full list of configurations data, including the angular momentum coupling.

3.7.1 Running of the Codes

The precise handling of these files will depend of the user's operating system. They are read or created via input or output channels as described below. The sequence needed for any calculations is as follows:

(a) Ensure that the initial/current set of radial function parameters is in channel 9 (e.g., ftn09).

(b) Run SETCIV3 and respond to the questions. This produces in:
channel 12, the basic input for CIV3;
channel 11, the input for GENCFG;
channel 10, the new set of radial functions for use in CIV3 if an optimization of radial functions is required.

(c) Run GENCFG. The only input is that now contained in channel 11. This produces in
channel 8, the configurations data for CIV3.

(d) Run CIV3. The input is in channel 12 (basic data), channel 9 or 10 (radial function parameters), and channel 8 (configurations data). No other data are needed. This produces in
channel 6, the results. They will appear on the screen or printer unless re-directed to a file;
channel 9, the optimized radial functions, when optimization has been undertaken.

Note that this sequence of commands leaves the current set of radial functions in channel 9, in *all* cases, ready for any further calculations.

3.7.2 Sample Radial Data

Although SETCIV3 will generate the input file for CIV3, some initial radial function data will be needed. We have provided, in correct format, the following radial data, from which a calculation can commence.

1. The Hartree–Fock ground-state functions for helium-like ions with nuclear charge $Z = 2, 3, \ldots, 10$, as given in [3.8], Table 3 (except for neutral He, which is taken from Table 1: the He data in Table 3 is incorrect).
2. 1s, 2s, 2p radial functions for beryllium-like ions (Z=5,6,7,8,9,10), being the HF functions of the $(1s^2 2s2p)^1P^o$ state. These are taken from [3.21] except for N IV (because of typographical errors for this ion in [3.21]) which instead is taken from [3.13].
3. The ground state $(1s^2 2s^2 2p^2)^3P$ of neutral carbon, given in [3.8], Table 7.

The formal specification of this data for case 3 is summarized in Table 3.2.

Table 3.2. Data specification for calculating the ground state $(1s^2 2s^2 2p^2)^3P$ of neutral carbon.

Format	Data	Parameters
A	2 2 8 6	l_{mp1}, l_{mp1}, NCOEFF, Z
A	2 2	n_1, n_2
A	2 2	n_1, n_2
A	1	NONCON
A	5	k for s orbitals
B	1 5.53875 1 9.52013 2 2.04126 2 5.30567 2 1.30552	I_{jnl}, ζ_{jnl} for s orbitals
C	0.89523 0.07720 0.00413 0.03954 -0.00123	c'_{jnl} for 1s
C	-0.18039 -0.02141 0.57726 -0.08643 0.50920	c'_{jnl} for 2s
A	4	k for p orbitals
B	2 1.44037 2 2.60786 2 0.96499 2 6.53286	I_{jnl}, ζ_{jnl} for p orbitals
C	0.56470 0.22955 0.26761 0.01016	c'_{jnl} for 2p

The formats are:

A `14I5`
B `5(I5,F9.5)`
C `8F9.5`

In the above, $l_{\mathrm{mp1}} = l_{\mathrm{max}} + 1$ where l_{max} is the largest l value of the HF orbitals; n_i are the maximum n values for each of these l values. Notice the line giving the n values is included twice. The data in the second of these lines will change as more orbitals are added. `NONCON = 1` if $k = n - l$, and `NONCON = 0` if $k > n - l$, for the non-HF orbitals (i.e., the ones you may add in the optimization process).

Other data taken from sources such as [3.8–13] can be set up in a similar way.

3.8 Examples

We give a sequence of three examples to illustrate the possible responses to the questions asked in running `SETCIV3`.

1. Example 1
   Starting from the 1s HF function of neutral helium, optimize a 2s radial function on the ground-state energy of helium. The 1s radial data is provided on the disk. A list of the questions and responses for `SETCIV3` is given below. The prompt > shows the responses to the questions.

```
 Do you want to optimize orbitals?
 Type 1 for yes, 0 for no
 > 1
 You have the following orbitals already:
  1s up to  1s
 Type in the n and l values of the orbital you want to optimize
 It can be one of these or a new one
 > 2 0
 You need to input initial estimates for the radial function of this
orbital.
 How many basis functions?  You need at least  2
 > 2
 Read in the  2 exponents
 > 2.0 1.0
 Read in the  2 powers of r.  The values must all be at least as big as  1
 > 1 2
 How do you want to vary the exponents?
 For a common value, type 1 else 2
 > 2
```

```
Which eigenvalue do you wish to minimize?
symmetry: 2S+1, 2L+1, parity (0 for even, 1 for odd)
> 1 1 0
Which eigenvalue? :
Type 1 for lowest, 2 for second lowest etc.
> 1
Do you wish to include all orbitals?
If so, type 1 else 0
> 1
This is the list of orbitals now available
 1   2
1s  2s
Specify the main configuration for each LS symmetry

First symmetry: for which 2S+1 =  1 2L+1 =  1 Parity = 0
How many subshells are occupied in the main configuration?
> 1
Which ones from the available list?
> 1
How many electrons in each?
> 2
How many of these subshells should remain filled and frozen?
> 0
```

The symmetry of the state is ^{1}S (parity even). The 2s orbital has two basis functions (the miminum allowed, $k = n - l$). The initial guesses for the two exponents (ζ_{jnl}) are 2.0 and 1.0; the I_{jnl} ("powers of r") are 1 and 2, respectively. (Note: although any integers can be used for I_{jnl}, we have found it convenient to use a sequence of consecutive integers, beginning with $I_{1nl} = l + 1$.)

To specify the configurations to be used, a basic configuration must be provided. SETCIV3 lists the orbitals/subshells available and asks which ones are occupied in this basic configuration, how many electrons are in each, and whether, in generating the configurations, any of the subshells should remain fixed. (For example, in considering Be-like ions, the user may wish to have all configurations contain 1s^2 – then the 1s subshell would remain fixed.) In the present case, the basic configuration is 1s^2. All the orbitals (we only have 1s and 2s) will be used to generate the configurations, which will then be 1s^2, 1s2s, and 2s^2.

The output file from CIV3 prints the data defining the type of calculation being undertaken, the initial radial data and the configurations. The eigenvalue and eigenvector corresponding to this initial guess are also printed. During the optimization process, the current radial function data is printed out periodically. The final, optimized data is also printed, together with the eigenvalue and eigenvector (which gives the $\{a_i\}$ in (3.46)) corresponding to this data. The optimal I_{jnl} are 1.59618 and 1.78873. The initial guess does not have to be very accurate.

2. Example 2
In this second example, we start with the optimized radial functions of Example 1 (1s and 2s) and optimize a 2p function on the energy of $(1s2p)^1P^o$. By way of illustrating a feature of SETCIV3, we will not include, in the configuration set, replacements by the 2s orbital. So, we have to specify the maximum n value for each l value (in this case, 1 and 2, corresponding to 1s and 2p). Hence the only configuration will be 1s2p. For simplicity, we choose just one basis function for the 2p radial function (so again $k = n - l$). The questions and responses in using SETCIV3 are shown below. The optimization using CIV3 proceeds as in Example 1.

```
Do you want to optimize orbitals?
Type 1 for yes, 0 for no
> 1
You have the following orbitals already:
 1s up to  2s
Type in the n and l values of the orbital you want to optimize
It can be one of these or a new one
> 2 1
You need to input initial estimates for the radial function of this
orbital.
How many basis functions? You need at least  1
> 1
Read in the  1 exponents
> 1.0
Read in the  1 powers of r.  The values must all be at least as big as  2
> 2
How do you want to vary the exponents?
For a common value, type 1 else 2
> 2
Which eigenvalue do you wish to minimize?
symmetry: 2S+1, 2L+1, parity (0 for even, 1 for odd)
> 1 3 1
Which eigenvalue? :
Type 1 for lowest, 2 for second lowest etc.
> 1
Do you wish to include all orbitals?
If so, type 1 else 0
> 0
What is the maximum l-value you wish to include?
> 1
```

```
For each l-value, type in the maximum n-value to be included
> 1 2
This is the list of orbitals now available
 1   2
1s  2p
Specify the main configuration for each LS symmetry

First symmetry: for which 2S+1 =  1 2L+1 =  3 Parity = 1
How many subshells are occupied in the main configuration?
> 2
Which ones from the available list?
> 1 2
How many electrons in each?
> 1 1
How many of these subshells should remain filled and frozen?
> 0
```

3. Example 3

   We now use these three orbitals (1s, 2s, 2p) to determine the oscillator strengths of the $(1s^2)^1S$ – $(1s2p)^1P^o$ transition in neutral helium. This time, no optimization is undertaken and the calculation of oscillator strengths and transition probabilities (both are given in the output) becomes possible. This time, we must specify the basic configuration for both states of the transition (i.e. $1s^2$ for ^1S, 1s2p for $^1P^o$). The questions and responses using SETCIV3 are as follows:

```
Do you want to optimize orbitals?
Type 1 for yes, 0 for no
> 0
Do you want to calculate transition probabilities?
Type 1 for yes, 0 for no
> 1
Which states?
First symmetry: 2S+1, 2L+1, parity (0 for even, 1 for odd)
> 1 1 0
Second symmetry: 2S+1, 2L+1, parity (0 for even, 1 for odd)
> 1 3 1
How many eigenvalues for each symmetry?
First symmetry:
> 1
Second symmetry:
> 1
Do you wish to include all orbitals?
If so, type 1 else 0
```

```
> 1
This is the list of orbitals now available
 1  2  3
1s  2s  2p
Specify the main configuration for each LS symmetry

First symmetry: for which 2S+1 =  1 2L+1 =  1 Parity = 0
How many subshells are occupied in the main configuration?
> 1
Which ones from the available list?
> 1
How many electrons in each?
> 2
How many of these subshells should remain filled and frozen?
> 0
This is the list of orbitals now available
 1  2  3
1s  2s  2p
Specify the main configuration for each LS symmetry

Second symmetry: for which 2S+1 =  1 2L+1 =  3 Parity = 1
How many subshells are occupied in the main configuration?
> 2
Which ones from the available list?
> 1 3
How many electrons in each?
> 1 1
How many of these subshells should remain filled and frozen?
> 0
```

All possible single and double replacements are included, giving the configurations $1s^2$, 1s2s, $2s^2$, and $2p^2$ for ^1S; 1s2p and 2s2p for $^1P^o$. The eigenvalues and eigenvectors of both states appear in the output from CIV3, as do the length and velocity forms of both the oscillator strength and the transition probability, for which the calculated value of ΔE has been used.

We also print out the contribution to the oscillator strengths from the different configurations. Using CI wave functions

$$\Psi_1 = \sum_i a_i \phi_i$$

and

$$\Psi_2 = \sum_i b_i \psi_j$$

then, for example, (3.73) becomes

$$\begin{aligned} f^{\mathrm{l}} &= \frac{2}{3}\Delta E \, |\langle \Psi_1 | \boldsymbol{r} | \Psi_2 \rangle|^2 \\ &= \frac{2}{3}\Delta E \left| \sum_i \sum_j a_i b_j \langle \phi_i | \boldsymbol{r} | \psi_j \rangle \right|^2 \equiv \left| \sum_i \sum_j F^{\mathrm{l}}_{ij} \right|^2 , \end{aligned}$$

where

$$F^{\mathrm{l}}_{ij} = \left(\frac{2}{3}\Delta E \right)^{\frac{1}{2}} a_i b_j \langle \phi_i | \boldsymbol{r} | \psi_j \rangle \, . \tag{3.75}$$

The values of i, j and F_{ij} are also given in the output to CIV3 if $|F^{\mathrm{l}}_{ij}|$ or $|F^{\mathrm{v}}_{ij}| \geq 0.001$. It enables the user to see which configurations are most significant in the calculation of the oscillator strengths.

This example is of course only illustrative. Much more CI would be needed for a correct result. For example, the 1s function of the HF ground state is quite different from the optimal 1s function of the $^1\mathrm{P^o}$ state. However, even in this simple approximation, f^{l} and f^{v} agree quite closely (0.253 and 0.245, respectively) and are also fairly close to the accurate value of 0.276 [3.22]. But notice that f^{l} and f^{v} agree more closely with each other than does either of them with the accurate result: agreement between f^{l} and f^{v} is, by itself, no guarantee for accuracy.

3.9 Summary

We have presented a discussion of the non-relativistic configuration interaction method, as it is used in the general atomic structure code CIV3 for the calculation of atomic energy levels and oscillator strengths. The wave functions obtained with this code can be adapted to serve as input data for the target description in the collision codes discussed in other chapters of this book.

3.10 Suggested Problems

3.10.1 He-Like Ions

1. Start with the 1s radial function of He I (put the data from the disk into channel 9), and by responding "0" to the first two questions in SETCIV3, determine the HF energy of He I. (The value should be −2.86168.)
2. Reproduce Table 3.1. To do this, you will need to go through the following stages:

(a) Start with the same radial data input as the previous exercise. Work through Example 1 of the previous section. This gives 2s. Then successively generate 3s, 4s, 5s, and 6s.

(b) Then generate 2p, ..., 6p; 3d, ..., 6d; 4f, 5f, 6f; 5g, 6g.
[Note: There is a limit of 40 on the number of configurations that can be included in an optimization. This limit is imposed to speed up the optimization process, but it can be lifted by extending the dimensions of the arrays in the `COMMON` block `ALTER3` in `CIV3` to $\frac{1}{2}n(n+1)$, where n is the desired number of configurations.]

In the optimization of the nl orbitals with $l > 0$, therefore, include only the configurations $1s^2$ plus those that can be constructed with the current l value ($= l_{max}$). To do this, respond "0" when asked if all the orbitals are to be included, and then set the maximum n value equal to unity for $l < l_{max}$. Hence, for p functions, l_{max}=1, $n_{max} = 1, n$ for $l = 0, 1$; for d functions, l_{max}=2 and $n_{max} = 1, 1, n$ for $l = 0, 1, 2$; etc.

(c) Once all the radial functions have been optimized (say up to 6g), re-run `SETCIV3` in each of the following cases, responding "0" to the first two questions. Determine
   i. The ground-state energy with all configurations included (l_{max}=4, up to n=6).
   ii. Repeat with l_{max}=3, up to n=6 for each l.
   iii. Repeat with l_{max}=2, up to n=6 for each l.
   iv. Repeat with l_{max}=1, up to n=6 for each l.
   v. Repeat with l_{max}=0, up to n=6 for each l.

3. Hence determine the contribution to the correlation energy of *each* of the summations in (3.52).

4. Repeat 1–3 for the other He-like ions for which HF radial function data is provided on the disk. What pattern do you notice about the results of 3?

3.10.2 Be-Like Ions

Evaluate the oscillator strengths of the $(1s^2 2s^2)^1S$ – $(1s^2 2s2p)^1P^o$ transition of Be-like ions ($5 \le Z \le 10$). The 1s, 2s, and 2p radial functions are provided on the disk.

Make the restriction that all configurations include $1s^2$. Although this is an approximation, it reflects the fact that these two electrons are predominantly in the region close to the nucleus (origin) while the other two electrons are generally much further out. It is these two outer electrons that we need to describe as accurately as possible.

In principle, we should use many configurations if we wish to obtain accurate oscillator strengths. But it turns out that, in order to understand which configurations provide the main contributions to the oscillator strengths, it is sufficient to limit the configurations to

^{1}S: $1s^22s^2$, $1s^22s3s$, $1s^23s^2$, $1s^22p^2$, $1s^22p3p$, $1s^23p^2$, $1s^23d^2$;
^{1}P^o: $1s^22s2p$, $1s^22s3p$, $1s^22p3s$, $1s^23s3p$, $1s^22p3d$, $1s^23p3d$.

Three new orbitals have been introduced: 3s, 3p, and 3d. As we remarked earlier, the same radial functions are used for both states: e.g., 3p in the ^{1}S configurations is the same as in the ^{1}P^o configurations. The configurations listed above are all those which can be formed from the orbitals 1s, 2s, 2p, 3s, 3p, and 3d with the assumption that each contains $1s^2$.

There are two parts to the calculation:

1. Determine the 3s, 3p, and 3d radial functions. These are obtained in two stages:
   (a) 3d is to be optimized on the lowest eigenvalue of the Hamiltonian matrix formed from the two ^{1}P^o configurations $1s^22s2p$ and $1s^22p3d$. This eigenvalue corresponds to the $(1s^22s2p)^1$P^o energy. The configuration $1s^22p3d$ represents a semi-internal correlation effect, and (in the absence of internal correlation for the ^{1}P^o state) is the major component in the wave function after the $1s^22s2p$ configuration.
   (a) 3s and 3p are to be optimized on the lowest eigenvalue of the ^{1}S Hamiltonian matrix.

   It is worth noting that although 3d is representing a semi-internal correlation effect in ^{1}P^o, both 3s and 3p have a dual role in the optimization (b) above. The 2s and 2p functions were obtained as HF functions for the $(1s^22s2p)^1$P^o state. If we were to compare with the HF functions for the $(1s^22s^2)^1$S state, we would find that the two 2s functions would be slightly different. Hence in determining 3s, we need it to represent both a correction to the 2s for the ^{1}S state (through the $1s^22s3s$ configuration) and the external correlation effect introduced by the configuration $1s^23s^2$. The configuration $1s^22p^2$ is an internal correlation effect, and so is important. But if 2p were to be determined by optimizing the lowest eigenvalue obtained with $1s^22s^2 + 1s^22p^2$, we would find that it would be somewhat different from the 2p we have from the ^{1}P^o state. The 3p function optimized as in (b) above both corrects for this difference and allows for further external correlation by including the configuration $1s^23p^2$.

2. Calculate oscillator strengths using configurations obtained as outlined below, and comment on the influence different correlation effects have on the oscillator strengths, particularly on the extent of the agreement between length and velocity forms.
   (a) All configurations possible from all six radial functions.
   (b) All configurations possible from 1s, 2s, 2p, and 3d.
   (c) All configurations possible from 1s, 2s, and 2p.
   (d) HF configurations only: $(1s^22s^2)^1$S and $(1s^22s2p)^1$P^o.

Some comparison may be made with the results given in [3.23], although the HF functions used in [3.23] were, in some cases, different from those provided here.

3.10.3 Ground State of Carbon

Starting with the HF orbitals of the $(1s^2 2s^2 2p^2)^3P$ ground state of carbon (from the disk), determine the correlation functions 3s, 3p, 3d, and 4f by optimizing the lowest ^3P energy with

(a) the $1s^2$ subshell held fixed;

(b) no subshells held fixed.

References

3.1 E.A. Hylleraas, Z. Phys. **54** (1929), 347–66.
3.2 C.L. Pekeris, Phys. Rev. **112** (1958), 1649–58; Y. Accad, C.L. Pekeris and B. Schiff, Phys. Rev. A. **4** (1971), 516–36.
3.3 O. Sinanoğlu, J. Chem. Phys. **36** (1962), 706–17.
3.4 D.R. Hartree, Proc. Camb. Phil. Soc. **24** (1927), 89–110; 111–32.
3.5 V. Fock, Z. Phys. **61** (1930), 126–48.
3.6 J.C. Slater, Phys. Rev. **35** (1930), 210.
3.7 C.F. Fischer, Comput. Phys. Commun. **1** (1969), 151–66; **4** (1972), 107–16.
3.8 E. Clementi and C. Roetti, Atom. Data Nucl. Data Tables **14** (1974), 177–478.
3.9 A.D. McLean and R.S. McLean, Atom. Data Nucl. Data Tables **26** (1981), 197–381.
3.10 J.G. Snijders, P. Vernooijs and E.J. Baerends, Atom. Data Nucl. Data Tables **26** (1981), 481–509.
3.11 T. Koga, Y. Seki, A.J. Thakkar and H. Tatewaki, J. Phys. B: Atom. Molec. Phys. **26** (1993), 2529–32.
3.12 A.W. Weiss, Astrophys. J. **138** (1963), 1262–76.
3.13 C.C.J. Roothaan and P.S. Kelly, Phys. Rev. **131** (1963), 1177–82.
3.14 E.A. Hylleraas and B. Undheim, Z. Phys. **65** (1930), 759–72.
3.15 A. Hibbert and C.F. Fischer, Comput. Phys. Commun. **64** (1991), 417–30.
3.16 A. Hibbert, Comput. Phys. Commun. **9** (1975), 141–72.
3.17 C.F. Fischer, Comput. Phys. Commun. **64** (1991), 431–54.
3.18 C. Bunge, Theo. Chim. Acta, **16** (1970), 126–44.
3.19 I. Oksüz and O. Sinanoğlu, Phys. Rev. **181** (1969), 42–53; 54–65.
3.20 A. Hibbert, P.C. Ojha and R.P Stafford, J. Phys. B: Atom. Molec. Phys. **25** (1992), 4427–32.
3.21 H. Tatewaki, H. Taketa and F. Sasaki, Int. J. Quan. Chem. **5** (1971), 335–57.
3.22 B. Schiff, Y. Accad and C.L. Pekeris, Phys. Rev. A. **4** (1971), 885–93.
3.23 P.G. Burke, A. Hibbert and W.D. Robb, J. Phys. B. **5** (1972), 37–43.

4. The Distorted-Wave Method for Elastic Scattering and Atomic Excitation

Don H. Madison [1] and Klaus Bartschat [2]

[1] Department of Physics and LAMR, University of Missouri-Rolla, Rolla, Missouri 65401, USA
[2] Department of Physics and Astronomy, Drake University, Des Moines, Iowa 50311, USA

Abstract

We present the first-order distorted-wave Born approximation (DWBA) for electron and positron scattering from neutral atoms. Explicit formulae are derived for elastic and inelastic scattering (excluding ionization) for quasi one-electron neutral targets. The computer code `DWBA1` for S–S and S–P transitions is described together with some sample input and output data.

4.1 Introduction

The calculation of reliable scattering amplitudes has been an issue of central importance in atomic physics since the first experiments were performed. While numerous works have been reported over the years using both classical and quantal approaches, most of the effort for electron and positron scattering has focused on the quantal methods. Within these approaches, essentially all the works fall either into the category of the close coupling approach (close coupling, R matrix, variational methods) or into the category of perturbation series expansions (Born series, Eikonal series, distorted-wave series, many-body theory, etc.). In the early days of atomic scattering experiments, only total cross sections were measured and most theoretical approaches gave results in reasonable agreement with the experimental data, at least for the higher energies. During this time, the first-order Born approximation became very popular due to its ease of calculation. In the 1960s, however, experimental techniques were improved and differential cross section measurements started to be reported. These experiments revealed that the first-order plane-wave Born approximation (PWB1) was valid only for small scattering angles and, further, that the angular range over which the PWB1 was valid decreased with increasing energy of the incident projectile. The reason that the PWB1 gave good results for the total cross sections lies in the fact that the major contribution to the total cross section came from very small scattering angles where the PWB1 was reliable. For applications that are sensitive to large-angle differential cross sections, a theoretical approach superior to the PWB1 is required. One of the most successful approaches has been the distorted-wave series. A primary advantage of this method lies in the fact that more of the important physical effects may be included in the leading terms of the perturbation series expansion such that the distorted-wave series converges faster than the Born series. Consequently, reasonably accurate results may be obtained from the first term of the distorted-wave series.

In this chapter, we will give a brief overview of the perturbation series approach. In particular, we will identify and describe the various types of distorted-wave series which can result from different types of distorting potentials and discuss how these expansions relate to the Born (plane-wave) series. Lastly, we will describe how the first term of the distorted-wave series may be numerically evaluated using partial-wave expansions, and we will describe a computer code which calculates first-order distorted-wave amplitudes.

4.2 Theory

The problem we will consider is the scattering of an electron or positron from an atom. For this case, the total hamiltonian for the system is expressed as

$$H = h_{\mathrm{a}} + T + V\,, \tag{4.1}$$

where h_{a} is the hamiltonian for an isolated atom, T is the hamiltonian for an isolated projectile (kinetic energy operator), and V is the interaction between the projectile and the atom. The initial-state full scattering wavefunction Ψ_{i} is a solution of Schrödinger's equation

$$(H - E)\,\Psi_{\mathrm{i}}^{+} = 0\,, \tag{4.2}$$

where the + superscript indicates the usual outgoing wave boundary conditions. If the projectile is an electron (which we shall label 0) and it experiences either an elastic or an inelastic collision with an N-electron atom, the exact $\mathbf{T}$ matrix in the two-potential approach is given by

$$\begin{aligned}\mathbf{T}_{\mathrm{fi}} = (N+1)\,&\langle \chi_{\mathrm{f}}^{-}(0)\psi_{\mathrm{f}}(1,\ldots,N)\,|V - U_{\mathrm{f}}|\,\mathcal{A}\Psi_{i}^{+}(0,\ldots,N)\rangle \\ &+ \langle \chi_{\mathrm{f}}^{-}(0)\psi_{\mathrm{f}}(1,\ldots,N)\,|U_{\mathrm{f}}|\,\psi_{\mathrm{i}}(1,\ldots,N)\beta_{\mathrm{i}}(0)\rangle\,.\end{aligned} \tag{4.3}$$

In (4.3), ψ_{i} and ψ_{f} are the properly antisymmetrized initial and final atomic wavefunctions for an isolated atom, which diagonalize the atomic hamiltonian h_{a} according to

$$\langle \psi_{n'}\,|h_{\mathrm{a}}|\,\psi_{n}\rangle = \varepsilon_{n}\delta_{n'n}. \tag{4.4}$$

Furthermore, β_{i} is an initial-state plane wave (eigenfunction for an isolated projectile) and $\mathcal{A}$ is the antisymmetrizing operator for the $N+1$ electrons.

If $\Psi_i^+(0,\ldots,N)$ is chosen to be a product of a projectile wavefunction (electron 0) times an antisymmetrized atomic wavefunction (electrons $1,...,N$), the antisymmetrization operator may be expressed as

$$\mathcal{A} = \frac{1}{N+1}\left(1 - \sum_{i=1}^{N} P_{i0}\right), \tag{4.5}$$

where P_{i0} is the operator that exchanges electrons 0 and i. The potential U_f in (4.3) is an arbitrary distorting potential for the projectile, which is used to calculate χ_f^- by solving the equation

$$(T + U_\mathrm{f} - E_\mathrm{f})\,\chi_\mathrm{f}^- = 0\,, \tag{4.6}$$

where the $-$ superscript designates incoming wave boundary conditions and E_f is the final-state energy of the projectile. Typically, U_f is chosen to be a spherically symmetric final-state approximation for V since it is the potential used to calculate the final-state wavefunction for the projectile. In principle, however, U_f can be any potential as long as χ_f fulfills the appropriate boundary conditions.

For inelastic scattering, the second term of (4.3) vanishes for orthogonal atomic wavefunctions since U_f depends only on the single coordinate of the projectile. For elastic scattering, on the other hand, the second term of (4.3) is the dominant term; in fact, it is generally the only contributing term since U_f is typically chosen such that the matrix elements of $V - U_\mathrm{f}$ vanish.

Finally, (4.3) may also be used for positron scattering if one sets $\mathcal{A} = 1/(N+1)$.

4.2.1 Distorted-Wave Series

Equation (4.3) cannot be evaluated without making some approximations due to the fact that Ψ_i^+ cannot be evaluated without making approximations. In the distorted-wave approach, Ψ_i^+ is expressed in terms of a product of an initial-state distorted wave χ_i^+ times an initial atomic wavefunction ψ_i. Then a power series expansion is made for the interaction which is assumed to be small. For an energy E_i of the incident projectile, the initial-state distorted wave is a solution of the Schrödinger equation,

$$(T + U_\mathrm{i} - E_\mathrm{i})\,\chi_\mathrm{i}^+ = 0\,, \tag{4.7}$$

for an arbitrary distorting potential U_i which vanishes asymptotically. Historically, U_i was chosen to be a spherical average of the initial-state projectile-atom interaction, but this is not required by the theory. In fact, this choice does not give the best agreement with experimental data, as will be discussed below.

The Lippmann–Schwinger solution for Ψ_i^+ in terms of χ_i^+ is given by

$$\Psi_\mathrm{i}^+ = \left[1 + G^+ (V - U_\mathrm{i})\right] \psi_\mathrm{i}\chi_\mathrm{i}^+\,, \tag{4.8}$$

where the full Green's function is

$$G^+ = (E - H + i\eta)^{-1}. \tag{4.9}$$

Note that the Lippmann–Schwinger wavefunction (4.8) is not properly antisymmetric for $N+1$ electrons but is assumed to be antisymmetric for the N atomic electrons. The operator $\mathcal{A}$ takes care of the complete antisymmetrization.

A series expansion for the Green's function may be obtained by introducing a third arbitrary potential U and the Green's function

$$g^+ = (E - h_a - T - U + i\eta)^{-1} \tag{4.10}$$

associated with this potential. The full Green's function G^+ may be expressed in terms of the distorted Green's function g^+ as

$$G^+ = g^+ + G^+(V - U)g^+. \tag{4.11}$$

Equation (4.11) can be iterated to obtain the series expansion for the full Green's function G^+ as

$$G^+ = g^+ + g^+(V-U)g^+ + g^+(V-U)g^+(V-U)g^+ \ldots \tag{4.12}$$

If the Lippmann–Schwinger solution (4.8) is used in (4.3) along with the Green's function expansion (4.12), we get

$$\mathbf{T}_{fi} = \mathbf{T}_1 + \mathbf{T}_2 + \mathbf{T}_3 + \ldots, \tag{4.13}$$

where

$$\begin{aligned}\mathbf{T}_1 = {} & (N+1)\Big\langle \chi_f^-(0)\psi_f(1,\ldots,N)\,|V - U_f|\,\mathcal{A}\psi_i(1,\ldots,N)\chi_i^+(0)\Big\rangle \\ & + \Big\langle \chi_f^-(0)\psi_f(1,\ldots,N)\,|U_f|\,\psi_i(1,\ldots,N)\beta_i(0)\Big\rangle,\end{aligned} \tag{4.14}$$

$$\begin{aligned}\mathbf{T}_2 = {} & (N+1)\Big\langle \chi_f^-(0)\psi_f(1,\ldots,N) \\ & \times \big|(V-U_f)\mathcal{A}g^+(V-U_i)\big|\,\psi_i(1,\ldots,N)\chi_i^+(0)\Big\rangle,\end{aligned} \tag{4.15}$$

and

$$\begin{aligned}\mathbf{T}_3 = {} & (N+1)\Big\langle \chi_f^-(0)\psi_f(1,\ldots,N) \\ & \times \big|(V-U_f)\mathcal{A}g^+(V-U)g^+(V-U_i)\big|\,\psi_i(1,\ldots,N)\chi_i^+(0)\Big\rangle.\end{aligned} \tag{4.16}$$

Equation (4.13) is the distorted-wave series for the $\mathbf{T}$ matrix whereas (4.14) is the first-order distorted-wave approximation (DWB1). Note that terminology can be troublesome since the distorted-wave approach contains three essentially arbitrary potentials which must be specified to characterize the particular type of "distorted-wave" calculation. Only two of these potentials, U_i and U_f, affect the first-order term while all three potentials affect the higher-order terms.

4.2.2 Discussion of Distorting Potentials U_i and U_f

Within the above framework, the three distorting potentials are arbitrary with the only constraint being that the asymptotic forms of the wavefunctions satisfy the appropriate boundary conditions. An intuitive feeling for how these potentials should be chosen can be obtained by expanding the distorted waves in the DWB1 amplitude in terms of plane waves. Such an expansion is readily achieved through the Lippmann–Schwinger solution for the distorted wave χ expressed in terms of plane waves β

$$\psi_\mathrm{i}\chi_\mathrm{i}^+ = \psi_\mathrm{i}\beta_\mathrm{i} + g_0^+ U_\mathrm{i}\psi_\mathrm{i}\chi_\mathrm{i}^+ \tag{4.17}$$

$$\psi_\mathrm{f}\chi_\mathrm{f}^- = \psi_\mathrm{f}\beta_\mathrm{f} + g_0^- U_\mathrm{f}\psi_\mathrm{f}\chi_\mathrm{f}^- \tag{4.18}$$

where

$$g_0^\pm = (E - h_\mathrm{a} - T \pm \mathrm{i}\eta)^{-1}. \tag{4.19}$$

If the Lippmann–Schwinger solutions (4.17) and (4.18) are iterated and the results used in the DWB1 $\mathbf{T}$ matrix, we obtain for the direct scattering amplitude

$$\begin{aligned} \mathbf{T}_1^\mathrm{direct} &= \langle \beta_\mathrm{f}\psi_\mathrm{f} | V | \psi_\mathrm{i}\beta_\mathrm{i}\rangle + \langle \beta_\mathrm{f}\psi_\mathrm{f} | (V - U_\mathrm{f}) g_0^+ U_\mathrm{i} | \psi_\mathrm{i}\beta_\mathrm{i}\rangle \\ &\quad + \langle \beta_\mathrm{f}\psi_\mathrm{f} | U_\mathrm{f} g_0^+ V | \psi_\mathrm{i}\beta_\mathrm{i}\rangle + \dots . \end{aligned} \tag{4.20}$$

Next we assume that the atomic states $\{|\psi_k\rangle,\ k = 1, \infty\}$ form a complete set such that

$$\sum_k |\psi_k\rangle \langle\psi_k| = 1\,, \tag{4.21}$$

and express V as

$$\sum_{j,k} |\psi_j\rangle V_{jk} \langle\psi_k|\,, \tag{4.22}$$

where

$$V_{jk} = \langle\psi_j | V | \psi_k\rangle\,. \tag{4.23}$$

If we use (4.22) and (4.23), the DWB1 direct $\mathbf{T}$ matrix becomes

$$\begin{aligned} \mathbf{T}_1^\mathrm{direct} &= \langle \beta_\mathrm{f} | V_\mathrm{fi} | \beta_\mathrm{i}\rangle + \langle \beta_\mathrm{f} | V_\mathrm{fi} g_0^+(E_\mathrm{i}) U_\mathrm{i} | \beta_\mathrm{i}\rangle \\ &\quad + \langle \beta_\mathrm{f} | U_\mathrm{f} g_0^+(E_\mathrm{f}) V_\mathrm{fi} | \beta_\mathrm{i}\rangle - \langle \beta_\mathrm{f} | U_\mathrm{f} g_0^+(E_\mathrm{i}) U_\mathrm{i} | \beta_\mathrm{i}\rangle \delta_\mathrm{if} + \dots , \end{aligned} \tag{4.24}$$

where

$$g_0^+(E_n) = (E_n - T + \mathrm{i}\eta)^{-1} \tag{4.25}$$

is a free-particle Green's function for projectile energy $E_n = E - \varepsilon_n$. Note that g_0^+ in (4.19) denotes an $(N+1)$-electron Green's function whereas $g_0^+(E_n)$ is a single-electron Green's function.

The standard plane-wave Born series may be obtained from the present formalism by setting $U_\mathrm{i} = U_\mathrm{f} = U = 0$. Hence, the plane-wave Born series for the **T** matrix (4.3) is given by

$$\mathbf{T}_{\mathrm{fi}}^{\mathrm{Born}} = \langle \beta_\mathrm{f} | V_{\mathrm{fi}} | \beta_\mathrm{i} \rangle + \sum_n \langle \beta_\mathrm{f} | V_{\mathrm{f}n} g_0^+(E_n) V_{n\mathrm{i}} | \beta_\mathrm{i} \rangle + \dots \tag{4.26}$$

If we compare (4.24) and (4.26), we see that the first-order distorted-wave amplitude contains second-order terms which are very similar to some of the second-order plane-wave Born (PWB2) terms. In fact, if $U_\mathrm{i} = V_{\mathrm{ii}}$ and $U_\mathrm{f} = V_{\mathrm{ff}}$, the DWB1 amplitude for excitation would contain the two terms in the PWB2 amplitude corresponding to the intermediate state being either the initial or final state. For the case of elastic scattering, the DWB1 contains the single PWB2 term for which the intermediate state equals the initial (= final) state. It is primarily for this reason that U_i has been chosen to be the initial-state distorting potential and U_f has been chosen to be the final-state distorting potential. However, this choice is reasonable only if those two terms completely dominate the second-order Born amplitude. It is now known from exact second-order calculations [4.1] that although these two terms are very important, particularly for large scattering angles, the other bound and continuum intermediate states also give significant contributions to the second Born amplitude. With this knowledge, it is clear that the best choice for U_i and U_f would be the one which simulates as much of the PWB2 amplitude (and even higher amplitudes) as possible. From separate orthogonality arguments [4.2–4] it can be seen that it is most desirable to choose $U_\mathrm{i} = U_\mathrm{f}$. By comparing DWB1 results with experiment, it was found that the choice of $U_\mathrm{i} = U_\mathrm{f} = V_{\mathrm{ff}}$ (final-state distorting potential) consistently gives the best results for excitation of P states in atomic hydrogen [4.5], all the inert gases [4.6,7], cadmium [4.8] and mercury [4.9]. Since this choice is typically far superior to the choice of $U_\mathrm{i} = V_{\mathrm{ii}}$ (initial-state potential) and $U_\mathrm{f} = V_{\mathrm{ff}}$ (final-state potential), one would conclude that the former choice simulates more of the second-order (and higher) Born terms than the latter choice, but the theoretical justification has not yet been given.

In a similar vein, it should be noted that the first-order many-body theory (FOMBT) is a first-order distorted-wave theory for which $U_\mathrm{i} = U_\mathrm{f} = V_{\mathrm{ii}}$ (initial-state distorting potential). These types of calculations are typically better than the $U_\mathrm{i} = V_{\mathrm{ii}}$, $U_\mathrm{f} = V_{\mathrm{ff}}$ calculations but not as good as the $U_\mathrm{i} = U_\mathrm{f} = V_{\mathrm{ff}}$ calculations. As a technical comment, we note that one does not actually use V_{ff}, but rather its spherical average, since spherical symmetry is required (the two are identical only for final S states).

4.3 First-Order Amplitudes

We shall now restrict our attention to scattering from hydrogen or other quasi one-electron neutral atoms. For this case the first-order distorted-wave amplitude (4.14) can be expressed in terms of a direct amplitude f^{dir} and an exchange amplitude f^{exch} as follows:

$$T_1^S = f^{\mathrm{dir}} + (-1)^S f^{\mathrm{exch}}, \tag{4.27}$$

with

$$\begin{aligned} f^{\mathrm{dir}} = &\left\langle \chi_{\mathrm{f}}^-(0)\psi_{\mathrm{f}}(1) \left| V \right| \psi_{\mathrm{i}}(1)\chi_{\mathrm{i}}^+(0) \right\rangle \\ &+ \left\langle \chi_{\mathrm{f}}^-(0) \left| U_{\mathrm{f}}(0) \right| \left[\beta_{\mathrm{i}}(0) - \chi_{\mathrm{i}}^+(0)\right] \right\rangle \delta_{\mathrm{fi}} \end{aligned} \tag{4.28}$$

and

$$\begin{aligned} f^{\mathrm{exch}} = &\left\langle \chi_{\mathrm{f}}^-(0)\psi_{\mathrm{f}}(1) \left| V \right| \psi_{\mathrm{i}}(0)\chi_{\mathrm{i}}^+(1) \right\rangle \\ &- \left\langle \chi_{\mathrm{f}}^-(0) \left| U_{\mathrm{f}}(0) \right| \psi_{\mathrm{i}}(0) \right\rangle \Big\langle \psi_{\mathrm{f}}(1) \mid \chi_{\mathrm{i}}^+(1) \Big\rangle. \end{aligned} \tag{4.29}$$

For the total spins $S = 0$ and 1, (4.27) gives the familiar singlet and triplet scattering amplitudes for electron scattering, respectively. For positron scattering, the exchange term g must be set to zero.

The second part of the exchange term (4.29) is the overlap integral between a distorted wave and the final bound state. The oscillatory nature of the distorted wave will tend to cause this integral to vanish, particularly when the distorted wave oscillates significantly more rapidly than the bound-state wavefunction, which is the case for intermediate to high projectile energies. As a consequence, we shall ignore this overlap term and approximate the exchange amplitude as

$$f^{\mathrm{exch}} = \left\langle \chi_{\mathrm{f}}^-(0)\psi_{\mathrm{f}}(1) \left| V \right| \psi_{\mathrm{i}}(0)\chi_{\mathrm{i}}^+(1) \right\rangle. \tag{4.30}$$

4.3.1 Inelastic Scattering

For scattering from atomic hydrogen the projectile–target interaction is given by (Rydberg energy units)

$$V(0,1) = -\frac{2Z_{\mathrm{p}}}{r_{01}} + \frac{2Z_{\mathrm{p}}Z_{\mathrm{N}}}{r_0}, \tag{4.31}$$

where Z_{p} is the charge of the projectile, Z_{N} (= +1 for hydrogen) is the charge of the nucleus, and r_{01} is the distance between the projectile and the atomic electron. If we use this interaction, the direct and exchange amplitudes for inelastic scattering to bound excited states (we do not consider ionization here) become

$$f^{\mathrm{dir}} = -2Z_{\mathrm{p}} \left\langle \chi_{\mathrm{f}}^-(0)\psi_{\mathrm{f}}(1) \left| \frac{1}{r_{01}} \right| \psi_{\mathrm{i}}(1)\chi_{\mathrm{i}}^+(0) \right\rangle \tag{4.32}$$

and

$$f^{\text{exch}} = -2Z_{\text{p}} \left\langle \chi_{\text{f}}^{-}(0)\psi_{\text{f}}(1) \left| \frac{1}{r_{01}} \right| \psi_{\text{i}}(0)\chi_{\text{i}}^{+}(1) \right\rangle + 2Z_{\text{p}}Z_{\text{N}} \left\langle \chi_{\text{f}}^{-}(0) \left| \frac{1}{r_{0}} \right| \psi_{\text{i}}(0) \right\rangle \langle \psi_{\text{f}}(1) \mid \chi_{\text{i}}^{+}(1) \rangle . \tag{4.33}$$

As was mentioned above, the overlap of a continuum wave and a bound wave, such as occurs in the second term of (4.33), is small for high energies. In fact, if all the distorted-waves and bound-state wavefunctions are eigenfunctions of the same distorting potential, this overlap is identically zero. Consequently, we shall also ignore the second term in (4.33) and set

$$f^{\text{exch}} = -2Z_{\text{p}} \left\langle \chi_{\text{f}}^{-}(0)\psi_{\text{f}}(1) \left| \frac{1}{r_{01}} \right| \psi_{\text{i}}(0)\chi_{\text{i}}^{+}(1) \right\rangle . \tag{4.34}$$

The inclusion of the second exchange term in (4.33) as well as of the second exchange term of (4.29) is left as an exercise for the interested reader.

4.3.2 Elastic Scattering

If one is interested in elastic scattering, it is possible either to use the present formalism and set the final state equal to the initial state or to calculate the results directly from elastic scattering phase shifts. Although the latter procedure is perhaps easier and more straightforward, we will use the former here so that elastic and inelastic amplitudes are treated in the same manner. For elastic scattering, $U_{\text{f}} = U_{\text{i}}$, and for scattering from an S state, $V_{\text{ii}} = U_{\text{i}}$. As a result, the terms containing distorted waves in the direct scattering amplitude f of (4.28) vanishes. If one is scattering elastically from an excited state which is not an S state, the spherically symmetric part of V_{ii} will equal U_{i} and this term will cancel. However, there will be a residual non-zero contribution to f^{dir} from the part of V_{ii} which is not spherically symmetric, but for present purposes we will ignore this term. Consequently, the direct amplitude for elastic scattering is

$$f^{\text{dir}} = \langle \chi_{\text{f}}^{-}(0) \left| V_{\text{ii}}(0) \right| \beta_{\text{i}}(0) \rangle . \tag{4.35}$$

The exchange term for elastic scattering is given by (4.34) with $\psi_{\text{f}} = \psi_{\text{i}}$ and χ_{f}^{-} an eigenfunction of the same potential as χ_{i}^{+}.

The amplitudes (4.32), (4.34), and (4.35) are single-particle direct and exchange amplitudes. For more complex atoms than hydrogen, the first-order amplitudes become more complicated due to multi-electron interactions and correlations. However, if the atomic wavefunction can be expressed either as a product of single-particle orthonormal wavefunctions or as a linear combination of such products, the first-order amplitude for any atom will eventually reduce to some factor times the basic direct amplitudes (4.32) or (4.35) and a different factor times the exchange amplitude (4.34) (for details we refer to Madison and Shelton [4.2]).

4.4 Partial-Wave Expansion of the T Matrix

The standard procedure for evaluating first-order amplitudes is to make partial-wave expansions for the continuum wavefunctions. Here, we shall assume that the continuum waves are normalized to $\delta(\boldsymbol{k}-\boldsymbol{k}')$ where $\boldsymbol{k}$ and $\boldsymbol{k}'$ are the wavevectors associated with these waves. For this normalization, the appropriate partial-wave expansion, with angular momenta ℓ_i and z components m_{ℓ_i}, for the distorted wave in the initial channel is given by

$$|\chi_{\mathrm{i}}^{+}\rangle = \sqrt{\frac{2}{\pi}}\frac{1}{k_{\mathrm{i}}r}\sum_{\ell_{\mathrm{i}}m_{\mathrm{i}}} \mathrm{i}^{\ell_{\mathrm{i}}}\chi_{\ell_i}(k_i,r)\, Y_{\ell_{\mathrm{i}}m_{\mathrm{i}}}(\hat{\boldsymbol{r}})Y^{*}_{\ell_{\mathrm{i}}m_{\mathrm{i}}}(\hat{\boldsymbol{k}}_{\mathrm{i}}), \tag{4.36}$$

where $Y_{\ell m}(\hat{\boldsymbol{r}})$ is a spherical harmonic for the angles associated with the direction of the $\boldsymbol{r} = r\hat{\boldsymbol{r}}$ vector, and $\boldsymbol{k}_i = k_i\hat{\boldsymbol{k}}_{\mathrm{i}}$ is the momentum of the incident projectile. The radial distorted wave is a solution of the Schrödinger equation

$$\left(\frac{\mathrm{d}^2}{\mathrm{d}r^2} - \frac{\ell(\ell+1)}{r^2} - U_{\mathrm{i}}(r) + k_{\mathrm{i}}^2\right)\chi_\ell(r) = 0. \tag{4.37}$$

In the asymptotic region, the radial distorted wave becomes

$$\lim_{r\to\infty}\chi_\ell(k_{\mathrm{i}},r) = j_\ell + B_\ell\left(-\eta_\ell + \mathrm{i}j_\ell\right) \tag{4.38}$$

where j_ℓ and η_ℓ are regular and irregular Riccatti–Bessel functions, and B_ℓ is a complex number related to the phase shift for elastic scattering (i.e., $B_\ell = \exp(\mathrm{i}\delta_\ell)\sin\delta_\ell$ where δ_ℓ is the elastic scattering phase shift). The asymptotic form (4.38) is appropriate for elastic scattering from and inelastic excitation of neutral atoms, but not for ionization. For scattering from ions, one must replace j_ℓ and η_ℓ with the corresponding Coulomb wavefunctions. The computer program provided here assumes scattering from neutral atoms, and the asymptotic wavefunctions are matched to Riccatti–Bessel functions.

The final-state distorted waves may be expanded similarly to (4.37), except that one must take the complex conjugate of the radial distorted wave to obtain the incoming-wave boundary conditions. This leads to

$$|\chi_{\mathrm{f}}^{-}\rangle = \sqrt{\frac{2}{\pi}}\frac{1}{k_{\mathrm{f}}r}\sum_{\ell_{\mathrm{f}}m_{\mathrm{f}}} \mathrm{i}^{\ell_{\mathrm{f}}}\chi^{*}_{\ell_f}(k_f,r)\, Y_{\ell_{\mathrm{f}}m_{\mathrm{f}}}(\hat{\boldsymbol{r}})Y^{*}_{\ell_{\mathrm{f}}m_{\mathrm{f}}}(\hat{\boldsymbol{k}}_{\mathrm{f}}). \tag{4.39}$$

The discrete atomic wavefunctions are expressed as

$$\psi_k(\boldsymbol{r}) = \frac{1}{r}P_{N_kL_k}(r)Y_{L_kM_k}(\hat{\boldsymbol{r}}) \tag{4.40}$$

and the electron–electron interaction is expanded as

$$-\frac{2Z_\mathrm{p}}{r_{01}} = -8\pi Z_\mathrm{p} \sum_{\ell m} \hat{\ell}^{-2} A_\ell(r_0,r_1) Y^*_{\ell m}(\hat{\boldsymbol{r}}_1) Y_{\ell m}(\hat{\boldsymbol{r}}_0)\,, \tag{4.41}$$

where

$$A_\ell(r_0,r_1) = r_<^\ell / r_>^{\ell+1}\,, \tag{4.42}$$

with $r_<$ $(r_>)$ = max (min) $\{r_0, r_1\}$ and $\hat{\ell} \equiv \sqrt{2\ell+1}$.

4.4.1 Inelastic Scattering

With these expansions, the first-order amplitude (4.27) for inelastic scattering can be expressed as

$$\begin{aligned} T_1^S = \sum_{\ell_\mathrm{i}\ell_\mathrm{f}J\mu} &\left[D^J_{\ell_\mathrm{i}\ell_\mathrm{f}} + (-1)^S E^J_{\ell_\mathrm{i}\ell_\mathrm{f}} \right] C(L_\mathrm{i}\ell_\mathrm{i}J; M_\mathrm{i}, \mu - M_\mathrm{i}, \mu) \\ &\times C(L_\mathrm{f}\ell_\mathrm{f}J; M_\mathrm{f}, \mu - M_\mathrm{f}, \mu)\, Y_{\ell_\mathrm{f},\mu-M_\mathrm{f}}(\hat{\boldsymbol{k}}_\mathrm{f})\, Y^*_{\ell_\mathrm{i},\mu-M_\mathrm{i}}(\hat{\boldsymbol{k}}_\mathrm{i}) \end{aligned} \tag{4.43}$$

where $L_\mathrm{i}M_\mathrm{i}$ $(L_\mathrm{f}M_\mathrm{f})$ are the angular momentum quantum numbers of the initial (final) atomic state. In (4.43), $C(\ell_1\ell_2\ell_3; m_1, m_2, m_3)$ is a Clebsch-Gordan coefficient,

$$D^J_{\ell_\mathrm{i}\ell_\mathrm{f}} = \frac{1}{k_\mathrm{i}k_\mathrm{f}} \int \chi_{\ell_f}(k_f, r_0) F_{\ell_\mathrm{i}\ell_\mathrm{f}J}(r_0) \chi_{\ell_i}(k_i, r_0)\, \mathrm{d}r_0 \tag{4.44}$$

and

$$E^J_{\ell_\mathrm{i}\ell_\mathrm{f}} = \frac{1}{k_\mathrm{i}k_\mathrm{f}} \int\!\!\int \chi_{\ell_f}(k_f, r_0) G_{\ell_\mathrm{i}\ell_\mathrm{f}J}(r_0, r_1) \chi_{\ell_i}(k_i, r_1)\, \mathrm{d}r_0 \mathrm{d}r_1 . \tag{4.45}$$

Furthermore, we have defined

$$\begin{aligned} F_{\ell_\mathrm{i}\ell_\mathrm{f}J}(r) = -\frac{4Z_\mathrm{p}}{\pi} \sum_\ell &\mathrm{i}^{\ell_\mathrm{i}-\ell_\mathrm{f}} (-1)^{J+L_\mathrm{f}+\ell_\mathrm{i}} \hat{\ell}_\mathrm{i} \hat{L}_\mathrm{i} W(\ell_\mathrm{i}\ell_\mathrm{f}L_\mathrm{i}L_\mathrm{f}; \ell J) \\ &\times C(L_\mathrm{i}\ell L_\mathrm{f}; 0,0,0) C(\ell_\mathrm{i}\ell\ell_\mathrm{f}; 0,0,0) I^\mathrm{dir}_{L_\mathrm{i}\ell L_\mathrm{f}}(r) \end{aligned} \tag{4.46}$$

with

$$I^\mathrm{dir}_{L_\mathrm{i}\ell L_\mathrm{f}}(r) = \int P_{N_\mathrm{f}L_\mathrm{f}}(r_1) A_\ell(r, r_1) P_{N_\mathrm{i}L_\mathrm{i}}(r_1) \mathrm{d}r_1\,, \tag{4.47}$$

and

$$\begin{aligned} G_{\ell_\mathrm{i}\ell_\mathrm{f}J}(r_0,r_1) = -\frac{4Z_\mathrm{p}}{\pi} \sum_\ell &\mathrm{i}^{\ell_\mathrm{i}-\ell_\mathrm{f}} (-1)^{\ell_\mathrm{f}+L_\mathrm{i}} \hat{\ell}_\mathrm{i} \hat{L}_\mathrm{i} W(L_\mathrm{i}\ell_\mathrm{f}\ell_\mathrm{i}L_\mathrm{f}; \ell J) \\ &\times C(L_\mathrm{i}\ell\ell_\mathrm{f}; 0,0,0) C(\ell_\mathrm{i}\ell L_\mathrm{f}; 0,0,0) I^\mathrm{exch}_{L_\mathrm{i}\ell L_\mathrm{f}}(r_0, r_1) \end{aligned} \tag{4.48}$$

with

$$I^{\text{exch}}_{L_i \ell L_f}(r_0, r_1) = P_{N_f L_f}(r_1) A_\ell(r_0, r_1) P_{N_i L_i}(r_0) . \tag{4.49}$$

It should be noted that the form of the amplitude (4.43) is a general one and not specific to distorted-wave amplitudes. Notice that the direct and exchange partial-wave amplitudes (4.44) and (4.45) are independent of the initial and final atomic magnetic sublevels and also independent of the coordinate system (here we do not specify any particular coordinate system). All the information about magnetic quantum numbers is contained in the Clebsch-Gordan coefficients and the spherical harmonics in (4.43).

4.4.2 Elastic Scattering

The partial-wave expressions for the elastic scattering amplitudes may be obtained in a similar fashion to that for the excitation amplitudes. The result for the direct scattering amplitude (4.35) is

$$f^{\text{dir}}_{\text{elastic}} = \sum_{J\mu} D^J Y_{J\mu}(\hat{\boldsymbol{k}}_f) Y^*_{J\mu}(\hat{\boldsymbol{k}}_i) , \tag{4.50}$$

where

$$D^J = \frac{2}{\pi} \frac{1}{k_i^2} \int \chi_J(k_f, r)\, V_{ii}(r)\, j_J(k_i, r)\, dr. \tag{4.51}$$

When expressions (4.50) and (4.51) are compared with the corresponding excitation results (4.43) and (4.44), it is seen that the elastic scattering results are a special case of the excitation results. In fact, if we set

$$L_i = L_f = 0 , \tag{4.52a}$$

$$F_{\ell_i \ell_f J}(r) = \frac{2}{\pi}\, V_{ii}(r) \delta_{\ell_i J} \delta_{\ell_f J} , \tag{4.52b}$$

$$\chi_J(k_i, r) = j_J(k_i, r) , \tag{4.52c}$$

the expression for the direct excitation amplitude reduces to the direct elastic scattering result. For the case of exchange, the amplitude (4.34) is of the same form for both elastic scattering and for excitation. Consequently, the partial-wave expansions of (4.43), (4.45) and (4.48) are also appropriate for elastic scattering if one simply sets $\psi_f = \psi_i$ and calculates χ_f using the same distorting potential as was used in the calculation of χ_i. The present formalism may be applied to elastic scattering from any atomic state, but the computer program supplied with this chapter was designed for elastic scattering from an S state. It must be modified if one wishes to calculate elastic scattering from a state with non-vanishing angular momentum.

4.4.3 Differential Cross Sections

The scattering amplitude of (4.43) may be used to calculate any physically observable quantity. This will be further discussed in Chap. 11 where the output from the scattering programs is used as input data for the calculation of various parameters. For completeness, the program for the present chapter will also calculate differential cross sections. With the continuum waves normalized to a deltafunction in $\boldsymbol{k}$, the singlet and triplet differential cross sections (in a_0^2/sr with a_0 as the Bohr radius) are given by

$$\sigma^S = 4\pi^4 \frac{k_\mathrm{f}}{k_\mathrm{i}} \left|T^S\right|^2 , \tag{4.53}$$

where T^S is the singlet or triplet scattering amplitude of (4.43). The differential cross section (4.53) will depend upon the initial and final angular momenta of the atom as well as the scattering angle.

For a practical calculation of differential cross sections, it is necessary to choose a coordinate system. The two most commonly used systems are one in which the z axis is parallel to the incident beam direction or one in which the z axis is perpendicular to the scattering plane. The former choice simplifies the angular momentum algebra somewhat and will be used here. Results may be transformed from one system to the other through a simple rotation as discussed in Chaps. 10 and 11. If one picks the z axis parallel to the incident beam direction and the y axis perpendicular to the scattering plane, the general scattering amplitude (4.43) takes the form

$$\begin{aligned} T^S = & \frac{1}{\sqrt{4\pi}} \sum_{\ell_\mathrm{i}\ell_\mathrm{f}J} \hat{\ell}_\mathrm{i} \left[D^J_{\ell_\mathrm{i}\ell_\mathrm{f}} + (-1)^S E^J_{\ell_\mathrm{i}\ell_\mathrm{f}} \right] C(L_\mathrm{i}\ell_\mathrm{i}J; M_\mathrm{i}, 0, M_\mathrm{i}) \\ & \times C(L_\mathrm{f}\ell_\mathrm{f}J; M_\mathrm{f}, M_\mathrm{i} - M_\mathrm{f}, M_\mathrm{i})\, Y_{\ell_\mathrm{f}, M_\mathrm{i} - M_\mathrm{f}}(\theta, 0) , \end{aligned} \tag{4.54}$$

where θ is the scattering angle. It is important to notice that the amplitudes $D^J_{\ell_\mathrm{i}\ell_\mathrm{f}}$ of (4.44) and $E^J_{\ell_\mathrm{i}\ell_\mathrm{f}}$ of (4.45) do not depend either on the magnetic quantum numbers or on the coordinate system.

Since the ground state of hydrogen and the hydrogen-like alkali atoms is an S state, it is convenient to set $L_\mathrm{i} = M_\mathrm{i} = 0$. For this case, (4.54) reduces to

$$\begin{aligned} T^S = & \frac{1}{\sqrt{4\pi}} \sum_{\ell_\mathrm{f}J} \hat{J} \left[D^J_{J\ell_\mathrm{f}} + (-1)^S E^J_{J\ell_\mathrm{f}} \right] \\ & \times\ C(L_\mathrm{f}\ell_\mathrm{f}J; M_\mathrm{f}, -M_\mathrm{f}, 0)\, Y_{\ell_\mathrm{f}, -M_\mathrm{f}}(\theta, 0). \end{aligned} \tag{4.55}$$

The scattering amplitude (4.55) and the resulting differential cross sections depend on the final angular momentum L_f and the magnetic sublevel M_f. The computer program will calculate both the magnetic sublevel cross sections and the cross section summed over these sublevels. As a final note, although the code for calculating the cross section assumes an initial S state and evaluates (4.55), the code for evaluating the $D^J_{\ell_i \ell_f}$ and the $E^J_{\ell_i \ell_f}$ amplitudes is written for any initial angular momentum. Consequently, the program is designed such that it may be generalized to more complicated scattering situations.

4.5 Computer Program

The computer program `DWBA1` supplied with this chapter will calculate the direct and exchange amplitudes for both elastic and inelastic scattering. The program is written for electron or positron scattering from hydrogen or any atom with one electron outside a closed shell configuration. For the case of scattering from hydrogen, the program will generate analytic hydrogenic wavefunctions and distorting potentials. For the case of scattering from an atom with a single electron outside a closed shell, one has to read in the initial and final atomic wavefunctions $P_{N_i L_i}(r)$ and $P_{N_f L_f}(r)$ of (4.40), the initial and final distorting potentials U_i and U_f (multiplied by r), and finally the radial mesh for all these functions. It should be noted that the program assumes that all four functions are on the same radial mesh and that the distorting potentials are in Rydberg energy units. In this section we will describe some of the important components of the program as well as the input data, and we will also present some sample output.

4.5.1 Radial Mesh

The radial mesh used by the program is an adaption of the Herman–Skillman mesh [4.10] which is a linear approximation to a logarithmic mesh. In the original Herman–Skillman mesh, a constant step size is used for forty points, then the step size is doubled for the next forty points. This process is repeated for each succeeding block of forty points. The problem with this procedure lies in the fact that the step size soon becomes too large for acceptable numerical accuracy. Consequently, we have modified this scheme for forming the radial mesh by doubling the step size until the largest numerically acceptable step size is reached and then by keeping the step size constant for all larger radii. The program contains an internal parameter defined as `NDS` which determines the point at which a constant step size will be used. The default value for this parameter is 7 which means that the original step size will be doubled six times and then kept constant thereafter.

The initial step size is determined by the size of the nuclear charge `ZNUC` as follows:

$$r_1 = 2.2133535 \times 10^{-3} \,/\, \mathtt{ZNUC}^{1/3} \tag{4.56}$$

and the maximum radius used in the calculation is

$$r_{\max} = 81\,880\, r_1 \,. \tag{4.57}$$

The internal parameter MAXF is the number of points required to reach the maximum radius. For the default mesh, this parameter is 2 720.

A finer mesh is required to solve the second-order differential equations than is necessary to perform the required integrals. Consequently, the program is set up to skip over points when performing the integrals. The internal parameter JUMP determines the number of points which are skipped in the numerical integration. The default value for JUMP is 2 which means that every second point is used in the integration process. This value for JUMP is *strongly* recommended. If JUMP is changed from 2, it must be even and 40/JUMP must be an integer.

The radial mesh used by the program is accurate for low-energy incident projectiles, but the numerical accuracy decreases with increasing projectile energy. As a rough rule-of-thumb, the default mesh used in the code is reasonably good up to about 150 eV incident energy but should not be used for energies greater than about 200 eV. If one tries to run the program for energies greater than 200 eV, it will stop immediately with an error message. However, higher energies can be calculated by making the mesh finer. This can be accomplished by decreasing the internal parameter NDS and by increasing the dimensions of the radial arrays. The next finer mesh may be obtained by decreasing NDS from 7 to 6 and by increasing MAXF from 2 720 to 5 240. In general, MAXF is related to NDS through

$$\mathtt{MAXF} = 40 \times \left(\mathtt{NDS} - 3 + 2^{13-\mathtt{NDS}}\right). \tag{4.58}$$

However, one must be very careful if NDS is reduced to less than 6 since the large number of points in the radial mesh can cause a significant accumulation error in the numerical solution of the differential equation.

4.5.2 Numerical Technique

The radial differential equation (4.37) is solved using Numerov's method. With this method, one chooses the first two points of the solution and these two points are used to form the third point. Once the third point is calculated, the second and third points are used to find the fourth point, and the process is repeated until the maximum radius is reached. To start the calculation, the first point (for $r = 0$) is set to zero and the second point is set to $r_1^{\ell+1}$ where r_1 is given by (4.56). Note that for large angular momentum, the second point can easily underflow on many computers. If this happens, this point is set equal to some small (non-zero) number. If the initial slope is too large, it is also likely that the solution will overflow before the maximum radius is reached. The program is written to watch for this possibility and scale the solution if this starts to happen. When the maximum radius is reached, the numerical solution will be an unnormalized form of the asymptotic solution (4.38). The normalization factor is found by matching the logarithmic derivative of the numerical solution to the logarithmic derivative of (4.38).

Before normalization, the numerical solution is real, but the normalization constant

$$N_\ell = a_l \exp(i\delta_\ell) \tag{4.59}$$

is complex. It can be shown that the phase δ_ℓ of this normalization constant is the elastic scattering phase shift. As a result, the real unnormalized numerical solution is multiplied by a_ℓ, and the routine returns this wavefunction along with the complex number $\exp(i\delta_\ell)$.

The computer program prints out the phase shift for each of the distorted waves. These phase shifts are not used any further but are nonetheless interesting since they indicate when the distorted waves become equivalent to plane waves (phase shift 0°). If the distorting potentials are set to zero, a PWB1 calculation is performed and all the phase shifts should vanish. For the radial mesh used in the program, the typical numerical error for most computers would produce phase shifts of the order of 0.2° in this case. Such errors are acceptable for the calculation of distorted waves, but more accurate phase shifts can be obtained with a finer radial mesh.

All the radial integrations are performed using the elementary Simpson's three-point method. While there are numerous higher-order methods which could produce equivalent accuracy with fewer mesh points, the present code is fairly easy to understand and provides reliable results in a reasonable amount of time. We note at this point that comprehensibility and efficiency of a computer program are sometimes difficult to achieve at the same time. Therefore, a compromise was attempted between recalculating some numbers and designing a more complicated flow structure with a large amount of storage space. It is left as an exercise to the interested reader to optimize the code for particular needs in regard to time and storage efficiency.

A simple, but illustrative example for the above point is the calculation of the exchange integral

$$\begin{aligned} I &= \int_0^\infty \mathrm{d}r' \int_0^\infty \mathrm{d}r\, \chi_{\ell_\mathrm{f}}(k_\mathrm{f}, r')\, P_{N_\mathrm{f} L_\mathrm{f}}(r)\, \frac{r_<^\ell}{r_>^{\ell+1}}\, \chi_{\ell_\mathrm{i}}(k_\mathrm{i}, r)\, P_{N_\mathrm{i} L_\mathrm{i}}(r') \\ &= \int_0^\infty \mathrm{d}r\, \chi_{\ell_\mathrm{i}}(k_\mathrm{i}, r)\, P_{N_\mathrm{f} L_\mathrm{f}}(r) \times \left\{ \frac{1}{r^{\ell+1}} \int_0^r \chi_{\ell_\mathrm{f}}(k_\mathrm{f}, r')\, r'^\ell\, P_{N_\mathrm{i} L_\mathrm{i}}(r')\, \mathrm{d}r' \right. \\ &\quad \left. + r^\ell \int_r^\infty \chi_{\ell_\mathrm{f}}(k_\mathrm{f}, r')\, \frac{1}{r'^{\ell+1}}\, P_{N_\mathrm{i} L_\mathrm{i}}(r')\, \mathrm{d}r' \right\} \end{aligned} \tag{4.60}$$

(cf Sect. 4.4). A straightforward evaluation of this double integral by setting up the integrand for every meshpoint (this is the "obvious" procedure) is extremely inefficient. However, the process can be speeded up significantly by rewriting (4.60) as

$$I = \int_0^\infty \mathrm{d}r\, \chi_{\ell_\mathrm{i}}(k_\mathrm{i}, r)\, P_{N_\mathrm{f} L_\mathrm{f}}(r) \left\{ \frac{1}{r^{\ell+1}} Y_1(r) + r^\ell Y_2(r) \right\}. \tag{4.61}$$

The function $Y_1(r)$ can now be evaluated in an outward integration (begin at $r = 0$) by using Simpson's three-point rule. However, this procedure will only give $Y_1(r)$ at every other meshpoint, i.e., on the JUMP $= 4$ mesh if JUMP $= 2$ (again, we strongly recommend not to change this value of JUMP !). Furthermore, it should be noted that $Y_1(r)$ will not change significantly after the meshpoint IPICUT (evaluated in the program) where the bound-state wavefunction $P_{N_i L_i}(r)$ is negligible. Beyond the same point IPICUT, the integral $Y_2(r)$ is essentially zero, and hence its values on the JUMP $= 4$ mesh can most easily be obtained by integrating backwards towards the origin (note that IPICUT must be even for Simpson's three-point rule). The final integral (4.61) can then be evaluated with the integrand given on the JUMP $= 4$ mesh, and again significant contributions will only be obtained from meshpoint numbers smaller than IPFCUT beyond which $P_{N_f L_f}(r)$ is negligible.

It should be noted that this procedure uses different mesh sizes in the calculations of the outer and inner integrals of (4.60). Although this feature may not be consistent with general numerical principles, the error is small ($\leq 1\%$) for practical applications of the present program. Improvements could be obtained, for example, by interpolating $Y_1(r)$ and $Y_2(r)$ back to the JUMP $= 2$ mesh or by averaging the result above and that obtained after a switch of the outer and inner integrals. The purpose of the present comment, however, was to demonstrate that the calculation of the double integral (4.60) can be reduced to the evaluation of three one-dimensional integrals, provided additional storage is used to hold temporary results on the JUMP $= 4$ mesh.

A further important problem deals with the calculation of the results for high ℓ values which are often needed to obtain converged cross sections and particularly to avoid wiggles in the results for differential cross sections at large scattering angles. However, it is undesirable to numerically evaluate very high ℓ values due to long execution times and large numerical errors.

A standard method for solving the above problem is to approximate the results for high ℓ values by analytic expressions. A typical example is the effective range formula of O'Malley *et al* [4.11]. This formula provides the **T**-matrix elements for elastic scattering in the presence of a long-range r^{-4} potential representing the polarization of the atomic charge cloud by the incident projectile. The present code is not set up to deal with such long-range potentials for elastic scattering, and we therefore do not include any analytic continuation of the results to high ℓ values for this case. However, only minor changes are necessary to include these effects, and these are again left as an exercise to the interested reader. We also note that the results for optically forbidden transitions usually converge very rapidly with increasing ℓ, and therefore no special procedure was included for the inelastic S–S transitions.

T-matrix results for large ℓ values are needed for the optically allowed S–P transitions, however. For this case, the semi-empirical procedure described below is very fast and generally produces results of sufficient accuracy, although a stringent theoretical justification has not been given so far. In detail, we proceed as follows. For $\ell \geq$ JFIT, we neglect the exchange matrix element $E^J_{\ell_i \ell_f}$ of (4.45) and take into account only the direct contribution $D^J_{\ell_i \ell_f}$ from (4.44) in (4.43). With $J = \ell_i \equiv \ell$ for an initial S-state of even parity and a final P-state of odd parity, we are then left with $\ell_f = \ell \pm 1$. We now assume that for $\ell \geq$ JFIT the direct matrix elements behave according to

$$D^\ell_{\ell,\ell\pm1} \equiv D^\pm_\ell = a \exp\{-b_\pm \ell\} \,. \tag{4.62}$$

Starting at $\ell =$ JFIT, we step ℓ by 5 and calculate $D^\pm_\ell$ numerically until $\ell =$ JFSTOP is reached. The parameters a and $b_\pm$ of (4.62) are then determined by a least-squares fit to the numerical values from JFIT to JFSTOP, and amplitudes for ℓ values greater than JFIT are calculated using (4.62) with the least-squares-fit parameters. Note that the appropriate values for JFIT and JFSTOP depend on the incident electron energy. The values that we found to work reasonably well are suggested in comment statements at the beginning of the program. If they were set in a region where (4.62) is not valid, one may get (unphysical) large-angle oscillations in the results for the cross section and the angular correlation parameters. Furthermore, the program will abort in the fitting procedure under the following two circumstances: (i) $D^\pm_\ell$ at JFIT+5 differs in sign from $D^\pm_\ell$ at JFIT, or (ii) the magnitude of $D^\pm_\ell$ at JFIT+5 is larger than that of $D^\pm_\ell$ at JFIT. If either one of these circumstances occurs, one must solve the problem by changing the value of JFIT and either increasing or decreasing its value until (4.62) is valid in the fitting range. Finally, the quality of the fit is determined and printed out. For all cases we have tested, the fit has typically been very good.

We will not go any further into the numerical details and just note that standard routines have been implemented to calculate, for example, the spherical Bessel functions (see also Chap. 9), the spherical harmonics and the angular momentum coefficients.

4.5.3 Program Input

The program requires seven lines of input which are read from tape unit IRD (the input and output units are set in a PARAMETER statement). All the input statements are format free, i.e., the data must be separated by either a comma or a space. All the input parameters are integers except for ENERGY, SMALL1, SMALL2, SMALL3 and SMALL4.

1. NI, LIA, NF, LFA, ENERGY

NI	principal quantum number for the initial state
LIA	angular momentum of the initial state
NF	principal quantum number for the final state
LFA	angular momentum for the final state

For elastic scattering set NI = NF and LIA = LFA.
The program is designed to treat analytically the $(1s)^2S^e$, $(2s)^2S^e$ and $(2p)^2P^o$ states of atomic hydrogen.

`ENERGY` energy of the incident projectile in eV

2. `NZP, NZNUC`

   `NZP` charge of the projectile (+1 or -1)

   `NZNUC` nuclear charge

3. `NUI, LUI`

   If the initial-state distorting potential U_i is evaluated analytically (`NZNUC` = 1), `NUI` is the N value and `LUI` is the L value for U_i. The program contains analytic formulae for potentials obtained from hydrogenic 1s, 2s and 2p orbitals. If `NUI` = `LUI` = 0, U_i will be set to zero and the initial-state wavefunctions will be plane waves.

4. `NUF, LUF`

   Same as 3 except for the final state.

5. `SMALL1, SMALL2, SMALL3, SMALL4`

   `SMALL1` value for which the atomic wavefunctions are small enough to be considered zero.

   `SMALL2` determines when the form factor of (4.47) may be considered zero. When the form factor becomes smaller than `SMALL2` * (maximum value of the form factor), it is set to zero for all larger radii.

   `SMALL3` if, for $\ell_f = \ell_i$ or $\ell_f = \ell_i - 1$, the magnitude of the exchange integral becomes smaller than `SMALL3` * (magnitude of the direct integral), the exchange contribution is neglected for higher ℓ values.

   `SMALL4` if the magnitude of the direct integral becomes smaller than `SMALL4` * (maximum of the direct integral for small ℓ values), the sum in ℓ is considered to have converged.

6. `NIPF, IPF(J), J = 1, NIPF`

   The program contains the option for printing various intermediate results. `NIPF` is the number of desired print options. Only if the number of the desired intermediate print option is included on this line, will it be printed. The possible print options are listed at the top of the computer program.

7. `ITAPUI, ITAPUF, ITPSII, ITPSIF, ITAPRM, ITAOUT`

   If U_i, U_f, $P_{N_i L_i}$ and $P_{N_f L_f}$ are read as input, they are each read from a different tape unit number. The assigned units should be

   `ITAPUI` r-multiplied initial-state distortion potential U_i (in Rydberg)

   `ITAPUF` r-multiplied final-state distortion potential U_f (in Rydberg)

   `ITPSII` initial-state atomic wavefunction $P_{N_i L_i}$

   `ITPSIF` final-state atomic wavefunction $P_{N_f L_f}$

   `ITAPRM` radial mesh

   In the mesh file, the first line should contain an integer indicating the number of radial points for the function, and the remaining lines should contain the function in the format `5E14.7`. The number of points in each file should be the same and equal to the number of points in the radial mesh. The wavefunctions and potentials which are read in are interpolated onto the radial mesh used by the computer program. If the maximum radius for the input radial mesh is less than $r_{\max}$ of (4.57), the wavefunctions and potentials are set to zero for radii larger than the maximum input r value.

`ITAOUT` output unit for the scattering amplitudes (see below).

8. `EI, EF` (only if `NZNUC` $\neq 1$)
   These are the ionization potentials for the initial and final states (in Rydberg) if the target is not atomic hydrogen.
9. `JFIT, JFSTOP` (only if `LFA` $= 1$)
   See text.

4.5.4 Sample Input and Output

The input file (read from unit `IRD` = 5) for 54.4 eV electron excitation of hydrogen from the $(1s)^2S^e$ to the $(2p)^2P^o$ state with the initial-state distorting potential being V_{ii} and the final-state distorting potential being V_{ff} would look like

```
  1     0     2     1     54.4
 -1     1
  1     0
  2     1
  1.0E-6     1.0E-6     1.0E-4     1.0E-5
  6    15    20    25    55    60    65
  0     0     0     0     0    16
 35    70
```

The print output is written to unit `IWRI` (set to 6 in the present program) and the scattering amplitudes (4.55) for the magnetic sublevels are written to unit 16. These amplitudes can then be used to calculate the various observables of interest (see Chap. 11). For comparison, we list in Tables 4.1 and 4.2 for a few scattering angles the singlet and triplet amplitudes (real and imaginary parts) as well as the differential cross section obtained with the above input data. The test run took approximately 10 seconds on the VAX 4000-600 computer installed at Drake University.

Table 4.1. Singlet and triplet amplitudes for electron impact excitation of the $(2p)^2P^o$ state of hydrogen for an incident energy of 54.4 eV.

	$T^{S=0}(M_f = 0)$		$T^{S=1}(M_f = 0)$		$T^{S=0}(M_f = 1)$		$T^{S=1}(M_f = 1)$	
Angle	Re	Im	Re	Im	Re	Im	Re	Im
0	-3.50E-02	3.55E-01	-2.52E-02	3.42E-01	0.00E 00	0.00E 00	0.00E 00	0.00E 00
10	-3.00E-02	7.89E-02	-2.05E-02	6.69E-02	-9.09E-03	9.96E-02	-7.22E-03	9.25E-02
20	-2.05E-02	1.42E-02	-1.19E-02	5.07E-03	-1.20E-02	4.59E-02	-8.90E-03	3.65E-02
30	-1.23E-02	4.97E-04	-5.20E-03	-5.28E-03	-1.01E-02	1.94E-02	-6.73E-03	1.17E-02
40	-6.97E-03	-3.13E-03	-1.35E-03	-6.03E-03	-6.90E-03	7.23E-03	-3.85E-03	2.24E-03
50	-3.62E-03	-3.95E-03	5.59E-04	-4.82E-03	-4.00E-03	1.91E-03	-1.63E-03	-8.49E-04
60	-1.48E-03	-3.90E-03	1.43E-03	-3.50E-03	-1.88E-03	-2.16E-04	-2.52E-04	-1.53E-03
70	-3.39E-05	-3.55E-03	1.82E-03	-2.45E-03	-4.91E-04	-9.17E-04	5.05E-04	-1.42E-03
80	9.80E-04	-3.11E-03	1.98E-03	-1.67E-03	3.54E-04	-9.96E-04	8.65E-04	-1.09E-03
90	1.72E-03	-2.66E-03	2.04E-03	-1.11E-03	8.22E-04	-8.44E-04	9.92E-04	-7.67E-04

Table 4.2. Differential cross sections for electron impact excitation of the $(2p)^2P^o$ state of hydrogen for an incident energy of 54.4 eV.

Angle	$\sigma(M_L = 0)$	$\sigma(M_L = 1)$	σ
0	4.22E 01	0.00E 00	4.22E 01
10	1.92E 00	3.15E 00	8.21E 00
20	9.88E-02	5.70E-01	1.24E-01
30	2.78E-02	9.00E-02	2.08E-01
40	1.52E-02	1.40E-02	4.32E-02
50	8.72E-03	2.62E-03	1.39E-02
60	5.29E-03	9.47E-04	7.19E-03
70	3.56E-03	6.91E-04	4.94E-03
80	2.70E-03	6.08E-04	3.92E-03
90	2.30E-03	5.36E-04	3.37E-03

4.6 Summary

In this chapter we have presented the distorted-wave approach to electron and positron scattering from atoms, together with a simplified computer code to calculate explicit numbers for elastic scattering from and excitation of quasi one-electron targets such as atomic hydrogen or the alkali atoms. The program has been developed particularly for the purpose of this book, which required a compromise between efficiency and comprehensibility. Although this code has never been used for scientific publications, various test runs have been performed to reproduce earlier results. The authors will be grateful for comments and suggestions and should be contacted in case of any problems.

4.7 Suggested Problems

Since the program provided with this chapter is only meant as a start-up code, there are numerous possibilities for further developments. On the other hand, some interesting problems can already be investigated with no or only minor changes to the program. The following list provides some suggestions, ordered approximately in a sequence of increasing complexity.

1. Compare the results for electron and positron scattering.
2. Compare the results obtained with different combinations of the initial-state and final-state distortion potentials U_{i} and U_{f}.
3. Extend the program to printout the amplitudes in a form that can be used by the program described in Chap. 11, or calculate some of the observables directly. Which of the combinations investigated under problem 2 agrees best with the available experimental data?
4. Include the neglected exchange terms in (4.29) and (4.33).
5. Extend the calculations to higher N and L levels (wavefunctions and potentials must be generated).
6. Study alkali targets by using either hydrogenic wavefunctions or orbitals obtained from a more sophisticated structure calculation. Describe the core electrons by a model potential.
7. Modify the program to treat scattering from positive ions. Compare your results for hydrogenic ions as a function of the nuclear charge.
8. Include model potentials to simulate electron exchange, charge cloud polarization and absorption (i.e., loss of flux into inelastic channels) in the calculation of the distorted waves. Note that long-range potentials can influence the matching procedure and the convergence with regard to the necessary number of ℓ values.
9. Extend the theory and the program to treat quasi two-electron systems.

Acknowledgments

The authors would like to thank K.M. DeVries for extensive checks of the computer code. This work has been supported, in part, by the National Science Foundation under grants PHY-8813799 and PHY-9014103.

References

4.1 D.H. Madison and K.H. Winters, J. Phys. B **20** (1987) 4173
4.2 D.H. Madison and W.N. Shelton, Phys. Rev. A **7** (1973) 499
4.3 D.H. Madison and W.N. Shelton, Phys. Rev. A **7** (1973) 514
4.4 D.H. Madison and E. Merzbacher in: *Atomic Inner-Shell Processes* Vol. I, ed. B. Crasemann (Academic Press, New York, 1975)
4.5 D.H. Madison and K.H. Winters, J. Phys. B **16** (1983) 4437
4.6 K. Bartschat and D.H. Madison, J. Phys. B **21** (1988) 153

4.7 K. Bartschat and D.H. Madison, J. Phys. B **20** (1987) 5839
4.8 D.H. Madison, K. Bartschat and R. Srivastava, J. Phys. B **24** (1991) 1839
4.9 K. Bartschat and D.H. Madison, J. Phys. B **20** (1987) 1609
4.10 F. Hermann and S. Skillman, *Atomic Structure Calculations* (Prentice–Hall, Englewood Cliffs, 1963)
4.11 T.F. O'Malley, L. Spruch and L. Rosenberg, J. Math. Phys. **2** (1961) 491

5. Distorted-Wave Methods for Ionization

Ian E. McCarthy[1] and Zhang Xixiang[2]

[1] Institute for Atomic Studies, The Flinders University of South Australia, Bedford Park, S.A. 5042, Australia
[2] Department of Modern Physics, The University of Science and Technology of China, Hefei, Anhui, China

Abstract

The distorted-wave Born (DWBA) and distorted-wave impulse (DWIA) approximations are discussed for electron–atom ionization. Explicit formulae are given in their computational form. A computer code for the ionization of s and p electrons is given. The code uses distorted radial wavefunctions and bound-state input in the form discussed for scattering in Chap. 4. Sample input and output data are given.

5.1 Introduction

The problem of the ionization of an atom by an incident electron is essentially a three-body problem, with Coulomb forces when the three bodies are far apart. The three bodies are the incident and target electrons and the residual ion. This problem cannot be solved in closed form. A satisfactory theoretical description of the differential cross section is arrived at by an iterative process involving close interaction between experiment and theory, where successive theoretical improvements are tested by corresponding experiments. Two theoretical methods that give good descriptions of experimental data in different ranges of validity are the distorted-wave Born (DWBA) and distorted-wave impulse (DWIA) approximations.

For a given incident energy E_0 and specified excitation level, electron scattering depends on one kinematic variable, the scattering angle. In contrast, ionization to a particular final ion state with separation energy ϵ_i depends on four kinematic variables, the energy E_{A} of one of the outgoing electrons (the energy E_{B} of the other electron is determined by energy conservation), the polar angles θ_{A}, θ_{B}, and the relative azimuthal angle ϕ of the outgoing electrons, defined with respect to the incident direction. The differential cross section for ionization is measured as a function of one kinematic variable, with the others fixed at particular values. This gives a rich diversity of possibilities, two of which are described here because they depend on quite different aspects of the reaction mechanism.

In *coplanar asymmetric* kinematics the energy of one of the outgoing electrons, conventionally called E_A, is fixed at a much larger value than that of the other energy E_B. The polar angle θ_A is fairly small (up to about 20° relative to the incident direction), the relative azimuthal angle ϕ is fixed at 0° or 180°, i.e., the incident and outgoing electron directions are coplanar, and the polar angle θ_B of the slow electron is varied. This kinematic arrangement was used in the first differential cross section measurements in 1969 by Ehrhardt *et al.* [5.1]. The magnitude K of the momentum transfer from the incident to the fast electron,

$$\boldsymbol{K} = \boldsymbol{k}_0 - \boldsymbol{k}_A \,, \tag{5.1}$$

is small in comparison with k_A. The ionizing collision between the two electrons is therefore a long-range glancing collision, which has a large cross section (recall, for example, that the differential cross section for Rutherford scattering is proportional to K^{-4}).

Recently the exact boundary condition for the problem of three charged particles was established by Brauner *et al.* [5.2]. It was used in an approximate calculation that gave an excellent description of coplanar asymmetric data for the atomic hydrogen target.

In *noncoplanar symmetric* kinematics both outgoing electrons have the same energy $E_A = E_B$, the polar angles are equal and usually chosen as $\theta_A = \theta_B = 45°$, and the relative azimuth ϕ is varied. In both kinematic cases the momentum transfer $\boldsymbol{K}$ is independent of the angle that is varied. In the latter case, however, its magnitude is almost equal to that of $\boldsymbol{k}_A$ and therefore assumes its largest value. Consequently, the two-electron collision is very close and results in rapid impulsive removal of the target electron. Under such conditions it is intuitively reasonable to consider the recoil momentum of the ion,

$$\boldsymbol{p} = \boldsymbol{k}_0 - \boldsymbol{k}_A - \boldsymbol{k}_B \,, \tag{5.2}$$

to be equal and opposite to the momentum $\boldsymbol{q}$ of the target electron in the bound state. The distribution of $\boldsymbol{p}$, which is measured in the experiment, will then give the momentum distribution of the target electron, which is the absolute square of its momentum-space wavefunction $|\langle \boldsymbol{q}|\psi_i\rangle|^2$. This intuition must be modified by the presence of distortion in the wavefunctions of the incident and outgoing electrons. Nevertheless the experiment is ideal for observation of the electronic structure of the target. This kinematic arrangement was used in the first observation of valence electronic structure in 1973 by Weigold *et al.* [5.3]. Since K is much larger than in the coplanar asymmetric case the cross section is much smaller.

The large-cross-section coplanar asymmetric case is very important for investigating approximations for the effect of ionization on scattering [5.4]. The direct observation of electronic structure afforded by the noncoplanar symmetric arrangement is the basis of the field of electron momentum spectroscopy of atoms, molecules and solids [5.5], which is fully understood in terms of the DWIA.

5.2 Theory

We consider the (e,2e) reaction as inelastic scattering from a one-electron orbital ψ_i of the target ground state to an ionized state of the target, characterized by the electron momentum $\boldsymbol{k}_\mathrm{B}$. We make a three-body model, treating the remainder of the target as an inert core. The Hamiltonian for the problem is

$$H = (K_1 + v_1) + (K_2 + v_2) + v_3. \tag{5.3}$$

The electron labelled 1 is the projectile, whereas the label 2 refers to the electron which is to be knocked out of the target. K_1 and K_2 are the respective kinetic energy operators, v_1 and v_2 are the potentials between the respective electrons and the ion core, and v_3 is the two-electron interaction potential. The core is treated as infinitely massive.

The electrons are in fact indistinguishable and obey the Pauli exclusion principle. We neglect spin-orbit coupling, which is a very good approximation for light target systems such as atomic hydrogen. In this case we have two independent Schrödinger equations, one for each value, 0 (singlet) and 1 (triplet), of the two-electron spin S. The effective Hamiltonian [5.6] is

$$H_S = (K_1 + v_1) + (K_2 + v_2) + V_S \tag{5.4}$$

with

$$V_S = v_3 + (-1)^S (H - E) P_\mathrm{r}, \tag{5.5}$$

where E is the total energy of the reaction and P_r is the space-exchange operator.

The differential cross section for the (e,2e) reaction is

$$\frac{\mathrm{d}^5\sigma}{\mathrm{d}\Omega_\mathrm{A}\mathrm{d}\Omega_\mathrm{B}\mathrm{d}E_\mathrm{A}} = (2\pi)^4 \frac{k_\mathrm{A} k_\mathrm{B}}{k_0} \frac{n_i}{4} \sum_S (2S+1) \left| T_S(\boldsymbol{k}_0, \boldsymbol{k}_\mathrm{A}, \boldsymbol{k}_\mathrm{B}) \right|^2 \tag{5.6}$$

where n_i is the number of equivalent electrons in the target orbital ψ_i and $\mathrm{d}\Omega_\mathrm{A,B} = \sin\theta_\mathrm{A,B} \mathrm{d}\theta_\mathrm{A,B} \mathrm{d}\phi_\mathrm{A,B}$. The **T**-matrix element for spin S is given by

$$T_S(\boldsymbol{k}_0, \boldsymbol{k}_\mathrm{A}, \boldsymbol{k}_\mathrm{B}) = \langle \chi^{(-)}(\boldsymbol{k}_\mathrm{A}) \chi^{(-)}(\boldsymbol{k}_\mathrm{B}) \mid V_S \mid \Psi_S^{(+)}(\boldsymbol{k}_0) \rangle, \tag{5.7}$$

where the distorted waves are solutions of the elastic scattering Schrödinger equations for the two electron–ion subsystems. These are given by

$$(E_\mathrm{A} - K_1 - v_1) \chi^{(-)}(\boldsymbol{k}_\mathrm{A}) = 0 \tag{5.8}$$

and

$$(E_{\rm B} - K_2 - v_2)\chi^{(-)}(\boldsymbol{k}_{\rm B}) = 0. \qquad (5.9)$$

We adopt the convention that the faster outgoing electron (labelled by the subscript "A") has coordinate $\boldsymbol{r}_1$, but we have already taken antisymmetry into account.

In effect we have said in (5.7) that the asymptotic form of the final three-body state is a product of two distorted waves with Coulomb boundary conditions. This is an assumption about the three-body boundary condition, which must be justified by comparison with experiment. In fact this choice has proved to be a good one. The exact boundary condition [5.2] multiplies the distorted waves by the phase factor for the interaction of two electrons in the absence of the core ion.

In considering the calculation of the **T**-matrix element we suppress the two-electron spin index S, since different spin states are independent. The wavefunction for the reaction $\Psi^{(+)}$ is expanded over the complete set of target eigenstates j, using for convenience a discrete notation for the continuum, which may be understood in the sense of enclosing the whole system in a box. Hence

$$|\,\Psi^{(+)}(\boldsymbol{k}_0)\rangle = \sum_j |\,ju_j^{(+)}(\boldsymbol{k}_j)\rangle. \qquad (5.10)$$

Substituting (5.10) into (5.7) we obtain for the (e,2e) amplitude

$$T(\boldsymbol{k}_0, \boldsymbol{k}_{\rm A}, \boldsymbol{k}_{\rm B}) = \sum_j \langle \chi^{(-)}(\boldsymbol{k}_{\rm A})\chi^{(-)}(\boldsymbol{k}_{\rm B}) \mid V \mid ju_j^{(+)}(\boldsymbol{k}_j)\rangle. \qquad (5.11)$$

This amplitude includes the ionization from an excited state of the target, which may be important when the excitation is on a resonance. Experience has shown that resonant excitation is not significant at energies that are at least several multiples of the ionization threshold. We therefore consider only the ground-state term of (5.11), defining the elastic distorted wave $\chi^{(+)}(\boldsymbol{k}_0)$ by

$$\chi^{(+)}(\boldsymbol{k}_0) = u_0^{(+)}(\boldsymbol{k}_0). \qquad (5.12)$$

The scattering wavefunctions $u_j^{(+)}(\boldsymbol{k}_j)$ are formally calculated by solving the set of coupled equations obtained by substituting the multichannel expansion (5.10) into the full Schrödinger equation. This yields

$$\sum_j [E_i^{(+)} - K_1 - V_{ij}]u_j^{(+)}(\boldsymbol{k}_j) = 0\,, \qquad (5.13)$$

where

$$E_i = E - \epsilon_i\,, \qquad (5.14a)$$
$$V_{ij} = \langle i \mid V \mid j\rangle\,, \qquad (5.14b)$$

and V is given by (5.5).

The elastic distorted wave $u_0^{(+)}(\boldsymbol{k})$ may be formally obtained by solving a one-body Schrödinger equation with an optical potential [5.6], which takes the other channels into account by means of a complex polarization term. Experience has shown that the first-order potential V_{00} is sufficient to describe ionization. The present program requires a real potential, which could include a real polarization part if desired. Polarization potentials are thoroughly discussed by Allen *et al.* [5.7].

The approximation to the (e,2e) amplitude obtained from all the above considerations is the distorted-wave Born approximation in which

$$T_S(\boldsymbol{k}_0, \boldsymbol{k}_{\rm A}, \boldsymbol{k}_{\rm B}) = \langle \chi^{(-)}(\boldsymbol{k}_{\rm A})\chi^{(-)}(\boldsymbol{k}_{\rm B}) \mid v_3(1 + (-1)^S)P_{\rm r} \mid \psi_i \chi^{(+)}(\boldsymbol{k}_0)\rangle. \tag{5.15}$$

Here the initial-state distorted wave is calculated in the ground-state average potential V_{00} and the final-state distorted waves are both calculated in the potential of the residual ion.

The distorted-wave impulse approximation is designed for the kinematic situation where the maximum possible momentum transfer from the incident to one of the outgoing electrons occurs. This is symmetric ionization where the electron–electron collision plays its most important role. Note that the DWBA treats the two-electron subsystem only in first order. If we neglect all other potentials in the problem we can treat the two-electron subsystem exactly by using the two-electron operator t_3. This gives the plane-wave impulse approximation where

$$T(\boldsymbol{k}_0, \boldsymbol{k}_{\rm A}, \boldsymbol{k}_{\rm B}) = \langle \boldsymbol{k}_{\rm A}\boldsymbol{k}_{\rm B} \mid t_3(E) \mid \psi_i \boldsymbol{k}_0\rangle. \tag{5.16}$$

We again treat the electrons as distinguishable pending antisymmetrization.

Introducing the momentum space representation of the orbital ψ_i, (5.16) becomes

$$T(\boldsymbol{k}_0, \boldsymbol{k}_{\rm A}, \boldsymbol{k}_{\rm B}) = \int \mathrm{d}^3\boldsymbol{q}\, \langle \boldsymbol{k}_{\rm A}\boldsymbol{k}_{\rm B} \mid t_3(E) \mid \boldsymbol{q}\boldsymbol{k}_0\rangle\langle \boldsymbol{q} \mid \psi_i\rangle. \tag{5.17}$$

Using the translational invariance of t_3 this reduces to

$$T(\boldsymbol{k}_0, \boldsymbol{k}_{\rm A}, \boldsymbol{k}_{\rm B}) = \langle \boldsymbol{k}' \mid t_3(k'^2) \mid \boldsymbol{k}\rangle\langle \boldsymbol{q} \mid \psi_i\rangle\,, \tag{5.18}$$

where

$$\boldsymbol{q} = \boldsymbol{k}_{\rm A} + \boldsymbol{k}_{\rm B} - \boldsymbol{k}_0\,, \tag{5.19a}$$

$$\boldsymbol{k}' = \frac{1}{2}(\boldsymbol{k}_{\rm A} - \boldsymbol{k}_{\rm B})\,, \tag{5.19b}$$

$$\boldsymbol{k} = \frac{1}{2}(\boldsymbol{k}_0 - \boldsymbol{q})\,. \tag{5.19c}$$

In (5.18) the two-electron **T**-matrix element is factorized out of the expression, and antisymmetrization is performed on this factor. It is "half-off-shell", since the relative momentum $\boldsymbol{k}'$ corresponds to the relative kinetic energy k'^2, but the relative momentum $\boldsymbol{k}$ does not. The half-off-shell Coulomb **T**-matrix element has been given by Ford [5.8]. The sum and average over two-electron spin states in the expression (5.6) for the differential cross section is performed, giving

$$\frac{\mathrm{d}^5\sigma}{\mathrm{d}\Omega_\mathrm{A}\mathrm{d}\Omega_\mathrm{B}\mathrm{d}E_\mathrm{A}} = (2\pi)^4 \frac{k_\mathrm{A}k_\mathrm{B}}{k_0} f_{\mathrm{ee}} n_i |M|^2 \tag{5.20}$$

where the electron–electron collision factor is

$$f_{\mathrm{ee}} = \frac{1}{(2\pi^2)^2} \frac{2\pi\eta}{\exp(2\pi\eta)-1} \left[\frac{1}{|\boldsymbol{k}-\boldsymbol{k}'|^4} - \frac{1}{|\boldsymbol{k}+\boldsymbol{k}'|^4} - \frac{1}{|\boldsymbol{k}-\boldsymbol{k}'|^2} \frac{1}{|\boldsymbol{k}+\boldsymbol{k}'|^2} \cos\left(\eta \ln \frac{|\boldsymbol{k}-\boldsymbol{k}'|^2}{|\boldsymbol{k}+\boldsymbol{k}'|^2} \right) \right] \tag{5.21}$$

with

$$\eta = 1/2k'. \tag{5.22}$$

The plane-wave overlap factor M is given by

$$M = \langle \boldsymbol{q} \mid \psi_i \rangle = \langle \boldsymbol{k}_\mathrm{A} \boldsymbol{k}_\mathrm{B} \mid \psi_i \boldsymbol{k}_0 \rangle . \tag{5.23}$$

We now re-introduce the electron–ion potentials into the approximation by substituting for the plane waves of (5.23) the same distorted waves as we have defined for the DWBA (5.15). The differential cross section for the DWIA is given by (5.20) with the distorted-wave overlap factor

$$M = \langle \chi^{(-)}(\boldsymbol{k}_\mathrm{A}) \chi^{(-)}(\boldsymbol{k}_\mathrm{B}) \mid \psi_i \chi^{(+)}(\boldsymbol{k}_\mathrm{A}) \rangle. \tag{5.24}$$

From the computational point of view the main difference between the DWBA and DWIA is that the former is a six-dimensional integral over the electron coordinates, which reduces to two-dimensional radial integrals on expanding the distorted waves in angular momentum states. The DWIA reduces to one-dimensional radial integrals.

5.3 Reduction of the (e,2e) Amplitudes to Computational Form

The expressions T of (5.15) and M of (5.24) are expressed in terms of functions of one variable by using the partial-wave expansion of the distorted wave

$$\langle \boldsymbol{r} \mid \chi^{(\pm)}(\boldsymbol{k}) \rangle = (2\pi)^{-3/2} (4\pi/kr) \times \sum_{LM} \mathrm{i}^{\pm L} \, \mathrm{e}^{\mathrm{i}\sigma_L} \chi_L(k,r) Y^*_{LM}(\hat{\boldsymbol{k}}) Y_{LM}(\hat{\boldsymbol{r}}) , \tag{5.25}$$

and the multipole expansion of the Coulomb potential

$$v_3(\boldsymbol{r}_1, \boldsymbol{r}_2) = \sum_{\lambda\mu} \frac{4\pi}{\hat{\lambda}^2} \frac{r_<^\lambda}{r_>^{\lambda+1}} Y_{\lambda\mu}^*(\hat{\boldsymbol{r}}_1) Y_{\lambda\mu}(\hat{\boldsymbol{r}}_2), \tag{5.26}$$

where we have used the abbreviation

$$\hat{\lambda} = (2\lambda + 1)^{1/2}. \tag{5.27}$$

Furthermore, $r_<$ and $r_>$ denote the lesser and greater of r_1 and r_2, respectively, and σ_L is the Coulomb phase shift. Finally, the orbital wavefunction of the knocked-out electron is

$$\langle \boldsymbol{r} \mid \psi_i \rangle = r^{-1} u_\ell(r) Y_{\ell m}(\hat{\boldsymbol{r}}). \tag{5.28}$$

The angular integrations are accomplished by using the Gaunt theorem

$$\int \mathrm{d}\hat{\boldsymbol{r}}\, Y_{LM}(\hat{\boldsymbol{r}}) Y_{L'M'}(\hat{\boldsymbol{r}}) Y_{L''M''}^*(\hat{\boldsymbol{r}}) = (4\pi)^{1/2} \hat{L}\hat{L}'\hat{L}'' \begin{pmatrix} L & L' & L'' \\ M & M' & M'' \end{pmatrix} \begin{pmatrix} L & L' & L'' \\ 0 & 0 & 0 \end{pmatrix}, \tag{5.29}$$

where the symbols in parentheses are Wigner $3j$-symbols. We choose $\hat{\boldsymbol{k}}_0$ to be the angular quantization axis, so that

$$Y_{LM}^*(\hat{\boldsymbol{k}}_0) = (4\pi)^{-1/2} \hat{L}, \tag{5.30a}$$

$$M = 0. \tag{5.30b}$$

5.3.1 Distorted-Wave Born Approximation: Direct Amplitude

We first consider the direct ionization amplitude

$$T_\mathrm{D}(\boldsymbol{k}_0, \boldsymbol{k}_\mathrm{A}, \boldsymbol{k}_\mathrm{B}) = \langle \chi^{(-)}(\boldsymbol{k}_\mathrm{A}) \chi^{(-)}(\boldsymbol{k}_\mathrm{B}) \mid v_3 \mid \psi_i \chi^{(+)}(\boldsymbol{k}_0) \rangle. \tag{5.31}$$

The association of the angular quantum numbers and the radial variables with the corresponding wavefunctions is described by Table 5.1. Note that we choose

$$k_\mathrm{A} \geq k_\mathrm{B}. \tag{5.32}$$

Arranging the order of summation for efficient computation, we have

$$\begin{aligned} T_\mathrm{D}(\boldsymbol{k}_0, \boldsymbol{k}_\mathrm{A}, \boldsymbol{k}_\mathrm{B}) &= (2\pi)^{-9/2} (4\pi)^{5/2} (k_0 k_\mathrm{A} k_\mathrm{B})^{-1} \\ &\times \sum_{L'L''M''} \Bigg[\sum_\lambda \begin{pmatrix} \ell & \lambda & L'' \\ 0 & 0 & 0 \end{pmatrix} \begin{pmatrix} \ell & \lambda & L'' \\ m & -M''-m & M'' \end{pmatrix} \\ &\times \sum_L \begin{pmatrix} L' & \lambda & L \\ 0 & 0 & 0 \end{pmatrix} \begin{pmatrix} L' & \lambda & L \\ m+M'' & -M''-m & 0 \end{pmatrix} \\ &\times R^{(\lambda)}_{L'L''\ell L}(k_0, k_\mathrm{A}, k_\mathrm{B}) \Bigg] Y_{L'm+M''}(\hat{\boldsymbol{k}}_\mathrm{A}) Y_{L''M''}(\hat{\boldsymbol{k}}_\mathrm{B}). \end{aligned} \tag{5.33}$$

Table 5.1. Association of angular quantum numbers and radial variables with wavefunctions.

function	angular momentum	radial variable
$\langle \boldsymbol{r}_1 \mid \chi^{(+)}(\boldsymbol{k}_0)\rangle$	$L,0$	r_1
$\langle \boldsymbol{r}_1 \mid \chi^{(-)}(\boldsymbol{k}_\mathrm{A})\rangle$	L',M'	r_1
$\langle \boldsymbol{r}_2 \mid \chi^{(-)}(\boldsymbol{k}_\mathrm{B})\rangle$	L'',M''	r_2
$\langle \boldsymbol{r}_2 \mid \psi_i\rangle$	ℓ,m	r_2

The direct radial integration factor is given by

$$R^{(\lambda)}_{L'L''\ell L}(k_0, k_\mathrm{A}, k_\mathrm{B}) = \mathrm{i}^{L-L'-L''} \mathrm{e}^{\mathrm{i}(\sigma_{L'}+\sigma_{L''})} N_L N_{L'} N_{L''} \hat{L}'\hat{L}''\hat{\ell}\hat{L}^2 \\ \times \int \mathrm{d}r_1 \int \mathrm{d}r_2\, u_{L'}(r_1) u_L(r_1) \frac{r_<^\lambda}{r_>^{\lambda+1}} u_{L''}(r_2) u_\ell(r_2)\,. \tag{5.34}$$

5.3.2 Radial Partial Waves

In (5.34) we have made use of the fact that the distorting potentials are real. This means that the partial waves $\chi_L(k,r)$ can be written in terms of a constant complex normalization factor N_L and a real function $u_L(k,r)$.

$$\chi_L(k,r) = N_L u_L(k,r). \tag{5.35}$$

They are obtained by solving the real radial Schrödinger equation up to an external matching point r_0, where it is matched to the external solution

$$\chi_L(k,r_0) = F_L(\eta, kr_0) + C_L[G_L(\eta, kr_0) + \mathrm{i}F_L(\eta, kr_0)]. \tag{5.36}$$

In (5.36) C_L is a complex number related to the phase shift, and F_L and G_L are respectively the regular and irregular Coulomb functions for the Coulomb parameter η, which is given in atomic units by

$$\eta = k^{-1}. \tag{5.37}$$

The normalization factor N_L is calculated by substituting (5.35) at r_0 in (5.36).

5.3.3 Exchange Amplitude

The exchange amplitude corresponding to (5.33) and (5.34) is obtained by exchanging $\boldsymbol{k}_\mathrm{A}$ and $\boldsymbol{k}_\mathrm{B}$ in Table 5.1. Its computational form is the same as (5.33) and (5.34) with the appropriate redefinitions of the momenta k_A and k_B, angles $\hat{\boldsymbol{k}}_\mathrm{A}$ and $\hat{\boldsymbol{k}}_\mathrm{B}$, and the corresponding wavefunctions. Note that for symmetric kinematics the direct and exchange amplitudes are identical.

5.3.4 Radial Integration

The radial integrals in (5.34) are of the form

$$R = \int_0^\infty dr_1 \int_0^\infty dr_2\ f(r_1)g(r_2)\frac{r_<^\lambda}{r_>^{\lambda+1}}, \tag{5.38}$$

where $f(r_1)$ oscillates with amplitude ≈ 1 for large r_1, and $g(r_2)$ contains the bound-state function $u_\ell(r_2)$ so that it decays exponentially and is negligible for $r > r_n$.

The integrals are computed by a quadrature rule at N points r_i with weights w_i. It is unnecessary to do the r_2 integration at all the points r_i. Instead the following procedure is adopted. We define

$$R = \sum_{i=1}^{n} w_i f_i \left(\frac{T_i}{r_i^{\lambda+1}} + S_i r_i^\lambda\right) + T_n \sum_{i=n}^{N} \frac{w_i f_i}{r_i^{\lambda+1}}, \tag{5.39}$$

where

$$T_i = T_{i-1} + w_i g_i r_i^\lambda; \qquad T_0 = 0; \tag{5.40a}$$

$$S_i = S_{i-1} - \frac{w_i g_i}{r_i^{\lambda+1}}; \qquad S_0 = \sum_{j=1}^{n} \frac{w_j g_j}{r_j^{\lambda+1}}. \tag{5.40b}$$

Note that the short range of $g(r_2)$ makes $T_i = T_n$ for $i > n$.

Another numerical aspect concerns the cases of small values of λ, which need some special treatment. The r_2 integral of (5.34) is

$$\begin{aligned} F(r_1) = {} & r_1^{-\lambda-1} \int_0^\infty dr_2\ r_2^\lambda u_{L''}(r_2) u_\ell(r_2) \\ & + \int_{r_1}^\infty dr_2 \left(\frac{r_1^\lambda}{r_2^{\lambda+1}} - \frac{r_2^\lambda}{r_1^{\lambda+1}}\right) u_{L''}(r_2) u_\ell(r_2). \end{aligned} \tag{5.41}$$

The second term of (5.41) is of short range in r_1, whereas the first term is a constant multiple of $r_1^{-\lambda-1}$. For $\lambda = 0$ the r_1 integration has parts that are equivalent to the momentum-space Coulomb potential, which is divergent for $\boldsymbol{k}_0 = \boldsymbol{k}_\mathrm{A}$. Although $\boldsymbol{k}_0$ can never be equal to $\boldsymbol{k}_\mathrm{A}$ in the real ionization problem, it is nevertheless useful that $u_{L''}(r_2)$ is orthogonal to $u_\ell(r_2)$ so that the first term of (5.41) vanishes identically for $\lambda = 0$. For the exchange term corresponding to (5.34) we again have orthogonality, since v_1 and v_2 are each electron–ion potentials.

For $\lambda = 1$ the r_1 integrand is an oscillating function multiplied by r_1^{-2}. The integral is definitely convergent but the integration range must be extended to $r_1 > 150\ a_0$, although the range of the bound-state function u_ℓ is only about 10 a_0 (with a_0 being the Bohr radius). The program extends the range of integration beyond the range of u_ℓ by a distance that decreases linearly with λ and is zero at a pre-chosen value of λ (usually 5).

5.3.5 Distorted-Wave Impulse Approximation

For the DWIA structure factor M of (5.24) the Coulomb potential $v_3(\boldsymbol{r}_1,\boldsymbol{r}_2)$ in (5.15) is replaced by $\delta(\boldsymbol{r}_1 - \boldsymbol{r}_2)$. The formalism in computational form is parallel to that of the DWBA except that v_3 of (5.26) is replaced by the multipole expansion of the delta function

$$\delta(\boldsymbol{r}_1 - \boldsymbol{r}_2) = \sum_{\lambda\mu} r_1^{-2}\delta(r_1 - r_2) Y^*_{\lambda\mu}(\hat{\boldsymbol{r}}_1) Y_{\lambda\mu}(\hat{\boldsymbol{r}}_2). \tag{5.42}$$

The radial integral (5.34) is replaced by the one-dimensional short-range integral

$$R^{(\lambda)}_{L'L''\ell L}(k_0, k_{\rm A}, k_{\rm B}) = {\rm i}^{L-L'-L''} {\rm e}^{{\rm i}(\sigma_{L'}+\sigma_{L''})} N_L N_{L'} N_{L''} (4\pi)^{-1} \hat{\lambda}^2 \hat{L}' \hat{L}'' \hat{\ell} \hat{L}^2 \times \int {\rm d}r \, u_{L'}(r) u_{L''}(r) u_\ell(r) u_L(r) \,. \tag{5.43}$$

There is no exchange term in this case, but antisymmetrization is performed in the electron–electron factor (5.21).

5.4 Computer Program

The essential part of the computer program is `SUBROUTINE DWBA`, which calculates differential cross sections for (e,2e) in the DWBA (if the option switch `IDWI` is 1) and also in the DWIA (if the option switch `IDWI` is 2). The driving program `DWE2E` reads the kinematic input and sets up option switches for different types of calculation. The input is described by comments in the program. `SUBROUTINE DWBA` requires radial distorted waves as well as a radial bound-state wavefunction. Example files are provided on the disk, but these functions could also be be supplied, for example, by the programs of Chaps. 2 and 4. The radial functions must be set up on a radial integration mesh, which again may be that of Chap. 4.

5.4.1 Input

The kinematic input consists of the kinetic energies of the incident, slow, and fast electrons, E_0, $E_{\rm B}$, and $E_{\rm A}$, and sets of angles which may be equally spaced, calculated from the number of angles, the starting angle, and the increment, or numerically specified by an extra input line if the increment is set negative. The meaning of the angle variables depends on the kinematic option chosen.

The distorted waves and associated quantities are labeled by 1, 2, and 3 for the incident, slow, and fast electrons, respectively. It is necessary to input the maximum number of partial waves used for each distorted wave. These are `LM1`, `LM2`, and `LM3`. In order to test the convergence of the partial-wave expansions, it is useful to have the option of setting all the distorting potentials to zero. The partial-wave calculation of the resulting plane-wave cross sections may be compared with the plane-wave cross sections calculated analytically. The various options are:

`IPW.EQ.0`	Distorted waves are calculated using potentials set up in a similar way to those discussed in Chap. 4.
`IPW.NE.0`	The potentials are set equal to zero. This option is used to check the convergence of the calculation in the radial variables and the partial-wave expansions.
`ICP.EQ.0`	Noncoplanar symmetric kinematics. Here the input angles are `THC(I)` $= \theta_A = \theta_B$ and `THD(I)` $= \phi$.
`ICP.NE.0`	Coplanar asymmetric kinematics. Here the input angles are `THC(I)` $= \theta_A$ and `THD(I)` $= \theta_B$.
`IEX.EQ.0`	Exchange amplitudes are omitted from the asymmetric calculation (they are equal to the direct amplitudes in the symmetric kinematic case). This approximation saves time and may be valid if E_B is small.
`IEX.NE.0`	Exchange amplitudes are included.
`IDWI.EQ.1`	The DWBA is calculated.
`IDWI.EQ.2`	The DWIA is calculated.

5.4.2 Calculation of the Differential Cross Section

The differential cross section for the DWBA is given by (5.6) with T_S calculated by (5.33) and (5.34). The differential cross section for the DWIA is calculated by (5.20) with M given by (5.33) and (5.43).

The mesh for the radial integrations is set up at `LIM` points `RDW(I)` with quadrature weights `WDW(I)`. The integrand of the r_1 integration for the DWBA is given by (5.41). It is proportional to $r_1^{-\lambda-1}$ for large r_1, except for λ=0 where the long-range part vanishes due to the orthogonality of the final-state distorted waves and the bound-state wavefunction. For λ=1, convergence requires `RDW(LIM)` >150 (atomic units are used throughout). However, for λ = `LAMN` ($\approx$5) convergence is determined (as for the DWIA) by the exponential tail of the radial bound-state function. The upper-limit index used for the DWBA integration is `LIMU`, which is reduced from `LIM` to `LIMT` in equal steps `LIMQ` for successive λ up to `LAMN`. The radial integration time for the DWBA is considerably shortened by using the method given by (5.39) and (5.40).

`SUBROUTINE DWBA` requires real radial partial waves $u_L(k,r)$ and complex normalization factors N_L defined by (5.25) and (5.35). For convenience the normalization factors are multiplied by the corresponding Coulomb phase factors $\exp(i\sigma_L)$, for which a subroutine is supplied. The use of real rather than complex arithmetic in the innermost loops (the radial integration) saves much computer time. Specifically, we choose the arrays

$$\text{U1(LB, I)} = u_L(k_0, r_i); \qquad \text{FNM1(LB)} = \mathrm{e}^{\mathrm{i}\sigma_L} N_L;$$
$$\text{U2(LD, I)} = u_{L''}(k_\mathrm{B}, r_i); \qquad \text{FNM2(LD)} = \mathrm{e}^{\mathrm{i}\sigma_{L''}} N_{L''};$$
$$\text{U3(LC, I)} = u_{L'}(k_\mathrm{A}, r_i); \qquad \text{FNM3(LC)} = \mathrm{e}^{\mathrm{i}\sigma_{L'}} N_{L'}.$$

Note that the loop indices LB, LD, and LC are positive integers so that they are one more than the corresponding radial quantum numbers L1, L2, and L3, respectively. For the m and M'' loops, however, the loop indices are the quantum numbers M1 and MD1. The bound-state orbital angular momentum quantum number is LA.

The bound-state data to be supplied are:

FNE	$= n_i$	electron multiplicity
LA	$= \ell$	orbital quantum number
WFN(I), I=1,LIM;	$= u_\ell(r_i)$	radial function

The spherical harmonics are tabulated in the arrays YLMC(LMC,MTH) and YLMD(LMD,MTH,NTH), corresponding to $Y_{L'M'}(\theta_\mathrm{A}, 0)$ and $Y_{L''M''}(\theta_\mathrm{B}, \phi)$, respectively. The indices LMC and LMD count the pairs of quantum numbers L and M, whereas MTH and NTH count the angles θ_A and θ_B or ϕ (depending on the kinematic option). The spherical harmonics are calculated from the associated Legendre polynomials $P_{LM}(\cos\theta)$ (SUBROUTINE PLM supplied) according to

$$Y_{\ell m}(\theta, \phi) = (-1)^m \left[\frac{(2\ell+1)(\ell-|m|)!}{4\pi(\ell+|m|)!}\right]^{1/2} P_{\ell m}(\cos\theta)\, \mathrm{e}^{\mathrm{i}m\phi}. \qquad (5.44)$$

5.4.3 Sample Input

The sample input is for a distorted-wave calculation (DWBA) of the coplanar asymmetric differential cross section for hydrogen with $E_0 = 100$ eV, $E_\mathrm{A} = 84.4$ eV, $E_\mathrm{B} = 2$ eV. There is one value of $\theta_\mathrm{A} = 8°$ and there are ten values of θ_B starting at 0° with increments of 10°.

The five lines of input on unit 9, file 'E2E.IN' are:

```
100.0   2.0  84.4         EIV, ESLOW, EFAST
    0     1     1     1   IPW, IDWI,  IEX,    ICOPS
    1   8.0   6.0         NTC, THCI,  THCINC
    8   0.0  10.0         NTD, THDI,  THDINC
   45    30    42         LM1, LM2,   LM3
```

The input is fully described by comments in the program. Additional input is required for the radial mesh, the distorting potentials and the bound-state wavefunction. As mentioned before, example files are provided on disk, but they can be replaced and/or extended by those generated with the programs of Chap. 4.

5.4.4 Sample Output

The output is printed on the terminal unit 6 and on unit 14, file '`E2E.OUT`'. Differential cross sections in atomic units are given in Table 5.2 for the analytic plane-wave Born approximation (PWBA), the partial-wave plane-wave Born approximation (PWBP), the distorted-wave Born approximationn (DWBA), the analytic plane-wave impulse approximation (PWIA), the partial-wave plane-wave impulse approximation (PWIP) and the distorted-wave impulse approximation (DWIA). In all cases, exchange amplitudes were included.

The distorted-wave cross sections of Table 5.2 were calculated using the Flinders potential and distorted-wave subroutines. Use of different subroutines is expected to give slightly different numbers. The convergence to the analytic plane-wave Born approximation can be improved by decreasing the step size in the radial integration mesh and by using more partial waves. The CPU time for the calculation by `SUBROUTINE DWBA` was approximately 30 minutes on a SUN4/280. This time is roughly proportional to $(\mathtt{LM1} + \mathtt{LM2} + \mathtt{LM3})^3$.

Table 5.2. Differential cross sections (in atomic units) for the (e,2e) case given by the sample input, using the Flinders distorting potential and distorted-wave subroutines with 1984 integration points and `RDW(LIM)` = 250. Column headings are explained in the text. Superscripts indicate powers of 10.

θ_B(deg)	PWBA	PWBP	DWBA	PWIA	PWIP	DWIA
0	2.024^1	1.983^1	9.078^{-1}	3.354^0	3.354^0	4.985^{-1}
10	2.364^1	2.324^1	1.078^0	3.984^0	3.985^0	2.803^{-1}
20	2.706^1	2.672^1	1.717^0	4.693^0	4.693^0	1.011^{-1}
30	3.016^1	2.991^1	2.854^0	5.432^0	5.432^0	6.853^{-3}
40	3.255^1	3.241^1	4.356^0	6.127^0	6.125^0	3.133^{-2}
50	3.386^1	3.386^1	5.919^0	6.685^0	6.682^0	1.760^{-1}
60	3.388^1	3.401^1	7.141^0	7.023^0	7.020^0	3.954^{-1}
70	3.259^1	3.286^1	7.674^0	7.090^0	7.088^0	6.058^{-1}

5.5 Summary

The essential subroutine for differential cross sections and a driving program giving the kinematic input and option switches has been described for DWBA and DWIA calculations of the (e,2e) reaction. Subroutines for calculating the radial partial waves for distorted (or plane) waves and for the radial bound-state wavefunction must be supplied.

Since there is one more distorted wave than in the scattering case treated in Chap. 4, the present calculations are much larger. The authors have tried to arrange the program for maximum computational efficiency but would welcome suggestions for improvements.

5.6 Suggested Problems

Useful approximations for two common kinematic arrangements can be computed with the existing program. The program can be used as the basis for different calculations. Some examples are given, ranging from slight rearrangements to full-scale research problems.

1. Calculate different kinematic arrangements such as fixed symmetric angles with variation of the energy balance between the slow and the fast electrons.
2. Include relativistic kinematics.
3. Include the possibility that the incident and outgoing electrons have their momenta defined by finite-size apertures as in a real experiment.
4. Replace the bound-state wavefunction by the overlap of target and ion states described by a configuration-interaction expansion. This eliminates the three-body approximation.
5. Devise a way to incorporate the exact boundary condition for three charged particles [5.2].
6. Calculate the double-differential cross section in which one electron is not observed.
7. Calculate the total ionization cross section.
8. Calculate the differential cross section in the case where one of the two-body subsystems is resonant.

Acknowledgments

This work was supported by the Australian Research Council. We are grateful to Professor D.H. Madison for help with numerical aspects.

References

5.1 H. Ehrhardt, M. Schulz, T. Tekaat and K. Willmann, Phys. Rev. Lett. **22** (1969) 89
5.2 M. Brauner, J.S. Briggs and H. Klar, J. Phys. B **22** (1989) 2265
5.3 E. Weigold, S.T. Hood and P.J.O. Teubner, Phys. Rev. Lett. **30** (1973) 475
5.4 I. Bray, D.H. Madison and I.E. McCarthy, Proc. XVI ICPEAC, New York (1989)
5.5 I.E. McCarthy and E. Weigold, Rep. Prog. Phys. **51** (1988) 299
5.6 I.E. McCarthy and A.T. Stelbovics, Phys. Rev. A **28** (1983) 2698
5.7 L.J. Allen, I. Bray and I.E. McCarthy, Phys. Rev. A **37** (1988) 49
5.8 W.F. Ford, Phys. Rev. B **133** (1964) 1616

6. The Close-Coupling Approximation

R.P. McEachran

Department of Physics and Astronomy, York University, Toronto, Canada M3J 1P3

Abstract

The close-coupling equations for positron–atom scattering are discussed. Detailed formulae are given for the special case of alkali-like target systems. A computer program, based upon an integral-equation approach combined with an asymptotic correction procedure, is presented together with some sample input and output data for positron scattering from lithium atoms.

6.1 Introduction

The close-coupling approximation has been a standard method for the treatment of low-energy scattering (elastic and inelastic) of positrons and electrons from atoms for many years. It is based upon an expansion of the total wavefunction of the system in terms of a sum of products of eigenstates of the atomic hamiltonian and unknown (to be determined) functions describing the motion of the projectile. These unknown functions or 'expansion coefficients' are then determined from the solution of a set of coupled integro-differential equations. This basic procedure was first proposed by Massey and Mohr [6.1]. In the late 1950s and early 1960s the close-coupling approximation was extensively tested on positron and electron scattering from the hydrogen atom ([6.2]–[6.9] and references cited therein). The specific details of the theory for electron (positron) scattering from hydrogen were developed by Percival and Seaton [6.10] and extended, in the case of positrons, to include positronium formation by Smith [6.11]. However, for atomic hydrogen the rate of convergence of the close-coupling method with respect to the inclusion of more and more target eigenstates was extremely slow (cf. [6.9]). This lack of convergence was due primarily to the fact that a few (or even many) eigenstates in the expansion did not adequately treat the long-range polarization effect [6.12]. This led to the original proposal of Burke and Schey [6.3] for the inclusion of pseudostates in the expansion in order to account for this effect.

On the other hand, for the alkali atoms, over 98% of their polarizabilities arises from just the resonance transition alone. In this regard the alkalis are then much more amenable to a close-coupling treatment than is atomic hydrogen. The close-coupling theory pertaining to the scattering of electrons by alkali-like systems was first presented by Salmona and Seaton [6.13]. Close-coupling calculations based upon these equations for electron scattering from the lighter alkalis have been carried out by Burke and Taylor [6.14] for lithium and by Salmona [6.15] and Moores and Norcross [6.16] for sodium. The R-matrix method (see Chap. 7 and [6.17]), which can be considered as an outgrowth of the close-coupling approach, has tended to supplant the latter method for electron scattering from atoms in general.

However, for low-energy positron scattering from the alkalis the close-coupling approximation has recently been used quite extensively and successfully [6.18–23] when compared with experiment [6.24–26]. In this chapter we will therefore discuss the close-coupling approximation within the framework of positron scattering from the alkali atoms. A brief derivation of the relevant equations and cross-section formulae, based upon the formalism established by Percival and Seaton [6.10], will be presented in Sect. 6.2. In Sect. 6.3 a numerical method, based upon an integral-equation formulation, is presented for the solution of the close-coupling equations. An asymptotic correction procedure appropriate for this approach is also given. In Sect. 6.4 a numerical procedure for the Born approximation is developed, and finally, in Sect. 6.5, the computer code is described together with some sample input and output.

6.2 Theory

As mentioned previously the theory for the close-coupling approximation for electron scattering from the alkali atoms was given by Salmona and Seaton [6.13]. Their derivation, as well as the one presented here, follows very closely the earlier work of Percival and Seaton [6.10] for atomic hydrogen. The close-coupling equations for positron scattering are, as will be shown, identical to those for electron scattering except that there are no exchange terms and the overall sign of all the potential terms is reversed.

6.2.1 The Close-Coupling Equations

We begin by constructing the total scattering wavefunction for the system consisting of an alkali atom and an incident positron. The wavefunction for an arbitrary valence state of the atom will be denoted by $\Psi_\nu(\boldsymbol{r}_{\mathrm{C}}\sigma_{\mathrm{C}}, \boldsymbol{r}_1\sigma_1)$ where $\boldsymbol{r}_{\mathrm{C}}$ and σ_{C} will specify collectively the space and spin coordinates of the core electrons respectively, i.e., $\boldsymbol{r}_{\mathrm{C}}\sigma_{\mathrm{C}} \equiv \boldsymbol{r}_i\sigma_i$, $i = 1, 2, \ldots, N$. Wherever possible $\boldsymbol{r}_1\sigma_1$ will be used to represent the space and spin coordinates of the valence electron. However, in order to avoid confusion with the designation of the coordinates for the *first* core electron, it will at times be necessary to use $\boldsymbol{r}_{N+1}\sigma_{N+1}$ rather than $\boldsymbol{r}_1\sigma_1$. Also the quantum number ν will initially represent collectively the quantum numbers of the valence electron which we will, in turn, denote by $n_1 l_1 m_1 m_{\mathrm{S}_1}$. Thus the wavefunction Ψ_ν is represented by the Slater determinant

$$\begin{aligned}\Psi_\nu(\boldsymbol{r}_{\mathrm{C}}\sigma_{\mathrm{C}},\boldsymbol{r}_1\sigma_1) &\equiv \Psi_\nu(\boldsymbol{r}_{\mathrm{C}}\sigma_{\mathrm{C}},\boldsymbol{r}_{\mathrm{N+1}}\sigma_{\mathrm{N+1}})\\ &= \frac{1}{\sqrt{(N+1)!}}\det\{\psi_1(\boldsymbol{r}_1,\sigma_1),\ldots,\psi_{N+1}(\boldsymbol{r}_{N+1},\sigma_{N+1})\},\end{aligned} \tag{6.1}$$

where the individual one-electron wavefunctions satisfy the orthonormality conditions

$$\langle\psi_i \mid \psi_j\rangle = \delta_{ij}\,. \tag{6.2}$$

The corresponding quantum numbers of the incident positron are, in turn, denoted by $kl_2m_2m_{\mathrm{S}_2}$ where k is the wavenumber of the incident positron. We next let γ collectively represent $n_1l_1m_1m_{\mathrm{S}_1}kl_2m_2m_{\mathrm{S}_2}$ and introduce the functions

$$\Psi_\gamma(\boldsymbol{r}_{\mathrm{C}}\sigma_{\mathrm{C}},\boldsymbol{r}_1\sigma_1,\hat{\boldsymbol{r}}_2\,\sigma_2) = \Psi_\nu(\boldsymbol{r}_{\mathrm{C}}\sigma_{\mathrm{C}},\boldsymbol{r}_1\sigma_1)\,Y_{l_2m_2}(\hat{\boldsymbol{r}}_2)\,\chi_{m_{\mathrm{S}_2}}(\sigma_2)\,, \tag{6.3}$$

where $Y_{l_2m_2}(\hat{\boldsymbol{r}}_2)$ and $\chi_{m_{\mathrm{S}_2}}(\sigma_2)$ represent the angular momentum and spin functions of the incident particle, respectively. Wherever possible $\boldsymbol{r}_2\sigma_2$ will be used to represent its space and spin coordinates. However, whenever it is necessary to use $\boldsymbol{r}_{N+1}\sigma_{N+1}$ for the valence electron then $\boldsymbol{x}\,\sigma_x$ will be used for the incident positron.

Since spin–orbit coupling is neglected, the total orbital and spin angular momentum quantum numbers $LSMM_{\mathrm{S}}$ are separately conserved. Consequently, calculations are simplified by employing, instead of γ, the alternative representation $\Gamma = n_1kl_1l_2LSMM_{\mathrm{S}}$. These two representations are connected by the unitary transformation

$$(\gamma \mid \Gamma') = \delta(n_1l_1l_2,n_1'l_1'l_2')\,C(l_1l_2L';m_1'm_2'M')\,C(\tfrac{1}{2}\tfrac{1}{2}S';m_{\mathrm{S}_1}'\,m_{\mathrm{S}_2}'\,M_{\mathrm{S}}'), \tag{6.4}$$

where the $C(j_1j_2j_3;m_1m_2m_3)$ are the usual Clebsch-Gordan coefficients as given by Rose [6.27]. We now define the functions

$$\Psi_\Gamma(\boldsymbol{r}_{\mathrm{C}}\sigma_{\mathrm{C}},\boldsymbol{r}_1\sigma_1,\hat{\boldsymbol{r}}_2\,\sigma_2) = \sum_\gamma (\gamma \mid \Gamma)\,\Psi_\gamma(\boldsymbol{r}_{\mathrm{C}}\sigma_{\mathrm{C}},\boldsymbol{r}_1\sigma_1,\hat{\boldsymbol{r}}_2\,\sigma_2)\,. \tag{6.5}$$

The summations over γ in (6.5) are therefore over just the magnetic orbital and spin quantum numbers $m_1m_2m_{\mathrm{S}_1}m_{\mathrm{S}_2}$.

The total wavefunction for the system is now taken to be of the form

$$\Psi(\boldsymbol{r}_{\mathrm{C}}\sigma_{\mathrm{C}},\boldsymbol{r}_1\sigma_1,\boldsymbol{r}_2\sigma_2) = \sum_\Gamma \Psi_\Gamma(\boldsymbol{r}_{\mathrm{C}}\sigma_{\mathrm{C}},\boldsymbol{r}_1\sigma_1,\hat{\boldsymbol{r}}_2\,\sigma_2)\,\frac{1}{r_2}\,F_\Gamma(r_2)\,. \tag{6.6}$$

The summations in (6.6) are, in principle, over all the quantum numbers represented by Γ, with the summation over k representing an integration over the continuum states of the atom. In practice, however, only a finite number of the *bound* atomic states can be included in the expansion (6.6) and consequently k will henceforth be omitted when we consider expansions over Γ.

The close-coupling equations are given by projecting the Schrödinger equation for Ψ onto Ψ_Γ, i.e.,

$$\langle \Psi_\Gamma(\boldsymbol{r}_C\sigma_C, \boldsymbol{r}_1\sigma_1, \hat{\boldsymbol{r}}_2\,\sigma_2) \mid H - E \mid \Psi(\boldsymbol{r}_C\sigma_C, \boldsymbol{r}_1\sigma_1, \boldsymbol{r}_2\sigma_2)\rangle = 0\,. \tag{6.7}$$

The total Hamiltonian, H, of the system can be expressed as

$$H = H_0 - \tfrac{1}{2}\nabla_2^2 + V\,, \tag{6.8}$$

where H_0 is the Hamiltonian of the atom and V is the interaction potential between the incident positron and the atomic electrons, i.e.,

$$V = V(\boldsymbol{r}_C, \boldsymbol{r}_1, \boldsymbol{r}_2) = V(\boldsymbol{r}_C, \boldsymbol{r}_{N+1}, \boldsymbol{x}) = \frac{Z}{x} - \sum_{i=1}^{N+1} \frac{1}{|\boldsymbol{r}_i - \boldsymbol{x}|}\,. \tag{6.9}$$

The total energy of the system is given by E where

$$E = E_0 + \tfrac{1}{2}k^2 \tag{6.10}$$

and E_0 is the total energy of the atom in its ground state and $\frac{1}{2}k^2$ is the kinetic energy of the incident positron. Thus we can write

$$H - E = H_0 - E_0 - \tfrac{1}{2}\nabla_2^2 + V - \tfrac{1}{2}k^2\,. \tag{6.11}$$

In the determination of the close-coupling equations it is assumed that

$$\langle \Psi_\Gamma \mid H_0 - E_0 \mid = \langle \Psi_\Gamma\,, \mid E_\nu - E_0 \mid\,, \tag{6.12}$$

where E_ν is the total energy of the atom in the state represented by Ψ_ν. Equation (6.12) would, of course, be true if Ψ_ν were an exact eigenstate of the atom. If Ψ_ν represents the ground state of the atom then the right-hand side of (6.12) becomes zero.

In a frozen-core approximation for an alkali atom E_ν satisfies [6.28]

$$E_\nu = E_{\rm ion} + \epsilon_\nu\,, \tag{6.13}$$

where $E_{\rm ion}$ is the total energy of the ion core. That is, $-\epsilon_\nu > 0$ is the ionization energy of the state represented by Ψ_ν and is also the eigenvalue parameter in the Hartree–Fock equation for the one-electron valence orbital $\psi_\nu(\boldsymbol{r}_1, \sigma_1)$. Equation (6.7) can thus be expressed as

$$\sum_{\Gamma'} \langle \Psi_\Gamma \mid -\tfrac{1}{2}\nabla_2^2 + V - \tfrac{1}{2}k_\nu^2 \mid \Psi_{\Gamma'}\rangle \frac{1}{r_2} F_{\Gamma'}(r_2) = 0 \tag{6.14}$$

or alternatively as

$$\sum_{\Gamma'}\left\langle\Psi_\Gamma\left|\left[\frac{\mathrm{d}^2}{\mathrm{d}r_2^2}-\frac{\ell_2'(\ell_2'+1)}{r_2^2}-2V+k_\nu^2\right]\right|\Psi_{\Gamma'}F_{\Gamma'}(r_2)\right\rangle=0\,, \tag{6.15}$$

where

$$\tfrac{1}{2}k_\nu^2=E-E_\nu=(E_0-E_\nu)+\tfrac{1}{2}k^2\,. \tag{6.16}$$

From (6.3) and (6.5) it follows that

$$\langle\Psi_\Gamma\mid \mathcal{O}p\mid\Psi_{\Gamma'}\rangle=\sum_\gamma(\gamma\mid\Gamma)\sum_{\gamma'}(\gamma'\mid\Gamma')\,\langle\Psi_\gamma\mid\mathcal{O}p\mid\Psi_{\gamma'}\rangle\,, \tag{6.17}$$

where $\mathcal{O}p$ is any operator. Now provided $\mathcal{O}p$ is symmetric in terms of the N+1 bound electrons, then, from the properties of Slater determinants [6.29], it can be shown that

$$\begin{aligned}&\langle\Psi_\gamma\mid\mathcal{O}p\mid\Psi_{\gamma'}\rangle\\ &\quad=\langle\{\psi_1(\boldsymbol{r}_1,\sigma_1),\ldots,\psi_\nu(\boldsymbol{r}_{N+1},\sigma_{N+1})\}Y_{l_2m_2}(\hat{\boldsymbol{x}})\chi_{m_{s2}}(\sigma_x)\mid\mathcal{O}p\mid\\ &\qquad\times\sum_{\mathrm{perm}}\{\psi_1(\boldsymbol{r}_1,\sigma_1),\ldots,\psi_{\nu'}(\boldsymbol{r}_{N+1},\sigma_{N+1}\}Y_{l_2'm_2'}(\hat{\boldsymbol{x}})\chi_{m_{s2}'}(\sigma_x)\rangle\,.\end{aligned} \tag{6.18}$$

From (6.17) it follows that the expression (6.18) needs to be evaluated only for the operators unity (1) and V. From (6.2) it follows immediately that

$$\langle\Psi_\gamma\mid\Psi_{\gamma'}\rangle=\delta(\gamma,\gamma')\,. \tag{6.19}$$

From (6.9) for V it then follows that

$$\langle\Psi_\gamma\mid V\mid\Psi_{\gamma'}\rangle=\frac{Z}{x}\,\delta(\gamma,\gamma')-\sum_{i=1}^{N+1}\left\langle\Psi_\gamma\left|\frac{1}{|\boldsymbol{r}_i-\boldsymbol{x}|}\right|\Psi_{\gamma'}\right\rangle \tag{6.20}$$

and from (6.18) that

$$\begin{aligned}&\sum_{i=1}^{N+1}\left\langle\Psi_\gamma\left|\frac{1}{|\boldsymbol{r}_i-\boldsymbol{x}|}\right|\Psi_{\gamma'}\right\rangle\\ &\quad=\left\langle Y_{l_2m_2}(\hat{\boldsymbol{x}})\chi_{m_{s2}}(\sigma_x)\right|\left\{\sum_{i=1}^{N}\left\langle\psi_i\left|\frac{1}{|\boldsymbol{r}_i-\boldsymbol{x}|}\right|\psi_i\right\rangle\langle\psi_\nu|\psi_{\nu'}\rangle\right.\\ &\quad+\left.\left\langle\psi_\nu\left|\frac{1}{|\boldsymbol{r}_{N+1}-\boldsymbol{x}|}\right|\psi_{\nu'}\right\rangle\right\}Y_{l_2'm_2'}(\hat{\boldsymbol{x}})\chi_{m_{s2}'}(\sigma_x)\Big\rangle\,.\end{aligned} \tag{6.21}$$

We now denote the quantum numbers represented by i, for an arbitrary core electron, by $nlmm_{\mathrm{s}}$ and represent these individual atomic core orbitals as well as the valence orbital by

$$\psi_i(\boldsymbol{r},\sigma) = \phi_{nlm}(\boldsymbol{r})\,\chi_{m_s}(\sigma) = \frac{1}{r}\,P_{nl}(r)\,Y_{lm}(\hat{\boldsymbol{r}})\,\chi_{m_s}(\sigma) \tag{6.22}$$

and

$$\psi_\nu(\boldsymbol{r},\sigma) = \phi_{n_1l_1m_1}(\boldsymbol{r})\,\chi_{m_{s_1}}(\sigma) = \frac{1}{r}\,P_{n_1l_1}(r)\,Y_{l_1m_1}(\hat{\boldsymbol{r}})\,\chi_{m_{s_1}}(\sigma)\,. \tag{6.23}$$

The reciprocal of the interparticle distance can be expressed in the usual manner as

$$\frac{1}{|\boldsymbol{r}-\boldsymbol{x}|} = \sum_\lambda \gamma_\lambda(r,x)\,P_\lambda(\hat{\boldsymbol{r}}\cdot\hat{\boldsymbol{x}})\,, \tag{6.24}$$

where

$$\gamma_\lambda(r,x) = \frac{r^\lambda}{x^{\lambda+1}} \quad \text{for} \quad r<x\,, \tag{6.25}$$

$$= \frac{x^\lambda}{r^{\lambda+1}} \quad \text{for} \quad r>x\,. \tag{6.26}$$

In the reduction of the matrix element involving ψ_i in (6.21) the summation over the spin coordinate σ gives unity and, since we are dealing with closed sub-shells in the atomic core, the summation over m_s yields a factor of 2. Thus (6.21) becomes

$$\begin{aligned}\sum_{i=1}^{N+1}\left\langle\Psi_\gamma\middle|\frac{1}{|\boldsymbol{r}_i-\boldsymbol{x}|}\middle|\Psi_{\gamma'}\right\rangle
&= 2\sum_{nlm}\sum_\lambda c_\lambda(lm,l_2m_2,lm,l_2'm_2')\,\frac{1}{x}\,y_\lambda(nl,nl;x)\,\delta(\nu,\nu')\\
&\times\;\delta(m_{\mathrm{s}_2},m'_{\mathrm{s}_2}) + \sum_\lambda c_\lambda(l_1m_1,l_2m_2,l_1'm_1',l_2'm_2')\\
&\times\;\frac{1}{x}\,y_\lambda(n_1l_1,n_1'l_1';x)\,\delta(m_{\mathrm{s}_1}m_{\mathrm{s}_2},m'_{\mathrm{s}_1}m'_{\mathrm{s}_2})\,,\end{aligned} \tag{6.27}$$

where

$$\begin{aligned}&c_\lambda(l_1m_1,l_2m_2,l_1'm_1',l_2'm_2')\\
&\quad=\big\langle Y_{l_1m_1}(\hat{\boldsymbol{r}})\,Y_{l_2m_2}(\hat{\boldsymbol{x}})\big|P_\lambda(\hat{\boldsymbol{r}}\cdot\hat{\boldsymbol{x}})\big|Y_{l_1'm_1'}(\hat{\boldsymbol{r}})\,Y_{l_2'm_2'}(\hat{\boldsymbol{x}})\big\rangle\end{aligned} \tag{6.28}$$

and

$$\frac{1}{x}\,y_\lambda(n_1l_1,n_1'l_1';x) = \int_0^\infty \mathrm{d}r\,\gamma_\lambda(r,x)\,P_{n_1l_1}(r)\,P_{n_1'l_1'}(r)\,. \tag{6.29}$$

The evaluation of the matrix element c_λ given by (6.26) can be performed straightforwardly with the aid of the following expression for $P_\lambda(\hat{\boldsymbol{r}}\cdot\hat{\boldsymbol{x}})$,

$$P_\lambda(\hat{\boldsymbol{r}}\cdot\hat{\boldsymbol{x}}) = \frac{4\pi}{2\lambda+1}\sum_\mu Y^*_{\lambda\mu}(\hat{\boldsymbol{r}})\,Y_{\lambda\mu}(\hat{\boldsymbol{x}}) \tag{6.30}$$

$$= \frac{4\pi}{2\lambda+1}\sum_\mu Y_{\lambda\mu}(\hat{\boldsymbol{r}})\,Y^*_{\lambda\mu}(\hat{\boldsymbol{x}})\,. \tag{6.31}$$

and the general result for the integration over three spherical harmonics [6.27], namely

$$\int \mathrm{d}\hat{\boldsymbol{r}}\, Y^*_{l_3m_3}(\hat{\boldsymbol{r}})\, Y_{l_2m_2}(\hat{\boldsymbol{r}})\, Y_{l_1m_1}(\hat{\boldsymbol{r}}) = \left[\frac{(2l_1+1)(2l_2+1)}{4\pi(2l_3+1)}\right]^{\frac{1}{2}} C(l_1l_2l_3;m_1m_2m_3)\, C(l_1l_2l_3;000)\,. \quad (6.32)$$

The summation over m in (6.25) then reduces to

$$\sum_m c_\lambda(lm, l_2m_2, lm, l_2'm_2') = (2l+1)\,\delta_{\lambda 0}\,\delta(l_2m_2, l_2'm_2') \quad (6.33)$$

and hence (6.25) takes the form

$$\sum_{i=1}^{N+1}\left\langle \Psi_\gamma \left| \frac{1}{|\boldsymbol{r}_i - \boldsymbol{x}|} \right| \Psi_{\gamma'}\right\rangle = V_{\mathrm{c}}(x)\delta(\gamma,\gamma') + \sum_\lambda c_\lambda(l_1m_1, l_2m_2, l_1'm_1', l_2'm_2') \times \frac{1}{x}\, y_\lambda(n_1l_1, n_1'l_1'; x)\,\delta(m_{\mathrm{s}_1}m_{\mathrm{s}_2}, m_{\mathrm{s}_1}'m_{\mathrm{s}_2}')\,, \quad (6.34)$$

where the atomic core potential $V_{\mathrm{c}}(x)$ is given by

$$V_{\mathrm{C}}(x) = \sum_{nl} q_{nl}\,\frac{1}{x}\, y_0(nl, nl; x) \quad (6.35)$$

with $q_{nl} = 2(2l+1)$. Thus the matrix element for V in (6.20) becomes

$$\langle \Psi_\gamma \mid V \mid \Psi_{\gamma'}\rangle = \left[\frac{Z}{x} - V_{\mathrm{C}}(x)\right]\delta(\gamma,\gamma') - \sum_\lambda c_\lambda(l_1m_1, l_2m_2, l_1'm_1', l_2'm_2') \times \frac{1}{x}\, y_\lambda(n_1l_1, n_1'l_1'; x)\,\delta(m_{\mathrm{s}_1}m_{\mathrm{s}_2}, m_{\mathrm{s}_1}'m_{\mathrm{s}_2}')\,. \quad (6.36)$$

We now consider the following two summations. First, from (6.4), (6.5), and (6.19) it readily follows that

$$\sum_{\Gamma'}\langle \Psi_\Gamma \mid \Psi_{\Gamma'}\rangle = \sum_{\Gamma'}\delta(\Gamma,\Gamma')\,. \quad (6.37)$$

Second, using (6.4), (6.5), and (6.36) it can be shown that

$$\sum_{\Gamma'}\langle \Psi_\Gamma \mid V \mid \Psi_{\Gamma'}\rangle = \sum_{\Gamma'}\sum_\gamma (\gamma \mid \Gamma)\sum_{\gamma'}(\gamma' \mid \Gamma')\langle \Psi_\gamma \mid V \mid \Psi_{\gamma'}\rangle$$
$$= \sum_{\Gamma'}\Big\{\delta(\Gamma,\Gamma')\left[\frac{Z}{x} - V_{\mathrm{C}}(x)\right] - \delta(SM_{\mathrm{S}}, S'M_{\mathrm{S}}') \times \sum_\lambda t_\lambda(l_1l_2LM, l_1'l_2'L'M')\,\frac{1}{x}\, y_\lambda(n_1l_1, n_1'l_1'; x)\Big\}\,, \quad (6.38)$$

where

$$t_\lambda(l_1 l_2 LM, l_1' l_2' L'M') = \sum_{m_1 m_2} \sum_{m_1' m_2'} C(l_1 l_2 L; m_1 m_2 M)\, C(l_1' l_2' L'; m_1' m_2' M') \times c_\lambda(l_1 m_1, l_2 m_2, l_1' m_1', l_2' m_2') . \tag{6.39}$$

From (6.26) as well as Eqs. (4.31) and (4.34) of Rose [6.27] c_λ can be expressed as

$$\begin{aligned} &c_\lambda(l_1 m_1, l_2 m_2, l_1' m_1', l_2' m_2') \\ &= \frac{(-1)^{m_2+m_1'}}{(2\lambda+1)^2} \left[(2l_1+1)(2l_1'+1)(2l_2+1)(2l_2'+1)\right]^{\frac{1}{2}} C(l_1 l_1' \lambda; 00) \\ &\times C(l_2 l_2' \lambda; 00)\, C(l_1 l_1' \lambda; -m_1, m_1')\, C(l_2 l_2' \lambda; -m_2, m_2')\, \delta(M, M'). \end{aligned} \tag{6.40}$$

After a straightforward but rather tedious calculation, (6.39) for t_λ may be expressed as

$$t_\lambda = \delta(LM, L'M')\, f_\lambda(l_1 l_2 l_1' l_2'; L) \tag{6.39$'$}$$

where

$$\begin{aligned} f_\lambda(l_1 l_2 l_1' l_2'; L) &= \frac{(-1)^{l_2+l_2'-L}}{2\lambda+1} C(l_1 l_1' \lambda; 00)\, C(l_2 l_2' \lambda; 00) \\ &\times \left[(2l_1+1)(2l_2+1)(2l_1'+1)(2l_2'+1)\right]^{\frac{1}{2}} W(l_1 l_2 l_1' l_2'; L\lambda) . \end{aligned} \tag{6.41}$$

Equation (6.38) can now be expressed as

$$\begin{aligned} \sum_{\Gamma'} \langle \Psi_\Gamma \mid V \mid \Psi_{\Gamma'} \rangle = \sum_{\Gamma'} \delta(LSMM_S, L'S'M'M_S') \Big\{ \delta(\nu, \nu') \left[\frac{Z}{r_2} - V_C(r_2)\right] \\ - \sum_\lambda f_\lambda(l_1 l_2 l_1' l_2'; L) \frac{1}{r_2} y_\lambda(n_1 l_1, n_1' l_1'; r_2) \Big\} , \end{aligned} \tag{6.42}$$

where the notation has been altered slightly so that now $\nu = n_1 l_1 l_2$ and $\nu' = n_1' l_1' l_2'$ and the independent variable has been changed from x to r_2. Upon substituting (6.37) and (6.42) into (6.15) and performing the remaining summation over Γ', the close-coupling equations are obtained, namely

$$\begin{aligned} \left\{ \frac{\mathrm{d}^2}{\mathrm{d}r_2^2} - \frac{\ell_2(\ell_2+1)}{r_2^2} - 2\left[\frac{Z}{r_2} - V_C(r_2)\right] + k_{n_1 l_1}^2 \right\} F_{\nu LS}(r_2) \\ = -2 \sum_{\nu'} V(\nu, \nu') F_{\nu' LS}(r_2) , \end{aligned} \tag{6.43}$$

where

$$V(\nu, \nu') = \sum_\lambda f_\lambda(l_1 l_2, l_1' l_2'; L) \frac{1}{r_2} y_\lambda(n_1 l_1, n_1' l_1'; r_2) . \tag{6.44}$$

Since both $l_1 + l_1' + \lambda$ and $l_2 + l_2' + \lambda$ must be even, or else f_λ will be zero, it follows that the phase factor $(-1)^{l_2+l_2'-L}$ in our definition of f_λ could also be given by $(-1)^{l_1+l_1'-L}$. If this were done then our definition of f_λ would be identical to that given by Percival and Seaton [6.10]. Also, if we put $V_C(r_2)$ to zero, set $Z = 1$, and change the overall sign of the potential terms, then (6.43) agrees with the corresponding result of Percival and Seaton for the partial-wave analysis for electron scattering from atomic hydrogen.

6.2.2 Boundary Conditions and Cross Sections

In this section we present the appropriate boundary conditions and cross-section formulae required in the solution of (6.43). If for a particular energy of the incident particle there are say, `NC` open channels, i.e., there are `NC` *different* values of ν for which $k_\nu > 0$ then `NC` linearly independent solutions of these coupled equations will be required. If the ν'th linearly independent solution for $F_{\nu LS}(r_2)$ is denoted by $F^{\nu'}_{\nu LS}(r_2)$ then the appropriate scattering boundary conditions are

$$F^{\nu'}_{\nu LS}(0) = 0 \tag{6.45a}$$

in all channels,

$$F^{\nu'}_{\nu LS}(r_2)_{r_2\to\infty} \sim \frac{1}{\sqrt{k_\nu}}\left[\delta(\nu,\nu')\sin\left(k_\nu r_2 - \frac{l\pi}{2}\right) + R^{LS}_{\nu\nu'}\cos\left(k_\nu r_2 - \frac{l\pi}{2}\right)\right] \tag{6.45b}$$

in the open channels, and

$$F^{\nu'}_{\nu LS}(r_2)_{r_2\to\infty} \sim C^{LS}_{\nu,\nu'}\ \exp(-|k_\nu|r_2) \tag{6.45c}$$

in the remaining closed channels. Here the numbers $R^{LS}_{\nu\nu'}$ are the corresponding elements of the reactance matrix or **R** matrix which, in turn, is related to the scattering matrix **S** and the transition matrix **T** according to

$$\mathbf{S} = [\mathbf{1} + \mathrm{i}\mathbf{R}]\,[\mathbf{1} - \mathrm{i}\mathbf{R}]^{-1} \tag{6.46}$$

and

$$\mathbf{T} = \mathbf{S} - \mathbf{1}\,. \tag{6.47}$$

The total cross section for the excitation of the atom from the state $n_1'l_1'$ to n_1l_1 is then given by (in units of πa_0^2)

$$Q(n_1'l_1' \to n_1l_1) = \sum_{LS}\sum_{l_2l_2'} \frac{(2L+1)(2S+1)}{4k^2_{\nu'}\,(2l_1'+1)}\,|\,T^{LS}_{\nu\nu'}\,|^2\,, \tag{6.48}$$

and the corresponding elastic differential cross section is given by (in units of $\pi a_0^2/\mathrm{sr}$)

$$\frac{\mathrm{d}\sigma}{\mathrm{d}\Omega}(n_1'l_1' \to n_1'l_1') = \sum_S \frac{(2S+1)}{16k^2_{\nu'}}\left|\sum_{Ll_2'}(2L+1)\,P_L(\cos\theta)\,T^{LS}_{\nu'\nu'}\right|^2. \tag{6.49}$$

For positron collisions it is not necessary to distinguish between singlet and triplet scattering and hence the summations over S in (6.48) and (6.49) can be carried out immediately, thereby simplifying these results.

6.3 The Numerical Solution of the Close-Coupling Equations

Here we consider a numerical procedure for solving the close-coupling equations (6.43) subject to the boundary conditions given in (6.45). The specific procedures to be developed here for the solution of these equations will be based upon their corresponding integral-equation formulation. Thus if the operator $K_{\nu LS}(r_2)$ is defined according to

$$K_{\nu LS}(r_2)\,F_{\nu LS}(r_2) = 2\Big[\frac{Z}{r_2} - V_{\mathrm{C}}(r_2)\Big]F_{\nu LS}(r_2) - 2\sum_{\nu''} V(\nu,\nu'')\,F_{\nu'' LS}(r_2) \tag{6.50}$$

then the equivalent integral-equation formulation of the close-coupling equations (6.43) is given by

$$F_{\nu LS}(r_2) = u_{l_2}(k_\nu r_2) - \frac{v_{l_2}(k_\nu r_2)}{k_\nu}\int_0^{r_2} \mathrm{d}r_1\, u_{l_2}(k_\nu r_1)\,K_{\nu LS}(r_1)F_{\nu LS}(r_1) + \frac{u_{l_2}(k_\nu r_2)}{k_\nu}\int_0^{r_2} \mathrm{d}r_1\, v_{l_2}(k_\nu r_1)\,K_{\nu LS}(r_1)F_{\nu LS}(r_1) \tag{6.51a}$$

in the open channels and

$$F_{\nu LS}(r_2) = -\frac{q_{l_2}(|k_\nu| r_2)}{|k_\nu|}\int_0^{r_2} \mathrm{d}r_1\, p_{l_2}(|k_\nu| r_1)\,K_{\nu LS}(r_1)F_{\nu LS}(r_1) - \frac{p_{l_2}(|k_\nu| r_2)}{|k_\nu|}\int_{r_2}^{\infty} \mathrm{d}r_1\, q_{l_2}(|k_\nu| r_1)\,K_{\nu LS}(r_1)F_{\nu LS}(r_1) \tag{6.51b}$$

in the closed channels.

The open-channel Green's functions $u_{l_2}(k_\nu r_2)$ and $v_{l_2}(k_\nu r_2)$ which occur in (6.51a) are defined by

$$u_{l_2}(k_\nu r_2) = \hat{\jmath}_{l_2}(k_\nu r_2) \tag{6.52a}$$

and

$$v_{l_2}(k_\nu r_2) = -\,\hat{n}_{l_2}(k_\nu r_2)\,, \tag{6.52b}$$

where $\hat{\jmath}_l(kr)$ and $\hat{n}_l(kr)$ are the Riccati-Bessel functions of the first and second kind, respectively. The properties of these Green's functions are such that their Wronskian, $W[u_{l_2}(k_\nu r_2), v_{l_2}(k_\nu r_2)]$, is $-k_\nu$ and their behaviour near the origin and asymptotically is as follows:

$$u_{l_2}(k_\nu r_2) = \frac{(k_\nu r_2)^{l_2+1}}{(2l_2+1)!!}\left[1 - \frac{(k_\nu r_2)^2}{2\,(2l_2+3)} + \cdots\right], \tag{6.53a}$$

$$v_{l_2}(k_\nu r_2) = \frac{(2l_2-1)!!}{(k_\nu r_2)^{l_2}}\left[1 - \frac{(k_\nu r_2)^2}{2\,(1-2l_2)} + \cdots\right], \tag{6.53b}$$

and

$$u_{l_2}(k_\nu r_2)_{r_2\to\infty} \sim \sin\!\left(k_\nu r_2 - \frac{l_2\pi}{2}\right) \tag{6.54a}$$

$$v_{l_2}(k_\nu r_2)_{r_2\to\infty} \sim \cos\!\left(k_\nu r_2 - \frac{l_2\pi}{2}\right). \tag{6.54b}$$

The closed-channel Green's functions $p_{l_2}(|k_\nu|r_2)$ and $q_{l_2}(|k_\nu|r_2)$ which occur in (6.51b) are defined by

$$p_{l_2}(|k_\nu|r_2) = \left(\frac{\pi|k_\nu|r_2}{2}\right)^{\frac{1}{2}} I_{l_2+\frac{1}{2}}(|k_\nu|r_2) \tag{6.55a}$$

and

$$q_{l_2}(|k_\nu|r_2) = \left(\frac{2|k_\nu|r_2}{\pi}\right)^{\frac{1}{2}} K_{l_2+\frac{1}{2}}(|k_\nu|r_2), \tag{6.55b}$$

where $I_{l_2}(|k_\nu|r_2)$ and $K_{l_2}(|k_\nu|r_2)$ are the modified Bessel functions of the first and third kind, respectively. The Wronskian of these Green's functions, namely $W[p_{l_2}(|k_\nu|r_2), q_{l_2}(|k_\nu|r_2)]$, is given by $-|k_\nu|$ and near the origin and asymptotically they behave according to

$$p_{l_2}(|k_\nu|r_2) = \frac{(|k_\nu|r_2)^{l_2+1}}{(2l_2+1)!!}\left[1 + \frac{(|k_\nu|r_2)^2}{2\,(2l_2+3)} + \cdots\right], \tag{6.56a}$$

$$q_{l_2}(|k_\nu|r_2) = \frac{(2l_2-1)!!}{(|k_\nu|r_2)^{l_2}} + O(r_2^{-l_2+1}), \tag{6.56b}$$

and

$$p_{l_2}(|k_\nu|r_2)_{r_2\to\infty} \sim \tfrac{1}{2}\exp(|k_\nu|r_2), \tag{6.57a}$$

$$q_{l_2}(|k_\nu|r_2)_{r_2\to\infty} \sim \exp(-|k_\nu|r_2). \tag{6.57b}$$

As was noted in Sect. 6.2.1, if there are `NC` open channels then `NC` linearly independent solutions of the coupled equations (6.51) will be required. If the ν'th linearly independent solution for $F_{\nu LS}(r_2)$ is denoted by $F^{\nu'}_{\nu LS}(r_2)$ then these particular functions can be obtained as *linear combinations* of the solutions to the equations

$$\mathcal{F}^{\nu'}_{\nu LS}(r_2) = \delta(\nu,\nu')u_{l_2}(k_\nu r_2) - \frac{v_{l_2}(k_\nu r_2)}{k_\nu}\int_0^{r_2} dr_1\, u_{l_2}(k_\nu r_1)\, K_{\nu LS}(r_1)\mathcal{F}^{\nu'}_{\nu LS}(r_1) + \frac{u_{l_2}(k_\nu r_2)}{k_\nu}\int_0^{r_2} dr_1\, v_{l_2}(k_\nu r_1)\, K_{\nu LS}(r_1)\mathcal{F}^{\nu'}_{\nu LS}(r_1) \quad (6.58a)$$

in the NC open channels and

$$\mathcal{F}^{\nu'}_{\nu LS}(r_2) = -\frac{q_{l_2}(|k_\nu|r_2)}{|k_\nu|}\int_0^{r_2} dr_1\, p_{l_2}(|k_\nu|r_1)\, K_{\nu LS}(r_1)\mathcal{F}^{\nu'}_{\nu LS}(r_1) - \frac{p_{l_2}(|k_\nu|r_2)}{|k_\nu|}\int_{r_2}^{\infty} dr_1\, q_{l_2}(|k_\nu|r_1)\, K_{\nu LS}(r_1)\mathcal{F}^{\nu'}_{\nu LS}(r_1) \quad (6.58b)$$

in the remaining closed channels.

The asymptotic form of the open-channel functions in (6.58a) is given by

$$\mathcal{F}^{\nu'}_{\nu LS}(r_2) \sim A^{LS}_{\nu\nu'} \sin\left(k_\nu r_2 - \frac{l_2\pi}{2}\right) + B^{LS}_{\nu\nu'} \cos\left(k_\nu r_2 - \frac{l_2\pi}{2}\right), \quad (6.59)$$

where

$$A^{LS}_{\nu\nu'} = \delta(\nu,\nu') + \frac{1}{k_\nu}\int_0^{\infty} dr_1\, v_{l_2}(k_\nu r_1)\, K_{\nu LS}(r_1)\mathcal{F}^{\nu'}_{\nu LS}(r_1), \quad (6.60a)$$

$$B^{LS}_{\nu\nu'} = -\frac{1}{k_\nu}\int_0^{\infty} dr_1\, u_{l_2}(k_\nu r_1)\, K_{\nu LS}(r_1)\mathcal{F}^{\nu'}_{\nu LS}(r_1). \quad (6.60b)$$

For each of the NC different values of ν', corresponding to the open channels, the required set of linearly independent solutions $F^{\nu'}_{\nu LS}(r_2)$ can be constructed from the open-channel solutions $\mathcal{F}^{\nu'}_{\nu LS}(r_2)$ according to

$$F^{\nu'}_{\nu LS}(r_2) = \sum_{\nu''} c^{\nu'}_{\nu''}\, \mathcal{F}^{\nu''}_{\nu LS}(r_2). \quad (6.61)$$

Each of the NC different sets of constants $c^{\nu'}_{\nu''}$ are then determined so that the $F^{\nu'}_{\nu LS}(r_2)$ satisfy the boundary conditions (6.45). Thus, for each value of ν' the $c^{\nu'}_{\nu''}$ are obtained from the solution of the linear algebraic equations

$$\mathbf{A}^{LS}\,\mathbf{c}^{\nu'} = \boldsymbol{k}^{\nu'} \quad (6.62)$$

where the column vector $\boldsymbol{k}^{\nu'}$ is a null vector except in the ν'th position where it has the element $1/\sqrt{k_{\nu'}}$. Then for each value of ν the elements in the ν'th column of the **R** matrix can be determined from

$$R^{LS}_{\nu\nu'} = \sqrt{k_\nu}\sum_{\nu''} c^{\nu'}_{\nu''}\, B^{LS}_{\nu\nu''}. \quad (6.63)$$

Once the **R** matrix has been obtained the transition matrix **T** can then be determined from (6.47). In principle, since the **T** matrix is complex, it would be necessary to resort to complex arithmetic. However, in practice it is usually easier to compute the real and imaginary parts of the **T** matrix separately, i.e.,

$$\Re e\,\mathbf{T} = -2\,\mathbf{R}^2\,[\mathbf{1} + \mathbf{R}^2]^{-1} \tag{6.64a}$$

and

$$\Im m\,\mathbf{T} = 2\,\mathbf{R}\,[\mathbf{1} + \mathbf{R}^2]^{-1}\,. \tag{6.64b}$$

In order to start the integration process for the solution of (6.58) it will first be necessary to determine power-series expansions about the origin for all the various terms. Once these have been obtained the integration of these equations can proceed quite simply using Simpson's 1/3 rule out into the asymptotic region where the short-range part of the overall potential vanishes. It is then necessary to develop an asymptotic correction method for extracting the elements of the **R** matrix when long-range potentials govern the asymptotic form of these equations.

6.3.1 The Power-Series Expansion About the Origin

The power-series expansion about the origin for the radial part of the bound-state Hartree–Fock wavefunctions is given by

$$P_{n_1 l_1}(r_2) = A_{n_1 l_1} r_2^{l_1+1}\Big[1 - \frac{Z}{l_1+1}\, r_2 + O(r_2^2)\Big] \tag{6.65}$$

and hence the potential terms of the form given in (6.44) can be expressed as

$$\begin{aligned} \frac{1}{r_2}\, y_\lambda(n_1 l_1, n_1' l_1'; r_2) &= d_\lambda(n_1 l_1, n_1' l_1')\, r_2^\lambda \\ &\quad - r_2^\lambda \int_0^{r_2} \mathrm{d}r_1\, \frac{1}{r_1^{\lambda+1}}\, P_{n_1 l_1}(r_1)\, P_{n_1' l_1'}(r_1) \\ &\quad + \frac{1}{r_2^{\lambda+1}} \int_0^{r_2} \mathrm{d}r_1\, r_1^\lambda\, P_{n_1 l_1}(r_1)\, P_{n_1' l_1'}(r_1) \\ &= d_\lambda(n_1 l_1, n_1' l_1')\, r_2^\lambda + O(r_2^{l_1 + l_1' + 2})\,, \end{aligned} \tag{6.66}$$

where

$$d_\lambda(n_1 l_1, n_1' l_1') = \int_0^\infty \mathrm{d}r_1 \frac{1}{r_1^{\lambda+1}} P_{n_1 l_1}(r_1) P_{n_1' l_1'}(r_1)\,. \tag{6.67}$$

Since $l_1 + l_1' = \max\lambda$ we note that $l_1 + l_1' + 2 \geq \lambda + 2$ and thus, in general, there is no term of order $\lambda + 1$ in (6.66).

Consequently, from (6.35) and (6.66) the nuclear plus core potentials have the following expansion about the origin

$$\left[\frac{Z}{r_2} - V_{\rm C}(r_2)\right] = \left[\frac{Z}{r_2} - d_0 + O(r_2^2)\right], \tag{6.68}$$

where

$$d_0 = \sum_{nl} q_{nl}\, d_0(nl, nl) = \sum_{nl} q_{nl}\, \langle nl \mid 1/r \mid nl \rangle. \tag{6.69}$$

If in all channels, open or closed, the following form is assumed near the origin for the series expansion of the linearly independent functions $\mathcal{F}^{\nu'}_{\nu LS}(r_2)$, namely

$$\mathcal{F}^{\nu'}_{\nu LS}(r_2) = a^{\nu'}_{\nu}\, r_2^{l_2+1} + b^{\nu'}_{\nu}\, r_2^{l_2+2} + c^{\nu'}_{\nu}\, r_2^{l_2+3} + \cdots, \tag{6.70}$$

then

$$\left[\frac{Z}{r_2} - V_{\rm C}(r_2)\right]\mathcal{F}^{\nu'}_{\nu LS}(r_2) = r_2^{l_2}\left[Za^{\nu'}_{\nu} + (Zb^{\nu'}_{\nu} - d_0 a^{\nu'}_{\nu})r_2 + \cdots\right] \tag{6.71}$$

and

$$\sum_{\nu''} V(\nu, \nu'')\mathcal{F}^{\nu'}_{\nu'' LS}(r_2) = \sum_{\nu''\lambda} f_\lambda(l_1 l_2, l_1'' l_2''; L)$$
$$\times \left[d_\lambda(n_1 l_1, n_1'' l_1'')\, r_2^\lambda + O(r_2^{l_1+l_1''+2})\right] \times \left[a^{\nu'}_{\nu''}\, r_2^{l_2''+1} + b^{\nu'}_{\nu''}\, r_2^{l_2''+2} + \cdots\right]. \tag{6.72}$$

From our definition of $f_\lambda(l_1 l_2, l_1' l_2'; L)$ given in (6.41) it is clear that this coefficient is zero unless l_2, l_2'', and λ satisfy the triangle inequality and the sum of these three numbers is even. For fixed l_2 we now consider all possible combinations of l_2'' and λ in the summations in (6.72) which satisfy the above two conditions. Since the sum over ν'' always contains the term $\nu'' = \nu$, for which $l_2'' = l_2$, and $\lambda = 0$, it follows that the minimum value of $\lambda + l_2''$ in (6.72) is l_2. Thus, in principle, the appropriate expansion of (6.72) would appear to be

$$\sum_{\nu''} V(\nu, \nu'')\mathcal{F}^{\nu'}_{\nu'' LS}(r_2) = \sum_{\nu''\lambda} m_\lambda(\nu, \nu'')\, r_2^{l_2+1}\Big[\delta(\lambda, l_2 - l_2'')$$
$$+ \delta(\lambda, l_2 + 2 - l_2'')\, r_2^2 + \cdots\Big]\left[a^{\nu'}_{\nu''} + b^{\nu'}_{\nu''}\, r_2 + \cdots\right], \tag{6.73}$$

provided all the $a^{\nu'}_{\nu}$ and $b^{\nu'}_{\nu}$ are non-zero. Here the coefficient $m_\lambda(\nu, \nu'')$ is given by

$$m_\lambda(\nu, \nu'') = f_\lambda(l_1 l_2, l_1'' l_2''; L)\, d_\lambda(n_1 l_1, n_1'' l_1'') \tag{6.74a}$$

and, in the particular case when $\nu'' = \nu$ and $\lambda = 0$, this reduces to

$$m_0(\nu, \nu) = d_0(n_1 l_1, n_1 l_1). \tag{6.74b}$$

Thus the power-series expansion of the nuclear plus core potential terms begins with a term of order $r_2^{l_2}$ whereas the corresponding expansions for the remaining potential terms begin with terms of order $r_2^{l_2+1}$. The next step is the substitution of (6.71) and (6.73) for the potential terms into the integral (6.58) to obtain the power-series expansions for the wavefunctions $\mathcal{F}^{\nu'}_{\nu LS}(r_2)$.

We shall first consider the power-series expansions for the open-channel wavefunctions as given by (6.58a). From (6.53) for the open-channel Green's functions and (6.71) for the nuclear plus core potential terms it follows that

$$\begin{aligned}
&-\frac{v_{l_2}(k_\nu r_2)}{k_\nu}\int_0^{r_2} \mathrm{d}r_1\, u_{l_2}(k_\nu r_1)\, 2\Big[\frac{Z}{r_1} - V_{\mathrm{C}}(r_1)\Big]\mathcal{F}^{\nu'}_{\nu LS}(r_1)\\
&\qquad + \frac{u_{l_2}(k_\nu r_2)}{k_\nu}\int_0^{r_2} \mathrm{d}r_1\, v_{l_2}(k_\nu r_1)\, 2\Big[\frac{Z}{r_1} - V_{\mathrm{C}}(r_1)\Big]\mathcal{F}^{\nu'}_{\nu LS}(r_1)\\
&= r_2^{l_2+2}\Big[\frac{Za^{\nu'}_\nu}{l_2+1} + \frac{Zb^{\nu'}_\nu - d_0 a^{\nu'}_\nu}{2l_2+3}\, r_2 + \cdots\Big]. \qquad (6.75)
\end{aligned}$$

The corresponding power-series expansions of the remaining potential as given by (6.73) will begin with terms of order $r_2^{l_2+3}$. It follows, therefore, that up to terms of order $r_2^{l_2+2}$, (6.58a) can be expressed as

$$a^{\nu'}_\nu r_2^{l_2+1} + b^{\nu'}_\nu r_2^{l_2+2} + \cdots = \frac{k_\nu^{l_2+1}}{(2l_2+1)!!}\,\delta(\nu,\nu')\, r_2^{l_2+1} + \frac{Za^{\nu'}_\nu}{l_2+1}\, r_2^{l_2+2} + \cdots \qquad (6.76)$$

Thus for the open-channel wavefunctions we have

$$a^{\nu'}_{\nu'} = \frac{k_{\nu'}^{l'_2+1}}{(2l'_2+1)!!} \quad \text{and} \quad b^{\nu'}_{\nu'} = \frac{Z}{l'_2+1}\, a^{\nu'}_{\nu'}\,, \qquad (6.77\text{a, b})$$

and

$$a^{\nu'}_\nu = b^{\nu'}_\nu = 0 \qquad \text{for } \nu \neq \nu'\,. \qquad (6.78\text{a, b})$$

Equation (6.58b) for the closed-channel wavefunctions can be re-written in the following manner:

$$\begin{aligned}
\mathcal{F}^{\nu'}_{\nu LS}(r_2) &= s^{\nu'}_\nu\, p_{l_2}(|k_\nu| r_2)\\
&\quad - \frac{q_{l_2}(|k_\nu| r_2)}{|k_\nu|}\int_0^{r_2} \mathrm{d}r_1\, p_{l_2}(|k_\nu| r_1)\, K_{\nu LS}(r_1)\mathcal{F}^{\nu'}_{\nu LS}(r_1)\\
&\quad + \frac{p_{l_2}(|k_\nu| r_2)}{|k_\nu|}\int_0^{r_2} \mathrm{d}r_1\, q_{l_2}(|k_\nu| r_1)\, K_{\nu LS}(r_1)\mathcal{F}^{\nu'}_{\nu LS}(r_1)\,, \qquad (6.79)
\end{aligned}$$

where

$$s^{\nu'}_\nu = -\frac{1}{|k_\nu|}\int_0^{\infty} \mathrm{d}r_1\, q_{l_2}(|k_\nu| r_1)\, K_{\nu LS}(r_1)\mathcal{F}^{\nu'}_{\nu LS}(r_1)\,. \qquad (6.80)$$

From (6.53) and (6.56) it is clear that the leading term in each of the Green's functions $p_{l_2}(r_2)$ and $q_{l_2}(r_2)$ is identical to that of $u_{l_2}(r_2)$ and $v_{l_2}(r_2)$, respectively, except that $|k_\nu|$ now replaces k_ν. Thus, up to terms of order $r_2^{l_2+2}$, (6.58b) can be expressed as

$$a_\nu^{\nu'} r_2^{l_2+1} + b_\nu^{\nu'} r_2^{l_2+2} + \cdots = \frac{|k_\nu|^{l_2+1}}{(2l_2+1)!!} s_\nu^{\nu'} r_2^{l_2+1} + \frac{Za_\nu^{\nu'}}{l_2+1} r_2^{l_2+2} + \cdots \quad (6.81)$$

and hence for all the closed-channel wavefunctions the coefficients $a_\nu^{\nu'}$ and $b_\nu^{\nu'}$ are given by

$$a_\nu^{\nu'} = \frac{|k_\nu|^{l_2+1}}{(2l_2+1)!!} s_\nu^{\nu'} \quad \text{and} \quad b_\nu^{\nu'} = \frac{Z}{l_2+1} a_\nu^{\nu'} . \quad (6.82\text{a, b})$$

Since these closed-channel functions have to be determined by an iterative procedure, the constants $s_\nu^{\nu'}$ and hence the coefficients $a_\nu^{\nu'}$ and $b_\nu^{\nu'}$ are in principle unknown. However, they can be estimated at each stage in the iteration cycle from the results of the previous iteration; initially they can be set to zero.

The coefficients $c_\nu^{\nu'}$ of $r_2^{l_2+3}$ for the open-channel wavefunctions can be determined in a similar manner. The corresponding results for the closed channels are not necessary since these wavefunctions must be determined iteratively and it is only necessary to consider the influence of, at most, the first two coefficients of the closed channels upon the series expansions for the open-channel wavefunctions. In addition to the free-wave term on the right-hand side of (6.58a) there are, in general, two more sources of terms of order $r_2^{l_2+3}$, namely, (6.71) for the nuclear plus core potential terms, and (6.73) for the remaining potential terms. In general, then, the coefficients $c_\nu^{\nu'}$ are given by

$$\begin{aligned} c_\nu^{\nu'} = {} & \frac{k_\nu^{l_2+1}}{(2l_2+1)!!} \left[\frac{-k_\nu^2}{2\,(2l_2+3)}\right] \delta(\nu,\nu') + \left[\frac{Zb_\nu^{\nu'} - d_0 a_\nu^{\nu'}}{2l_2+3}\right] - \delta(\lambda, l_2 - l_2') \\ & \times m_\lambda(\nu,\nu') \frac{a_{\nu'}^{\nu'}}{2l_2+3} - {\sum_{\nu''}}' \delta(\lambda, l_2 - l_2'')\, m_\lambda(\nu,\nu'') \frac{a_{\nu''}^{\nu'}}{2l_2+3} . \end{aligned} \quad (6.83)$$

From (6.74a) the case when $\nu = \nu'$ yields

$$\begin{aligned} c_{\nu'}^{\nu'} = {} & \frac{1}{2(2l_2'+3)} \big[2(Zb_{\nu'}^{\nu'} \\ & - \tilde{d}_0 a_{\nu'}^{\nu'}) - k_{\nu'}^2 a_{\nu'}^{\nu'}\big] - {\sum_{\nu''}}' \delta(\lambda, l_2' - l_2'')\, m_\lambda(\nu',\nu'') \frac{a_{\nu''}^{\nu'}}{2l_2'+3} , \end{aligned} \quad (6.84)$$

where

$$\tilde{d}_0 = d_0 + d_0(n_1 l_1, n_1 l_1) . \quad (6.85)$$

When $\nu \neq \nu'$ there are several different cases to be considered. Since, for the open channels, $a_\nu^{\nu'} = b_\nu^{\nu'} = 0$ for $\nu \neq \nu'$ it follows that the second term in (6.83) for $c_\nu^{\nu'}$ vanishes. The third term also vanishes unless $l_2 \geq l_2'$, whereas the fourth term is only present if there are closed channels and even then it may or may not be non-zero depending upon the particular value of l_2 and the various values of the l_2''. Thus for $\nu \neq \nu'$, and $l_2 \geq l_2'$ we have

$$c_\nu^{\nu'} = -\,\delta(\lambda, l_2 - l_2')\, m_\lambda(\nu,\nu')\,\frac{a_{\nu'}^{\nu'}}{2l_2+3}$$
$$-\sum_{\nu''}{}'\delta(\lambda, l_2 - l_2'')\, m_\lambda(\nu,\nu'')\,\frac{a_{\nu''}^{\nu'}}{2l_2+3} \tag{6.86}$$

and for $\nu \neq \nu'$, and $l_2 < l_2'$

$$c_\nu^{\nu'} = -\sum_{\nu''}{}'\delta(\lambda, l_2 - l_2'')\, m_\lambda(\nu,\nu'')\,\frac{a_{\nu''}^{\nu'}}{2l_2+3}\,, \tag{6.87}$$

where the prime on the summation signs indicates that the summation is to be taken over only the closed channels. As noted above it is possible that the coefficient $c_\nu^{\nu'}$, as given by (6.87), will be zero and in this case it can be shown from (6.75) that the next term, say $d_\nu^{\nu'}$, in the series expansion will also be zero. If there are no closed channels then the leading term in the series expansion will arise from the term $\nu'' = \nu'$ in (6.72). When $l_2 < l_2'$ the minimum value of λ will be $l_2' - l_2$ and hence the minimum value of $\lambda + l_2'$ will be $2l_2' - l_2$. Equation (6.73) should then be expressed as

$$\sum_{\nu''} V(\nu,\nu'')\,\mathcal{F}_{\nu''LS}^{\nu'}(r_2)$$
$$= \sum_{\lambda} \delta(\lambda, l_2' - l_2) m_\lambda(\nu,\nu')\, r_2^{2l_2'-l_2+1}\left[a_{\nu'}^{\nu'} + b_{\nu'}^{\nu'} r_2 + \cdots\right]. \tag{6.88}$$

It can be shown that the leading term in the series expansion of $\mathcal{F}_{\nu LS}^{\nu'}(r_2)$ is then given by

$$-\,\delta(\lambda, l_2' - l_2)\, m_\lambda(\nu,\nu')\,\frac{a_{\nu'}^{\nu'}}{(2l_2'+3)(l_2'-l_2+1)}\, r_2^{2l_2'-l_2+3}\,. \tag{6.89}$$

However, if there are closed channels then it is necessary to determine at which order they would contribute, if at all.

The above analysis has so far assumed that the $d_\lambda(n_1l_1, n_1'l_1')$ are never zero. For the alkali atoms this will be true in general. However, for hydrogen where the energy levels, for a fixed principal quantum number, are degenerate some of the d_λ are zero [e.g., $d_1(2s, 2p)$ and $d_1(3s, 3p)$] and hence the order of the leading coefficient is two higher than that given in the above analysis.

Finally we consider the power-series expansions for the integrals which arise in (6.58a) for the open channels. When $\nu = \nu'$ it follows from (6.71) and (6.72) that

$$
\begin{aligned}
I^{\nu'}_{\nu' LS}(r_2) &= \int_0^{r_2} \mathrm{d}r_1\, u_{l'_2}(k_{\nu'} r_1)\, K_{\nu' LS}(r_1) \mathcal{F}^{\nu'}_{\nu' LS}(r_1) \\
&= \frac{2k_{\nu'}^{l'_2+1}}{(2l'_2+1)!!}\, r_2^{2l'_2+2} \Big\{ \frac{Za^{\nu'}_{\nu'}}{2l'_2+2} + \frac{1}{2l'_2+3} \Big[Zb^{\nu'}_{\nu'} - \tilde{d}_0 a^{\nu'}_{\nu'} \\
&\quad - {\sum_{\nu''}}' \delta(\lambda, l'_2 - l''_2)\, m_\lambda(\nu', \nu'')\, a^{\nu'}_{\nu''} \Big] r_2 + \cdots \Big\} . \qquad (6.90)
\end{aligned}
$$

On the other hand when $\nu \neq \nu'$ and $l_2 \geq l'_2$ then the $c^{\nu'}_{\nu}$, as given by (6.86), are always non-zero and hence

$$
\begin{aligned}
I^{\nu'}_{\nu LS}(r_2) &= \frac{2k_{\nu}^{l_2+1}}{(2l_2+1)!!}\, r_2^{2l_2+3} \Big\{ c^{\nu'}_{\nu} + \frac{1}{2l_2+4} \Big[Zc^{\nu'}_{\nu} \\
&\quad - \delta(\lambda, l_2 - l'_2)\, m_\lambda(\nu, \nu')\, b^{\nu'}_{\nu'} \\
&\quad - {\sum_{\nu''}}' \delta(\lambda, l_2 - l''_2)\, m_\lambda(\nu, \nu'')\, b^{\nu'}_{\nu''} \Big] r_2 + \cdots \Big\} . \qquad (6.91)
\end{aligned}
$$

When $\nu \neq \nu'$ and $l_2 < l'_2$ there are two cases. First, if the $c^{\nu'}_{\nu}$, as given by (6.87), are non-zero then

$$
\begin{aligned}
I^{\nu'}_{\nu LS}(r_2) &= \frac{2k_{\nu}^{l_2+1}}{(2l_2+1)!!}\, r_2^{2l_2+3} \Big\{ c^{\nu'}_{\nu} + \frac{1}{2l_2+4} \Big[Zc^{\nu'}_{\nu} \\
&\quad - {\sum_{\nu''}}' \delta(\lambda, l_2 - l''_2)\, m_\lambda(\nu, \nu'')\, b^{\nu'}_{\nu''} \Big] r_2 + \cdots \Big\} , \qquad (6.92)
\end{aligned}
$$

and second if, for positron scattering, the $c^{\nu'}_{\nu}$ are zero then from (6.88)

$$
I^{\nu'}_{\nu LS}(r_2) = -\frac{2k_{\nu}^{l_2+1}}{(2l_2+1)!!}\, \delta(\lambda, l'_2 - l_2)\, m_\lambda(\nu, \nu')\, \frac{a^{\nu'}_{\nu'}}{(2l'_2+3)}\, r_2^{2l'_2+3} + \cdots \qquad (6.93)
$$

Similarly, if

$$
J^{\nu'}_{\nu LS}(r_2) = \int_0^{r_2} \mathrm{d}r_1\, v_{l'_2}(k_{\nu'} r_1)\, K_{\nu LS}(r_1) \mathcal{F}^{\nu'}_{\nu' LS}(r_1) , \qquad (6.94)
$$

then for $\nu = \nu'$ it follows that

$$
\begin{aligned}
J^{\nu'}_{\nu' LS}(r_2) &= \frac{2(2l'_2-1)!!}{k_{\nu'}^{l'_2}}\, r_2 \Big\{ Za^{\nu'}_{\nu'} + \tfrac{1}{2} \Big[Zb^{\nu'}_{\nu'} - \tilde{d}_0 a^{\nu'}_{\nu'} \\
&\quad - {\sum_{\nu''}}' \delta(\lambda, l'_2 - l''_2)\, m_\lambda(\nu', \nu'')\, a^{\nu'}_{\nu''} \Big] r_2 + \cdots \Big\} , \qquad (6.95)
\end{aligned}
$$

and for $\nu \neq \nu'$ and $l_2 \geq l'_2$ that

$$
\begin{aligned}
J^{\nu'}_{\nu LS}(r_2) &= \frac{2(2l_2-1)!!}{k_{\nu}^{l_2}}\, r_2^2 \Big\{ \frac{2l_2+3}{2}\, c^{\nu'}_{\nu} \\
&\quad + \tfrac{1}{3} \Big[Zc^{\nu'}_{\nu} - \delta(\lambda, l_2 - l'_2)\, m_\lambda(\nu, \nu')\, b^{\nu'}_{\nu'} \\
&\quad - {\sum_{\nu''}}' \delta(\lambda, l_2 - l''_2) m_\lambda(\nu, \nu'') b^{\nu'}_{\nu''} \Big] r_2 + \cdots \Big\} . \qquad (6.96)
\end{aligned}
$$

When $\nu \neq \nu'$ and $l_2 < l_2'$ then, if the $c_\nu^{\nu'}$, as given by (6.87), are non-zero we have

$$J_{\nu LS}^{\nu'}(r_2) = \frac{2(2l_2-1)!!}{k_\nu^{l_2}}\, r_2^2 \Big\{ \frac{2l_2+3}{2}\, c_\nu^{\nu'} + \tfrac{1}{3}\Big[Z c_\nu^{\nu'} - \sum_{\nu''}{}' \delta(\lambda, l_2 - l_2'')\, m_\lambda(\nu,\nu'')\, b_{\nu''}^{\nu'} \Big] r_2 + \cdots \Big\}, \tag{6.97}$$

and if the $c_\nu^{\nu'}$ are zero then

$$J_{\nu LS}^{\nu'}(r_2) = -\frac{(2l_2-1)!!}{k_\nu^{l_2}}\, \delta(\lambda, l_2' - l_2)\, m_\lambda(\nu,\nu') \times \frac{a_{\nu'}^{\nu'}}{(l_2' - l_2 + 1)}\, r_2^{2l_2' - 2l_2 + 2} + \cdots . \tag{6.98}$$

We also note from (6.77a) and (6.94) that when $\nu = \nu'$ then

$$\lim_{r_1 \to 0} v_{l_2'}(k_{\nu'} r_1)\, K_{\nu' LS}(r_1) \mathcal{F}_{\nu' LS}^{\nu'}(r_1) = \frac{2Z k_{\nu'}}{2l_2' + 1} . \tag{6.99}$$

The corresponding limit of all the other integrands above is zero.

6.3.2 Asymptotic Correction Procedures

Whenever the close-coupling expansion (6.6) contains atomic bound states other than s orbitals, then some of the terms in the potential $V(\nu,\nu')$ will behave asymptotically as $1/r_2^\lambda$ as $r_2 \to \infty$ with $\lambda \geq 2$. In these instances it is either necessary to integrate (6.58a) extremely far out into the asymptotic region or to develop an asymptotic correction procedure which accurately approximates the contributions to the $A_{\nu\nu'}^{LS}$ and $B_{\nu\nu'}^{LS}$ [(6.60a,b)] from the asymptotic region. In this latter approach it is usually only necessary to consider the asymptotic corrections to $A_{\nu\nu'}^{LS}$ and $B_{\nu\nu'}^{LS}$ due to the NC open-channels functions. Thus we begin by re-writing (6.58a) in the following form

$$\mathcal{F}_{\nu LS}^{\nu'}(r_2) = A_{\nu\nu'}^{LS}(r_2)\, u_{l_2}(k_\nu r_2) + B_{\nu\nu'}^{LS}\, v_{l_2}(k_\nu r_2), \tag{6.100}$$

where

$$A_{\nu\nu'}^{LS}(r_2) = \delta(\nu,\nu') + \frac{1}{k_\nu} \int_0^{r_2} \mathrm{d}r_1\, v_{l_2}(k_\nu r_1)\, K_{\nu LS}(r_1) \mathcal{F}_{\nu LS}^{\nu'}(r_1), \tag{6.101a}$$

and

$$B_{\nu\nu'}^{LS}(r_2) = -\frac{1}{k_\nu} \int_0^{r_2} \mathrm{d}r_1\, u_{l_2}(k_\nu r_1)\, K_{\nu LS}(r_1) \mathcal{F}_{\nu LS}^{\nu'}(r_1). \tag{6.101b}$$

If we now define the quantities

$$\Delta A^{LS}_{\nu\nu'}(r_2) = \frac{1}{k_\nu}\int_{r_2}^{\infty} \mathrm{d}r_1\, v_{l_2}(k_\nu r_1)\, K_{\nu LS}(r_1)\mathcal{F}^{\nu'}_{\nu LS}(r_1) \tag{6.102a}$$

and

$$\Delta B^{LS}_{\nu\nu'}(r_2) = -\frac{1}{k_\nu}\int_{r_2}^{\infty} \mathrm{d}r_1\, u_{l_2}(k_\nu r_1)\, K_{\nu LS}(r_1)\mathcal{F}^{\nu'}_{\nu LS}(r_1)\,, \tag{6.102b}$$

then from (6.60a,b) it follows that

$$A^{LS}_{\nu\nu'} = A^{LS}_{\nu\nu'}(r_2) + \Delta A^{LS}_{\nu\nu'}(r_2) \tag{6.103a}$$

and

$$B^{LS}_{\nu\nu'} = B^{LS}_{\nu\nu'}(r_2) + \Delta B^{LS}_{\nu\nu'}(r_2)\,. \tag{6.103b}$$

From (6.66) the asymptotic form of the potential terms is given by

$$\lim_{r_2\to\infty} \frac{1}{r_2}\, y_\lambda(n_1l_1, n_1'l_1'; r_2) = \frac{e_\lambda(n_1l_1, n_1'l_1')}{r_2^{\lambda+1}}\,, \tag{6.104}$$

where

$$e_\lambda(n_1l_1, n_1'l_1') = \int_0^\infty \mathrm{d}r_1\, r_1^\lambda\, P_{n_1l_1}(r_1)\, P_{n_1'l_1'}(r_1)\,, \tag{6.105a}$$

and hence

$$e_0(n_1l_1, n_1'l_1') = \delta(n_1l_1, n_1'l_1')\,. \tag{6.105b}$$

Thus in any scattering channel (open or closed) represented by νLS, the sum over the potential terms, $V(\nu,\nu')$, in (6.43) will always contain the term

$$\frac{1}{r_2}\, y_0(n_1l_1, n_1l_1; r_2)$$

which multiplies the function $\mathcal{F}^{\nu'}_{\nu LS}(r_2)$ of that channel and which behaves asymptotically as $1/r_2$. This potential term can then always be combined with

$$\frac{Z}{r_2} - V_c(r_2)$$

to produce a net overall potential which multiplies this channel function and which decays exponentially as $r_2 \to \infty$.

By analogy with (6.50) we now define

$$\mathcal{K}_{\nu LS}(r_2)\, F_{\nu LS}(r_2) = -2\sum_{\nu''}{}' V(\nu,\nu'')\, F_{\nu'' LS}(r_2)\,, \tag{6.106}$$

where $\sum'$ means that the term with $\nu'' = \nu$ and $\lambda = 0$ has been omitted from the sum. If r_2 is sufficiently large, say $r_2 \geq r_{\mathrm{M}}$, so that all the potential terms which decay exponentially are vanishingly small, then it follows from (6.104) that

$$\mathcal{K}_{\nu LS}(r_2)\mathcal{F}^{\nu'}_{\nu LS}(r_2) = -2\sum_{\nu''}{}' f_\lambda(l_1l_2,l''_1l''_2;L)\,e_\lambda(n_1l_1,n''_1l''_1)\,\frac{1}{r_2^{\lambda+1}}\,\mathcal{F}^{\nu'}_{\nu''LS}(r_2)\,. \tag{6.107}$$

Thus for $r_2 \geq r_{\mathrm{M}}$, $\Delta A^{LS}_{\nu\nu'}(r_2)$ and $\Delta B^{LS}_{\nu\nu'}(r_2)$ can be expressed as

$$\Delta A^{LS}_{\nu\nu'}(r_2) = -\frac{2}{k_\nu}\sum_{\nu''}{}'\sum_{\lambda} f_\lambda(l_1l_2,l''_1l''_2;L)\,e_\lambda(n_1l_1,n''_1l''_1) \times \int_{r_2}^{\infty} \mathrm{d}r_1\, v_{l_2}(k_\nu r_1)\,\frac{1}{r_1^{\lambda+1}} \times \left[A^{LS}_{\nu''\nu'}(r_1)\,u_{l''_2}(k_{\nu''}r_1) + B^{LS}_{\nu''\nu'}(r_1)\,v_{l''_2}(k_{\nu''}r_1)\right] \tag{6.108a}$$

and

$$\Delta B^{LS}_{\nu\nu'}(r_2) = \frac{2}{k_\nu}\sum_{\nu''}{}'\sum_{\lambda} f_\lambda(l_1l_2,l''_1l''_2;L)\,e_\lambda(n_1l_1,n''_1l''_1) \times \int_{r_2}^{\infty} \mathrm{d}r_1\, u_{l_2}(k_\nu r_1)\,\frac{1}{r_1^{\lambda+1}} \times \left[A^{LS}_{\nu''\nu'}(r_1)\,u_{l''_2}(k_{\nu''}r_1) + B^{LS}_{\nu''\nu'}(r_1)\,v_{l''_2}(k_{\nu''}r_1)\right]. \tag{6.108b}$$

So far (6.108a,b) are exact provided $r_2 \geq r_{\mathrm{M}}$. However, if it is assumed that for $r_1 \geq r_{\mathrm{M}}$ both $A^{LS}_{\nu''\nu'}(r_1)$ and $B^{LS}_{\nu''\nu'}(r_1)$ vary sufficiently slowly so that they can be approximated in (6.108a,b) by their respective values at the lower limit of integration, then these equations can be approximated according to

$$\Delta A^{LS}_{\nu\nu'}(r_2) \simeq -\frac{2}{k_\nu}\sum_{\nu''}{}'\sum_{\lambda} f_\lambda(l_1l_2,l''_1l''_2;L)\,e_\lambda(n_1l_1,n''_1l''_1) \times \left[A^{LS}_{\nu''\nu'}(r_2)\,\mathcal{W}_{\nu\nu''}(r_2) + B^{LS}_{\nu''\nu'}(r_2)\,\mathcal{V}_{\nu\nu''}(r_2)\right] \tag{6.109a}$$

and

$$\Delta B^{LS}_{\nu\nu'}(r_2) \simeq \frac{2}{k_\nu}\sum_{\nu''}{}'\sum_{\lambda} f_\lambda(l_1l_2,l''_1l''_2;L)\,e_\lambda(n_1l_1,n''_1l''_1) \times \left[A^{LS}_{\nu''\nu'}(r_2)\,\mathcal{U}_{\nu\nu''}(r_2) + B^{LS}_{\nu''\nu'}(r_2)\,\mathcal{W}_{\nu''\nu}(r_2)\right], \tag{6.109b}$$

where

$$\mathcal{U}_{\nu\nu''}(r_2) = \int_{r_2}^{\infty} \mathrm{d}r_1\,\frac{1}{r_1^{\lambda+1}}\,u_{l_2}(k_\nu r_1)\,u_{l''_2}(k_{\nu''}r_1)\,, \tag{6.110a}$$

$$\mathcal{V}_{\nu\nu''}(r_2) = \int_{r_2}^{\infty} \mathrm{d}r_1\,\frac{1}{r_1^{\lambda+1}}\,v_{l_2}(k_\nu r_1)\,v_{l''_2}(k_{\nu''}r_1)\,, \tag{6.110b}$$

$$\mathcal{W}_{\nu\nu''}(r_2) = \int_{r_2}^{\infty} \mathrm{d}r_1\,\frac{1}{r_1^{\lambda+1}}\,v_{l_2}(k_\nu r_1)\,u_{l''_2}(k_{\nu''}r_1)\,, \tag{6.110c}$$

and

$$\mathcal{W}_{\nu''\nu}(r_2) = \int_{r_2}^{\infty} \mathrm{d}r_1\,\frac{1}{r_1^{\lambda+1}}\,u_{l_2}(k_\nu r_1)\,v_{l''_2}(k_{\nu''}r_1)\,. \tag{6.110d}$$

Provided l_2 and l_2'' are not too large these integrals can be evaluated directly by the method given in McEachran and Stauffer [6.30]. An alternative approach is to combine the Green's functions $u_{l_2}(k_\nu r_2)$ and $v_{l_2}(k_\nu r_2)$ to form complex functions which are proportional to the spherical Bessel functions of the third kind, i.e.,

$$h_{l_2}^{\pm}(k_\nu r_2) = v_{l_2}(k_\nu r_2) \pm \mathrm{i}\, u_{l_2}(k_\nu r_2). \tag{6.111}$$

Equations (6.110a–d) can then be re-written in the form

$$\mathcal{U}_{\nu\nu''}(r_2) = -\frac{1}{2}\,\Re e\left[Q_{\nu\nu''}(r_2) - S_{\nu\nu''}(r_2)\right], \tag{6.112a}$$

$$\mathcal{V}_{\nu\nu''}(r_2) = \frac{1}{2}\,\Re e\left[Q_{\nu\nu''}(r_2) + S_{\nu\nu''}(r_2)\right], \tag{6.112b}$$

$$\mathcal{W}_{\nu\nu''}(r_2) = \frac{1}{2}\,\Im m\left[Q_{\nu\nu''}(r_2) - S_{\nu\nu''}(r_2)\right], \tag{6.112c}$$

and

$$\mathcal{W}_{\nu''\nu}(r_2) = \frac{1}{2}\,\Im m\left[Q_{\nu\nu''}(r_2) + S_{\nu\nu''}(r_2)\right], \tag{6.112d}$$

where

$$Q_{\nu\nu''}(r_2) = \int_{r_2}^{\infty} \mathrm{d}r_1\, \frac{1}{r_1^{\lambda+1}}\, h_{l_2}^{+}(k_\nu r_1)\, h_{l_2''}^{+}(k_{\nu''} r_1) \tag{6.113}$$

and

$$S_{\nu\nu''}(r_2) = \int_{r_2}^{\infty} \mathrm{d}r_1\, \frac{1}{r_1^{\lambda+1}}\, h_{l_2}^{+}(k_\nu r_1)\, h_{l_2''}^{-}(k_{\nu''} r_1)\,. \tag{6.114}$$

It is now convenient to express the Bessel functions $h_l^{\pm}(z)$ as

$$h_l^{\pm}(z) = H_l^{\pm}(z)\,\mathrm{e}^{\pm \mathrm{i}z}\,, \tag{6.115}$$

with

$$H_0^{\pm}(z) = 1 \qquad \text{and} \qquad H_1^{\pm}(z) = \frac{1}{z} \mp \mathrm{i}\,, \tag{6.116a, b}$$

and

$$H_l^{\pm}(z) = \frac{2l-1}{z}\, H_{l-1}^{\pm}(z) - H_{l-2}^{\pm}(z) \quad \text{for } l \geq 2\,, \tag{6.117}$$

where z is, in general, a complex variable. It should be noted, however, if z is real then $H_l^{-}(z)$ is the complex conjugate of $H_l^{+}(z)$. Equations (6.113) and (6.114) can then be re-written as

$$Q_{\nu\nu''}(r_2) = \int_{r_2}^{\infty} \mathrm{d}r_1\, \frac{1}{r_1^{\lambda+1}}\, H_{l_2}^{+}(k_\nu r_1)\, H_{l_2''}^{+}(k_{\nu''} r_1)\, \exp\{\mathrm{i}(k_\nu + k_{\nu''}) r_1\} \tag{6.118}$$

and

$$S_{\nu\nu''}(r_2) = \int_{r_2}^{\infty} \mathrm{d}r_1\, \frac{1}{r_1^{\lambda+1}}\, H_{l_2}^{+}(k_\nu r_1)\, H_{l_2''}^{-}(k_{\nu''} r_1)\, \exp\{\mathrm{i}(k_\nu - k_{\nu''}) r_1\}\,. \tag{6.119}$$

The integrands in the integrals of (6.118) and (6.119) not only decay slowly but oscillate infinitely. This makes the accurate numerical evaluation of these integrals along the real axis particularly difficult, especially if either $k_\nu + k_{\nu''}$ or $k_\nu - k_{\nu''}$ is small in magnitude. These difficulties can in part be overcome by adopting the method proposed by Sil *et al.* [6.31] and deforming the integration contour into the complex plane. Equation (6.113) for $Q_{\nu\nu''}(r_2)$ then becomes

$$Q_{\nu\nu''}(r_2) = \mathrm{i}\, r_2^{-\lambda} \exp\{\mathrm{i}(k_\nu + k_{\nu''})r_2\} \int_0^\infty \mathrm{d}\eta \,\frac{1}{(1+\mathrm{i}\eta)^{\lambda+1}}\, H_{l_2}^{+}[k_\nu r_2(1+\mathrm{i}\eta)] \times H_{l_2''}^{+}[k_{\nu''} r_2(1+\mathrm{i}\eta)] \exp\{-(k_\nu + k_{\nu''})r_2\eta\}\,, \tag{6.120}$$

which can in turn be evaluated by complex Gauss-Laguerre quadrature. Provided $k_\nu \neq k_{\nu''}$, then for $k_\nu > k_{\nu''}$, $S_{\nu\nu''}(r_2)$ can be determined from

$$S_{\nu\nu''}(r_2) = \mathrm{i}\, r_2^{-\lambda} \exp\{\mathrm{i}(k_\nu - k_{\nu''})r_2\} \int_0^\infty \mathrm{d}\eta \,\frac{1}{(1+\mathrm{i}\eta)^{\lambda+1}}\, H_{l_2}^{+}[k_\nu r_2(1+\mathrm{i}\eta)] \times H_{l_2''}^{-}[k_{\nu''} r_2(1+\mathrm{i}\eta)] \exp\{-(k_\nu - k_{\nu''})r_2\eta\}\,, \tag{6.121}$$

and when $k_\nu < k_{\nu''}$, $S^*_{\nu\nu''}(r_2)$ can be evaluated, i.e.,

$$S^*_{\nu\nu''}(r_2) = \mathrm{i}\, r_2^{-\lambda} \exp\{\mathrm{i}(k_{\nu''} - k_\nu)r_2\} \int_0^\infty \mathrm{d}\eta \,\frac{1}{(1+\mathrm{i}\eta)^{\lambda+1}}\, H_{l_2}^{-}[k_\nu r_2(1+\mathrm{i}\eta)] \times H_{l_2''}^{+}[k_{\nu''} r_2(1+\mathrm{i}\eta)] \exp\{-(k_{\nu''} - k_\nu)r_2\eta\}\,. \tag{6.122}$$

Finally if $k_\nu = k_{\nu''}$ then

$$S_{\nu\nu''}(r_2) = \int_{r_2}^\infty \mathrm{d}r_1 \,\frac{1}{r_1^{\lambda+1}}\, H_{l_2}^{+}(k_\nu r_1)\, H_{l_2''}^{-}(k_\nu r_1)\,. \tag{6.123}$$

It should be noted that if $l_2 = l_2''$ then $S_{\nu\nu''}(r_2)$ is real. In order to evaluate the integral in (6.123) the change of variable $r_1 = 1/s$ is made and hence

$$S_{\nu\nu''}(r_2) = \int_0^{1/r_2} \mathrm{d}s\, s^{\lambda-1}\, H_{l_2}^{+}(k_\nu/s)\, H_{l_2''}^{-}(k_\nu/s)\,, \tag{6.124}$$

which can be evaluated, in general, by complex Gauss-Legendre quadrature.

6.4 The Born Approximation

Whenever the total angular momentum, L, is sufficiently large, such that the overall interaction is sufficiently weak, the Born approximation can be used to compute the **R** matrix directly. This approximation can be developed by noting that the integral-equation equivalent of the coupled differential equations [cf. (6.43)] can also be expressed as

$$F_{\nu LS}(r_2) = u_{l_2}(k_\nu r_2) - \frac{v_{l_2}(k_\nu r_2)}{k_\nu} \int_0^{r_2} \mathrm{d}r_1\, u_{l_2}(k_\nu r_1)\, K_{\nu LS}(r_1)\, F_{\nu LS}(r_1)$$
$$- \frac{u_{l_2}(k_\nu r_2)}{k_\nu} \int_{r_2}^{\infty} \mathrm{d}r_1\, v_{l_2}(k_\nu r_1)\, K_{\nu LS}(r_1)\, F_{\nu LS}(r_1) \qquad (6.125)$$

in the open channels, where $K_{\nu LS}(r_1)\, F_{\nu LS}(r_1)$ is given, as before, by (6.50). This particular form for the integral equations is not as convenient computationally, however, because the second integral on the right-hand side precludes any possibility of integrating these equations directly out from the origin. Nonetheless, this form for the integral equation will enable us to obtain an explicit expression for each individual element of the **R** matrix.

Once again, if the ν'th linearly independent solution for $F_{\nu LS}(r_2)$ is denoted by $F^{\nu'}_{\nu LS}(r_2)$ then these particular functions can be obtained, in principle, as linear combinations of the solutions to the equation

$$\mathcal{F}^{\nu'}_{\nu LS}(r_2) = \delta(\nu, \nu')\, u_{l_2}(k_\nu r_2)$$
$$- \frac{v_{l_2}(k_\nu r_2)}{k_\nu} \int_0^{r_2} \mathrm{d}r_1\, u_{l_2}(k_\nu r_1)\, K_{\nu LS}(r_1)\, \mathcal{F}^{\nu'}_{\nu LS}(r_1)$$
$$- \frac{u_{l_2}(k_\nu r_2)}{k_\nu} \int_{r_2}^{\infty} \mathrm{d}r_1\, v_{l_2}(k_\nu r_1)\, K_{\nu LS}(r_1)\, \mathcal{F}^{\nu'}_{\nu LS}(r_1)\,. \qquad (6.126)$$

The asymptotic form of these open-channel functions is then given by

$$\mathcal{F}^{\nu'}_{\nu LS}(r_2) \sim \delta(\nu, \nu')\, \sin\Big(k_\nu r_2 - \frac{l_2 \pi}{2}\Big) + B^{LS}_{\nu\nu'}\, \cos\Big(k_\nu r_2 - \frac{l_2 \pi}{2}\Big) \qquad (6.127)$$

where

$$B^{LS}_{\nu\nu'} = -\frac{1}{k_\nu} \int_0^{\infty} \mathrm{d}r_1\, u_{l_2}(k_\nu r_1)\, K_{\nu LS}(r_1)\, \mathcal{F}^{\nu'}_{\nu LS}(r_1)\,. \qquad (6.128)$$

It should be noted that the numerical values of the coefficients, $B^{LS}_{\nu\nu'}$, in (6.128) are different from those in (6.101b) since the $\mathcal{F}^{\nu'}_{\nu LS}(r_1)$ are now different.

The set of linearly independent solutions $F^{\nu'}_{\nu LS}(r_2)$ can once again be expressed, in principle, as the following linear combination of the $\mathcal{F}^{\nu''}_{\nu LS}(r_2)$:

$$F^{\nu'}_{\nu LS}(r_2) = \sum_{\nu''} c^{\nu'}_{\nu''}\, \mathcal{F}^{\nu''}_{\nu LS}(r_2)\,. \qquad (6.129)$$

Then, from (6.45b) and (6.127) it follows that

$$c^{\nu'}_{\nu'} = \frac{1}{\sqrt{k_{\nu'}}} \quad \text{for } \nu = \nu', \qquad (6.130a)$$

$$c^{\nu'}_{\nu} = 0 \quad \text{for } \nu \neq \nu'\,. \qquad (6.130b)$$

Consequently, (6.129) reduces to just

$$F^{\nu'}_{\nu LS}(r_2) = \frac{1}{\sqrt{k_{\nu'}}}\, \mathcal{F}^{\nu'}_{\nu LS}(r_2)\,, \tag{6.131}$$

and the individual elements of the **R** matrix are given simply by

$$\begin{aligned} R^{LS}_{\nu\nu'} &= \sqrt{\frac{k_\nu}{k_{\nu'}}}\, B^{LS}_{\nu\nu'} \\ &= -\frac{1}{\sqrt{k_\nu k_{\nu'}}} \int_0^\infty \mathrm{d}r_1\, u_{l_2}(k_\nu r_1)\, K_{\nu LS}(r_1)\, \mathcal{F}^{\nu'}_{\nu LS}(r_1)\,. \end{aligned} \tag{6.132}$$

This expression is exact, provided the $\mathcal{F}^{\nu'}_{\nu LS}(r_2)$ are solutions of (6.126).

In the Born approximation it is assumed that the total angular momentum, L, of the system is sufficiently large, or alternatively, the interaction is so weak, that we can simply take $\mathcal{F}^{\nu'}_{\nu LS}(r_2)$ to be

$$\mathcal{F}^{\nu'}_{\nu LS}(r_2) = \delta(\nu,\nu')\, u_{l_2}(k_\nu r_2)\,. \tag{6.133}$$

That is, the $\mathcal{F}^{\nu'}_{\nu LS}(r_2)$ given by (6.133) are the solutions of (6.126) when all the potential terms are set to zero. From (6.50), $K_{\nu LS}(r_1)\, \mathcal{F}^{\nu'}_{\nu LS}(r_1)$ reduces to

$$\begin{aligned} K_{\nu LS}(r_1)\, \mathcal{F}^{\nu'}_{\nu LS}(r_1) = 2\epsilon &\left[\frac{Z}{r_1} - V_{\mathrm{C}}(r_1)\right] \delta(\nu,\nu')\, u_{l_2}(k_\nu r_1) \\ &- 2\epsilon\, V(\nu,\nu')\, u_{l'_2}(k_{\nu'} r_1)\,, \end{aligned} \tag{6.134}$$

where $V(\nu,\nu')$ is given by (6.44). From (6.132) and (6.134) it then follows that the individual **R**-matrix elements, within the Born approximation, are given by

$$R^{\mathrm{B},LS}_{\nu\nu} = -\frac{2\epsilon}{k_\nu} \int_0^\infty \mathrm{d}r_1\, u_{l_2}(k_\nu r_1) \left\{\left[\frac{Z}{r_1} - V_{\mathrm{C}}(r_1)\right] - V(\nu,\nu)\right\} u_{l_2}(k_\nu r_1)\,, \tag{6.135}$$

and for $\nu \neq \nu'$

$$R^{\mathrm{B},LS}_{\nu\nu'} = \frac{2\epsilon}{\sqrt{k_\nu k_{\nu'}}} \int_0^\infty \mathrm{d}r_1\, u_{l_2}(k_\nu r_1)\, V(\nu,\nu')\, u_{l'_2}(k_{\nu'} r_1)\,. \tag{6.136}$$

As was discussed in Sect. 6.3.2, an asymptotic correction procedure is necessary to accurately approximate the contribution to $B^{LS}_{\nu\nu'}$ from the asymptotic region whenever the coupled equations (6.126) contain potential terms $V(\nu,\nu')$ which behave asymptotically as $1/r_2^\lambda$ as $r_2 \to \infty$ with $\lambda \geq 2$. As before, we can write

$$B^{LS}_{\nu\nu'} = B^{LS}_{\nu\nu'}(r_{\mathrm{M}}) + \Delta B^{LS}_{\nu\nu'}(r_{\mathrm{M}})\,, \tag{6.137}$$

where

$$B^{LS}_{\nu\nu'}(r_{\mathrm{M}}) = -\frac{1}{k_\nu} \int_0^{r_{\mathrm{M}}} \mathrm{d}r_1\, u_{l_2}(k_\nu r_1)\, K_{\nu LS}(r_1)\, \mathcal{F}^{\nu'}_{\nu LS}(r_1) \tag{6.138}$$

and

$$\Delta B^{LS}_{\nu\nu'}(r_{\rm M}) = -\frac{1}{k_\nu}\int_{r_{\rm M}}^{\infty} {\rm d}r_1\, u_{l_2}(k_\nu r_1)\, K_{\nu LS}(r_1)\,\mathcal{F}^{\nu'}_{\nu LS}(r_1)\,, \tag{6.139}$$

with $K_{\nu LS}(r_1)\,\mathcal{F}^{\nu'}_{\nu LS}(r_1)$ given by (6.134). Provided $r_{\rm M}$ is sufficiently large, so that all the potential terms which decay exponentially are vanishingly small, then $\mathcal{K}_{\nu LS}(r_1)\,\mathcal{F}^{\nu'}_{\nu LS}(r_1)$ can be expressed as

$$\begin{aligned}\mathcal{K}_{\nu LS}(r_1)\,\mathcal{F}^{\nu'}_{\nu LS}(r_1)&\\ &= -2\epsilon\sum_\lambda{}' f_\lambda(l_1l_2,l_1'l_2';L)\,e_\lambda(n_1l_1,n_1'l_1')\,\frac{1}{r_1^{\lambda+1}}\,u_{l_2'}(k_{\nu'}r_1)\,,\end{aligned} \tag{6.140}$$

where $\sum'$ means that when $\nu'=\nu$ the term with $\lambda=0$ has been omitted from the sum. Consequently

$$\Delta B^{LS}_{\nu\nu'}(r_{\rm M}) = \frac{2\epsilon}{k_\nu}\sum_\lambda{}' f_\lambda(l_1l_2,l_1'l_2';L)\,e_\lambda(n_1l_1,n_1'l_1')\,\mathcal{U}_{\nu\nu'}(r_{\rm M})\,, \tag{6.141}$$

where $\mathcal{U}_{\nu\nu'}(r_{\rm M})$ is given by (6.110a). This latter integral can be determined by numerical quadrature in the same manner as described in Sect. 6.3.2. It should be noted that (6.141) for $\Delta B^{LS}_{\nu\nu'}$ is *exact* whereas when (6.58a) is solved, the expression for $\Delta B^{LS}_{\nu\nu'}$ [cf. (6.109b)] is only an approximation. Thus, the individual elements of the **R** matrix, within the Born approximation, are given by

$$\begin{aligned}R^{{\rm B},LS}_{\nu\nu} &= -\frac{2\epsilon}{k_\nu}\int_0^{r_{\rm M}} {\rm d}r_1\, u_{l_2}(k_\nu r_1)\left\{\left[\frac{Z}{r_1}-V_{\rm C}(r_1)\right]-V(\nu,\nu)\right\}u_{l_2}(k_\nu r_1)\\ &\quad + \Delta R^{{\rm B},LS}_{\nu\nu}(r_{\rm M})\,,\end{aligned} \tag{6.142}$$

and for $\nu\neq\nu'$

$$R^{{\rm B},LS}_{\nu\nu'} = \frac{2\epsilon}{\sqrt{k_\nu k_{\nu'}}}\int_0^{r_{\rm M}} {\rm d}r_1\, u_{l_2}(k_\nu r_1)\,V(\nu,\nu')\,u_{l_2'}(k_{\nu'}r_1) + \Delta R^{{\rm B},LS}_{\nu\nu'}(r_{\rm M})\,, \tag{6.143}$$

where, in general

$$\Delta R^{{\rm B},LS}_{\nu\nu'}(r_{\rm M}) = \frac{2\epsilon}{\sqrt{k_\nu k_{\nu'}}}\sum_\lambda{}' f_\lambda(l_1l_2,l_1'l_2';L)\,e_\lambda(n_1l_1,n_1'l_1')\,\mathcal{U}_{\nu\nu'}(r_{\rm M})\,. \tag{6.144}$$

Equations (6.142) and (6.143) conform to the definition of the Born approximation as given by Moores and Norcross [6.16]. In practice the Born approximation is often only required for the resonance transition in the alkalis and here it works quite well. On the other hand, the Born approximation is particularly unreliable in estimating the elastic cross section since the leading element in the Born **R** matrix, namely $R^{{\rm B},LS}_{11}$, does not contain any of the long-range interactions. Consequently, if the Born **R**-matrix elements are to be used to determine any scattering parameters other than the resonance cross section then it is usually better to employ effective-range theory and determine $R_{11}^{\mathrm{B},LS}$ by means of the so-called Ali-Fraser formula [6.32].

6.5 Computer Program

The computer program first reads the input data, including the core potential (6.35) and the radial valence orbitals (6.22b). It then sets up and solves the appropriate integral equations (6.58a,b) and determines the asymptotic constants $A_{\nu\nu'}^{LS}$ and $B_{\nu\nu'}^{LS}$ given by (6.60a,b). The program then calculates the **R** matrix and **T** matrix from (6.63) and (6.64a,b), and finally determines the elastic and excitation cross sections from (6.48). Alternatively, the Born approximation can be used to compute the **R**-matrix elements directly from (6.142–144). The program is written for positron scattering from the alkali atoms as well as atomic hydrogen. Nonetheless, the input data can be setup so that the program will also do electron scattering but *without exchange*; this could be useful when either the incident angular momentum or energy (or both) are sufficiently large that exchange is relatively unimportant. The program assumes that the core potential and the valence atomic orbitals have been determined in separate atomic-structure calculations. In particular, that the core orbitals, used to determine the core potential, have been determined from a fully varied Hartree–Fock (or equivalent) calculation of the corresponding alkali$^+$ ion and that the valence orbitals have been then calculated within a *frozen-core* framework. In the following sections some of the more important aspects of the program are described; some sample input and output data are also provided.

6.5.1 The Radial Meshes

The program assumes that the core and valence atomic orbitals have been determined on a logarithmic mesh [6.33] which is defined according to

$$r_i = \exp[\rho_0 + (i-1)h] \quad \text{for } i = 1, 2, \ldots, n\,. \tag{6.145}$$

Appropriate values for the constants ρ_0 and h are -4.0 and $1/16$ respectively; n, the number of points in the logarithmic mesh, will vary according to the particular atom and orbital. Furthermore, the program assumes that the same values of ρ_0 and h have been used for all orbitals, core and valence, and that these orbitals are available in the form [6.33]

$$\overline{P}_{nl}(r) = r^{-\frac{1}{2}}\, P_{nl}(r)\,. \tag{6.146}$$

The program uses a '3-interval' equidistant mesh for the scattering calculation; the step-size in each of these three intervals is h, as given above, divided by 2 raised to an integer power. The smallest interval is in the inner atomic region near the origin and is $2\,a_0$ (with a_0 being the Bohr radius) in length, the next largest interval is $4\,a_0$ in length, and, finally, the largest interval is in the outer atomic and 'asymptotic' regions. In general, these interval sizes need to be decreased with increasing angular momentum L and increasing nuclear charge Z. For lithium, a step size of $1/32$ is sufficient in all three intervals for small angular momentum whereas for rubidium step sizes of $1/1024$, $1/256$, and $1/64$ for large angular momentum, say $L \approx 30$, are required. However, for the Born approximation, a step size of $1/32$ is sufficient in all three intervals, independent of the nuclear charge Z.

The program computes all of the necessary valence potential functions (6.44) on the logarithmic mesh and then interpolates them as well as the core potential (6.35) onto the scattering mesh by means of Aitken's method.

6.5.2 Numerical Procedures

The integration of the integral equations (6.58a,b) is performed using Simpson's 1/3 rule. For each value of ν' the integration of all the open-channel functions proceeds outwardly from the origin to a point $r_{\rm M}$ [see text just after (6.106)]. If there are any closed channels then these functions are determined iteratively, while the open-channel functions remain fixed. Once these closed-channel wavefunctions have been determined they are, in turn, held fixed while the open-channel wavefunctions are determined once more; in the initial determination of the open-channel functions the closed-channel functions are set to zero. This overall iterative procedure is continued until convergence is achieved. It should be noted that, first, the entire procedure is only iterative when the energy of the incident positron is such that there are closed channels present in the close-coupling expansion (6.6) and, second, like any iterative scheme it can *diverge.* In order to either speed up convergence or help to overcome these divergence difficulties the corresponding wavefunctions from consecutive iterations may be averaged by means of an acceleration parameter which can either be fixed or determined dynamically [6.34].

Once the wavefunctions $\mathcal{F}^{\nu'}_{\nu LS}(r)$ in all channels, open and closed, have been determined out to $r_{\rm M}$ for *all* values of ν' the program makes an initial estimate of the various elastic and inelastic cross sections. The open-channel wavefunctions are then integrated out an additional number of atomic units (specified in the input data) and another estimate of the scattering cross sections is made. This procedure is continued until the cross sections from one stage agree with those from the previous stage to within specified tolerances. Whenever required, the asymptotic procedure outlined in Sect. 6.3.2 is used in determining these estimates for the cross sections.

In the Born approximation the **R**-matrix elements are computed directly using Simpson's 1/3 rule out to $r_{\rm M}$, together with the asymptotic procedure outlined in Sect. 6.3.2. For very large values of the angular momentum L it is often necessary to increase the effective value of $r_{\rm M}$; this can be accomplished through the parameter NDEL in the input data. For the values of the constants ρ_0 and h given in Sect. 6.5.1 the choice of NDEL equal to 1, 2, or 3 will increase r_M by approximately 28%, 65%, or 112%, respectively.

6.5.3 Convergence Criteria

There are four different convergence tests each of which is controlled by a separate subroutine. If the close-coupling expansion contains only atomic S states then there are no long-range potentials; all the $V(\nu,\nu'')$ in (6.106) decay exponentially. In this case it is possible, during the outward integration of an open-channel wavefunction, to achieve convergence of the integrals for $A^{LS}_{\nu\nu'}$ and $B^{LS}_{\nu\nu'}$ while still inside the atomic region, i.e., prior to the point r_M. The subroutine CONVG1 makes both an absolute and a relative test for the convergence of the ratio of these integrals. When there are closed channels the subroutine CONVG2 tests for convergence of the corresponding wavefunctions by monitoring successive iterations for the convergence of the two integrals in (6.58b). Once again both a relative and an absolute test are made. Also, when there are closed channels, it is necessary to repeat the outward integration of the open-channel functions each time a new set of converged closed-channel wavefunctions is obtained. The subroutine CONVG3 monitors the convergence of the ratio of the integrals in (6.60a,b) between successive determinations of the open-channel wavefunctions. The convergence tests here closely parallel those in the subroutine CONVG1.

In the asymptotic region the program calculates, at specified radial points, the elastic and inelastic cross sections as given by (6.48). The convergence of these cross sections, between successive determinations, is monitored by the subroutine CONVG4.

6.5.4 Program Input

All the data is read from the unit IREAD=5 by the subroutine INPUT; the form of the data is controlled by FORMAT statements. A brief description of each of the various input records and blocks of records follows.

Record 1.

- NO — Maximum number of points in the logarithmic bound-state integration mesh; normally 180.
- HH — Logarithmic integration interval h [(6.145)] for the bound-state wavefunctions; normally 0.0625.
- Z — Nuclear charge of the atom.
- RHO — Starting value ρ_0 [(6.145)] on the logarithmic bound-state mesh; normally -4.0.
- EPSILN — Charge of the incident particle; $+1$ for positrons, -1 for electrons.

Record 2.

- NCORE Equals 0 for hydrogen and 1 for the alkalis.
- NDX1 Number of integration points in the atomic region between tests for convergence.
- NDX2 Number of integration points in the asymptotic region between tests for convergence.
- NMAX Maximum number of times permitted to test for convergence in the asymptotic region.
- ITER Maximum number of iterations permitted to determine all of the closed-channel wavefunctions as well as the maximum number of iterations to determine any individual closed-channel wavefunction.
- IH1 H1=HH/IH1 is the scattering integration interval in the outer atomic and asymptotic regions.
- IH2 H2=H1/IH2 is the scattering integration interval in the middle atomic region.
- IH3 H3=H2/IH3 is the scattering integration interval in the inner atomic region.
  Note: IH1, IH2, and IH3 should be powers of 2.

Record 3.

- ISYM Parameter to control symmetrization of the **R** matrix; if ISYM=0 then the **R** matrix will not be symmetrized before computing the cross sections.
- IAVE1 Parameter to control the averaging of the current iteration of the open-channel wavefunctions with the previous iteration. If IAVE1=1 then AVER=0.5 in the SOLVE1 subroutine; else AVER=0.0.
- IAVE2 Parameter to control the averaging of the current iteration of the closed-channel wavefunctions with the previous iteration. If IAVE2=1 then AVER=0.5 in the SOLVE2 subroutine. If IAVE2=2 then AVER is determined dynamically; else AVER=0.0.
- IAVE3 Parameter to control the averaging of the *converged* closed-channel wavefunctions from one iteration cycle with the *converged* results from the previous cycle. If IAVE3=1 the functions are averaged.

Record 4.

- TOLER1 Tolerance for the subroutine CONVG1; normally 1.0×10^{-5}.
- TOLER2 Tolerance for the subroutine CONVG1; normally 1.0×10^{-7}.
- TOLER3 Tolerance for the subroutine CONVG3; normally 1.0×10^{-5}.
- TOLER4 Tolerance for the subroutine CONVG3; normally 1.0×10^{-7}.
- TOLER5 Tolerance for the subroutine CONVG4; normally 2.0×10^{-5}.
- TOLER6 Tolerance for the subroutine CONVG4; normally 2.0×10^{-6}.

Record 5.

- NLEG Number of mesh points in Gauss-Legendre integration; normally 16 or 32.

NLAG Order of the Laguerre polynomial used in the Gauss-Laguerre integration; normally 16 or 32.

LAGMAX Number of mesh points in Gauss-Laguerre integration; normally the same as NLAG.

NDEL Parameter to extend the effective value of r_M.

Record 6.

NK Number of linear-momentum values of the incident particle.

LMIN The minimum value of the total angular momentum L of the system.

LMAX The maximum value of the total angular momentum L of the system.

LBORN If L is greater than LBORN the Born approximation is used.

ALD Dipole polarizability of the atom.

ALQ Quadrupole polarizability of the atom.

Record 7.

E Values of the linear momentum of the incident particle.

Block 1, card 1.

M Number of logarithmic mesh points in the static core potential.

CO Value of the constant d_0 in (6.68) and (6.69).

Block 1, remaining cards.

The remaining cards in this block contain the static core potential on the logarithmic mesh under the specified format of 5F16.9.
Note: Block 1 is not required for the hydrogen atom.

Block 2, card 1.

NS Number of atomic states included in the eigenfunction expansion.

NFORM Parameter to specify the format of the atomic wavefunctions. If NFORM=1 the format is 7F11.7; if NFORM=2 the format is 5F16.9.

Block 2, card 2.

EL Hollerith label specifying $n_1 l_1$ of the first atomic wavefunction.

N1 Value of n_1 of the first atomic wavefunction.

L1 Value of l_1 of the first atomic wavefunction.
Note: Block 2, card 2 is repeated for each of the NS atomic wavefunctions.

Block 3, card 1.

ATOM Hollerith label specifying the atomic symbol for the atom.

M Number of mesh points in the first atomic wavefunction.

ENL Value of $-\epsilon_\nu$ for the first atomic wavefunction [cf. (6.13)] in Rydberg energy units.

Block 3, remaining cards.

The remaining cards in this block contain the first atomic wavefunction on the logarithmic mesh under the specified format.
Note: Block 3 is repeated for each of the NS atomic wavefunctions.

Next to last card

`HEAD` Hollerith heading to be printed on the top of all the output pages; it should describe the atomic states included in the calculation.

Last card

`PRNT` Logical variable: if `TRUE` then the subroutine `PRINT3` will write the **R** matrix, **T** matrix and the cross-section matrix to unit `IWRITE` each time they are determined, whereas if `FALSE` only the final converged matrices will be written.

6.5.5 Sample Input and Output

The following are the input data for a two-state (2s-2p) close-coupling calculation for e^+–Li scattering when the total angular momentum of the system is $L = 0$ and the incident positron has a linear momentum of 1.0 atomic unit. In this case both channels are open.

```
180       .0625      3.0000     -4.0000      1.0000
   1    32  160    40    40     2     1     1
   1     1     2     1
1.00D-05  1.00D-07  1.00D-05  1.00D-07  2.00D-05  2.00D-06
  16    16    16     0
1  0   0 20       164.00        0.00
   1.0000
 ATOM LI+          116                         CO=   5.374839
```

[Records containing the static potential of Li^+ are inserted here.]

```
2  1
2S     2     0
2P     2     1
```

[Records containing the 2s and 2p valence orbitals of Li are inserted here.]

```
IN A 2S-2P CLOSE COUPLING APPROXIMATION WITH HF WAVEFUNCTS
T
```

All the output is written onto two units, namely, `IWRITE=6` and `IPNCH=7`. The output written on each page of the unit `IWRITE` consists of a ‘heading’ followed by the current estimates of the **R** matrix, the **T** matrix, and the various elastic and inelastic cross sections. The following is the partial output on the unit `IWRITE` for this case; it consists of just the first estimates and the final converged values for the above mentioned quantities.

```
PHASE SHIFTS/CROSS SECTIONS FOR E+ LI SCATTERING IN A 2S-2P CC-APPROX
INCIDENT WAVE NUMBER K =  1.0000 ( 13.61 EV) AND THE TOTAL  L = 0
INTEGRATION INTERVALS =  0.031250000  0.031250000 AND 0.031250000
PRIMARY INTEGRATION RANGE: 1312 POINTS, SECONDARY INTEGRATION RANGE: 160
CONVERGENCE TOLERANCES = 1.0D-05 1.0D-07 1.0D-05 1.0D-07 2.0D-05 2.0D-06
GAUSSIAN INTEGRATION PARAMETERS: NLEG = 16 NLAG = 16 LAGMAX = 16 NDEL = 0
R-MATRIX WAS SYMMETRIZED
R MATRIX AT R =  41.00000
 -1.086955    0.353134
  0.353134   -0.764165
REAL PART OF THE T MATRIX
 -1.027278    0.372142
  0.372142   -0.687113
IMAG PART OF THE T MATRIX
 -0.925889    0.059123
  0.059123   -0.871846
 CROSS SECTIONS
          2S           2P
 2S    1.912571     0.141985
 2P    0.054736     0.475036

PHASE SHIFTS/CROSS SECTIONS FOR E+ LI SCATTERING IN A 2S-2P CC-APPROX
INCIDENT WAVE NUMBER K =  1.0000 ( 13.61 EV) AND TOTAL L = 0
INTEGRATION INTERVALS =  0.031250000  0.031250000 AND 0.031250000
PRIMARY INTEGRATION RANGE: 1312 POINTS, SECONDARY INTEGRATION RANGE: 160
CONVERGENCE TOLERANCES = 1.0D-05 1.0D-07 1.0D-05 1.0D-07 2.0D-05 2.0D-06
GAUSSIAN INTEGRATION PARAMETERS: NLEG = 16 NLAG = 16 LAGMAX = 16 NDEL = 0
R-MATRIX WAS SYMMETRIZED
R MATRIX AT R =  71.00000
 -1.086499    0.353162
  0.353162   -0.764568
REAL PART OF THE T MATRIX
 -1.026850    0.372187
  0.372187   -0.687577
IMAG PART OF THE T MATRIX
 -0.925884    0.059118
  0.059118   -0.871994
CROSS SECTIONS
         2S           2P
2S    1.911682     0.142018
2P    0.054749     0.475381
CALCULATION CONVERGED
```

6.6 Summary

We have summarized the basic equations of the close-coupling approximation for electron–atom and positron–atom scattering. As an example, the formalism was applied to positron scattering from alkali-like targets. A computer code to study these collision systems is provided on the disk.

6.7 Suggested Problems

1. Use the computer code to perform 2-state close-coupling calculations for positron–lithium scattering for various collision energies between 1 eV and 100 eV.
2. Investigate the validity of the Born approximation for the **R**-matrix elements for various collision energies and partial wave angular momenta.
3. Extend the calculations to other alkali targets.
4. Perform DWBA calculations and compare the results with those from the 2-state close-coupling approximation.
5. Calculate scattering amplitudes, differential cross sections and other angle-differential observables (see also Chaps. 10 and 11).
6. Investigate the effects of including more than just two states, in particular d orbitals.

Acknowledgments

This work was supported, in part, by the National Science and Engineering Research Council of Canada.

References

6.1 H.S.W. Massey and C.B.O. Mohr, Proc. Roy. Soc. (London) A **136** (1932) 289
6.2 K. Smith, R.P. McEachran and P.A. Fraser, Phys. Rev. **125** (1962) 553
6.3 P.G. Burke and H.M. Schey, Phys. Rev. **126** (1962) 147
6.4 P.G. Burke and H.M. Schey, Phys. Rev. **126** (1962) 163
6.5 P.G. Burke and K. Smith, Rev. Mod. Phys. **34** (1962) 458
6.6 P.G. Burke, H.M. Schey and K. Smith, Phys. Rev. **129** (1963) 1258
6.7 R.P. McEachran and P.A. Fraser, Proc. Phys. Soc. **82** (1963) 1038
6.8 W.J. Cody, J. Lawson, Sir Harrie Massey and K. Smith, Proc. Roy. Soc. (London) A **278** (1964) 479
6.9 R.P. McEachran and P.A. Fraser, Proc. Phys. Soc. **86** (1965) 369
6.10 I.C. Percival and M.J. Seaton, Proc. Cam. Phil. Soc. **53** (1957) 654
6.11 K. Smith, Proc. Phys. Soc. **78** (1961) 549
6.12 L. Castillejo, I.C. Percival and M.J. Seaton, Proc. Roy. Soc. (London) A **254** (1960) 259

6.13 A. Salmona and M.J. Seaton, Proc. Phys. Soc. **77** (1961) 617
6.14 P.G. Burke and A.J. Taylor, J. Phys. B **2** (1969) 869
6.15 A. Salmona, C. R. Acad. Sci. Paris **260** (1965) 2434
6.16 D.L. Moores and D.W. Norcross, J. Phys. B **5** (1972) 1482
6.17 P.G. Burke and W.D. Robb, Adv. At. Molec. Phys. **11** (1975) 143
6.18 P. Khan, S. Dutta and A.S. Ghosh, J. Phys. B **20** (1987) 2927
6.19 K.P. Sarkar, M. Basu and A.S. Ghosh, J. Phys. B **21** (1988) 1649
6.20 S.J. Ward, M. Horbatsch, R.P. McEachran and A.D. Stauffer, *Atomic Physics with Positrons* ed. J.W. Humberston and E.A.G. Armour (Plenum, London, 1988) p.265
6.21 S.J. Ward, M. Horbatsch, R.P. McEachran and A.D. Stauffer, J. Phys. B **21** (1988) L611
6.22 S.J. Ward, M. Horbatsch, R.P. McEachran and A.D. Stauffer, Nucl. Instrum. Methods B **42** (1989) 472
6.23 S.J. Ward, M. Horbatsch, R.P. McEachran and A.D. Stauffer, J. Phys. B **22** (1989) 1845
6.24 T.S. Stein, R.D. Gomez, Y.-F. Hsieh, W.E. Kauppila, C.K. Kwan and Y.J. Wan, Phys. Rev. Lett. **55** (1985) 488
6.25 T.S. Stein, M.S. Dababneh, W.E. Kauppila, C.K. Kwan, Y.J. Wan, *Atomic Physics with Positrons* ed. J.W. Humberston and E.A.G. Armour (Plenum, London, 1988) p.251
6.26 C.K. Kwan, W.E. Kauppila, R.A. Likaszew, S.P. Parikh, T.S. Stein, Y.J. Wan and M.S. Dababneh, Phys. Rev. A **44** (1991) 1620
6.27 M.E. Rose, *Elementary Theory of Angular Momentum* (Wiley, New York, 1957)
6.28 R.P. McEachran, C.E. Tull and M. Cohen, Can. J. Phys. **46** (1968) 2675
6.29 J.C. Slater, *Quantum Theory of Atomic Structure* Vol. I (McGraw-Hill, New York, 1960)
6.30 R.P. McEachran and A.D. Stauffer, J. Phys. B **16** (1983) 255
6.31 N.C. Sil and M.A. Crees, M.J. Seaton, J. Phys. B **17** (1984) 1
6.32 M.K. Ali and P.A. Fraser, J. Phys. B **10** (1977) 3091
6.33 C. Froese Fischer, *The Hartree–Fock Method for Atoms* (Wiley, New York, 1977)
6.34 R.P. McEachran, A.G. Ryman and A.D. Stauffer, J. Phys. B **11** (1978) 551

7. The R-Matrix Method

P.G. Burke and M.P. Scott

Department of Applied Mathematics and Theoretical Physics,
The Queen's University of Belfast, Belfast BT7 1NN, Northern Ireland

Abstract

We introduce the R-matrix method as applied to electron–atom and electron–ion scattering, and in particular we discuss the specific two-electron problem of electron–hydrogen-like ion scattering. A computer code for the calculation of cross sections for electron–hydrogen-like ion scattering is described together with some sample input and output data.

7.1 Introduction

The R-matrix method was introduced in nuclear physics in an analysis of resonance reactions by Wigner [7.1] and Wigner and Eisenbud [7.2]. It was subsequently used by many workers in nuclear physics and comprehensive reviews of the theory were written by Lane and Thomas [7.3] and by Breit [7.4].

Over the last twenty years the R-matrix method has been extensively developed as an *ab initio* approach for calculating a broad range of atomic and molecular structure and collision processes. These processes include: electron–atom and electron–ion collisions [7.5]; atomic polarizabilities [7.6]; atomic photoionization [7.7]; attachment energies of negative ions [7.8]; electron–molecule collisions [7.9]; atom–molecule reactive collisions [7.10]; free–free transitions [7.11]; dissociative attachment [7.12]; charge transfer collisions [7.13]; molecular photoionization [7.14]; positron–atom collisions [7.15]; positron–molecule collisions [7.16]; and electron impact ionization of atoms and ions [7.17].

In this chapter we illustrate the R-matrix method by considering its application to electron–hydrogen-like ion scattering. In Sect. 7.2 we summarize the general theory of the R-matrix method and consider its application to the specified problem in Sect. 7.3. In Sect. 7.4 we discuss the computational procedure for the solution of this problem and give details of the computer codes in Sect. 7.5. A summary is given in Sect. 7.6 and suggested problems in Sect. 7.7.

7.2 General R-Matrix Theory

In this section we introduce the basic concepts and equations of the R-matrix method. We consider the collision process

$$e^- + A_i \rightarrow e^- + A_j \tag{7.1}$$

between an electron (e^-) and an atom (A), which undergoes a transition from an initial state (i) to a final state (j).

The R-matrix method proceeds by partitioning configuration space into two regions by a sphere of radius $r = a$, where r is the relative coordinate of the scattered electron. The radius a of the sphere is chosen so that the sphere just envelopes the atomic target states and possible pseudostates of interest in the calculation. It follows that when the scattered electron is in the internal region, electron exchange and correlation effects between the scattered electron and the target are important and the electron–atom complex behaves very much as a bound state. Consequently, in this region a configuration interaction expansion of the total wavefunction similar to that used in atomic structure calculations is appropriate. On the other hand, when the scattered electron is in the external region, exchange with the target atom can be neglected and the scattered electron moves in the weak long-range multipole potential of the target. The wavefunction in this region is then often accurately represented by an asymptotic expansion or by a perturbation approach.

We consider the solution of the Schrödinger equation for the wavefunction Ψ_E at a total energy E in the internal region. We write this equation as

$$(H + L - E)\Psi_E = L\Psi_E \,, \tag{7.2}$$

where H is the electron–atom Hamiltonian and L is a surface operator introduced by Bloch [7.18] which ensures that $H + L$ is hermitian in the internal region. The Bloch operator is defined by

$$L = \frac{1}{2} \sum_i \sum_p |\bar{\phi}_i\rangle \delta(r_p - a) \left(\frac{\mathrm{d}}{\mathrm{d}r_p} - \frac{b}{r_p} \right) \langle \bar{\phi}_i | \,, \tag{7.3}$$

where $\bar{\phi}_i$ are channel functions constructed by coupling the target atom eigenstates and pseudostates of interest, ϕ_i, to the spin-angle function describing the scattered electron, and b is an arbitrary constant which may be channel dependent but is usually chosen to be zero. Finally, the summations in (7.3) are over all channels and over all the permutations of the scattered and target electrons. Equation (7.2) can now be formally solved, yielding

$$\Psi_E = (H + L - E)^{-1} L\Psi_E \,. \tag{7.4}$$

In order to obtain an explicit representation of the inverse operator in the equation we introduce an energy independent basis Ψ_k in the internal region. We then write

$$\Psi_k = \sum_q \chi_q \beta_{qk} \,. \tag{7.5}$$

The set of functions χ_q consists of products of channel functions and radial functions describing the scattered electron with the addition of some quadratically integrable functions which vanish on the boundary of the internal region and which are included to allow for short-range correlation effects between the scattered electron and the target atom. The coefficients β_{qk} can be obtained by diagonalizing $H + L$ according to

$$\langle \psi_k | H + L | \psi_k' \rangle = E_k \delta_{kk'} \,, \tag{7.6}$$

where the integrals in this equation are carried out over the internal region. It follows that (7.4) can be rewritten as

$$\Psi_E = \sum_k \frac{|\psi_k\rangle \langle \psi_k | L | \Psi_E \rangle}{E_k - E} \,. \tag{7.7}$$

We project this equation onto the channel function $\bar{\phi}_i$ and evaluate it on the boundary $r = a$ giving

$$F_i(a) = \sum_j R_{ij}(E) \left(a \frac{\mathrm{d}F_j}{\mathrm{d}r} - bF_j \right)_{|r=a} \tag{7.8}$$

where we have introduced the **R** matrix with elements

$$R_{ij} = \frac{1}{2a} \sum_k \frac{\omega_{ik}\omega_{jk}}{E_k - E} \,. \tag{7.9}$$

The reduced radial functions are defined by

$$r^{-1} F_j(r) = \langle \bar{\phi}_j | \Psi_E \rangle \,, \tag{7.10}$$

and the reduced width amplitudes are

$$r^{-1}\omega_{ik} = \langle \bar{\phi}_i | \psi_k \rangle_{|r=a} = \sum_q \langle \bar{\phi}_i | \chi_q \rangle_{|r=a} \beta_{qk} \,. \tag{7.11}$$

Equations (7.8) and (7.9) are the basic equations of the R-matrix method. The **R** matrix is obtained at all energies by a single diagonalization of $H + L$ in (7.6) to obtain the eigenenergies E_k and the corresponding eigensolutions ψ_k. From these eigensolutions the reduced width amplitudes are given immediately.

The final step is to solve the collision problem in the external region where $r > a$. In this region we expand the total wavefunction as

$$\Psi_E = \sum_i \bar{\phi}_i r^{-1} F_i(r) . \tag{7.12}$$

Substituting this expansion into the Schrödinger equation and projecting onto the channel functions $\bar{\phi}_i$ yields a set of coupled differential equations for the functions F_i:

$$\left(\frac{\mathrm{d}^2}{\mathrm{d}r^2} - \frac{\ell_i(\ell_i+1)}{r^2} + \frac{2Z}{r} + k_i^2\right) F_i(r) = 2\sum_{j=1}^{n} V_{ij}(r) F_j(r) ;$$

$$i = 1, n;\ r \geq a , \tag{7.13}$$

where n is the number of coupled channels retained in (7.12), ℓ_i are the channel angular momenta, k_i^2 are the channel energies, and the potential matrix elements V_{ij} are given by

$$V_{ij}(r) = \langle \bar{\phi}_i \,|\sum_{k=1}^{N} \frac{1}{r_{k,N+1}} |\, \bar{\phi}_j \rangle . \tag{7.14}$$

Equation (7.13) can be solved by one of a number of standard methods giving n_a linearly independent solutions where n_a is the number of open channels $(k_i^2 \geq 0)$ at the energy of interest. We can choose these solutions to satisfy the asymptotic boundary conditions

$$F_{ij} \underset{r\to\infty}{\sim} k_i^{-1/2} \left(\sin\theta_i \delta_{ij} + \cos\theta_i K_{ij}\right) \quad \text{for open channels} , \tag{7.15a}$$

$$F_{ij} \underset{r\to\infty}{\sim} 0 \quad \text{for closed channels} . \tag{7.15b}$$

The second index j on the F_{ij} distinguishes the linearly independent solutions, the number of which equals n_a, and

$$\theta_i = k_i r - \frac{1}{2}\ell_i \pi - \eta_i \ln 2k_i r + \sigma_i , \tag{7.16a}$$

$$\eta_i = -\frac{Z-N}{k_i} , \tag{7.16b}$$

$$\sigma_i = \arg\Gamma\left(\ell_i + 1 + i\eta_i\right) . \tag{7.16c}$$

where Z is the charge on the nucleus. Substituting these solutions evaluated at $r = a$ into (7.8) enables the $n_a \times n_a$ **K**-matrix elements K_{ij} to be determined in terms of the $n \times n$ **R**-matrix elements R_{ij}. The $n_a \times n_a$ **S** matrix is then given by the matrix equation

$$\mathbf{S} = [\mathbf{1} + \mathrm{i}\mathbf{K}]\,[\mathbf{1} - \mathrm{i}\mathbf{K}]^{-1} \tag{7.17}$$

and the cross section then follows in the usual way (see, for example, Mott and Massey [7.19]).

7.3 Electron–Hydrogen-like Ion Scattering

In this section we illustrate the general theory described in the previous section by applying it to electron–hydrogen-like ion scattering.

The energy independent basis functions in (7.5) adopted in the Queen's University non-relativistic programs are defined by

$$\chi^{LS\Pi}_{n_1\ell_1 n_2\ell_2}(\hat{\boldsymbol{r}}_1,\hat{\boldsymbol{r}}_2) = \frac{1}{\sqrt{2}}\Big[\frac{1}{r_1}u_{n_1\ell_1}(r_1)\frac{1}{r_2}u_{n_2\ell_2}(r_2)\mathcal{Y}_{\ell_1\ell_2 LM_L}(\hat{\boldsymbol{r}}_1,\hat{\boldsymbol{r}}_2) + (-1)^{L+S+\ell_1+\ell_2}\frac{1}{r_1}u_{n_2\ell_2}(r_1)\frac{1}{r_2}u_{n_1\ell_1}(r_2)\mathcal{Y}_{\ell_2\ell_1 LM_L}(\hat{\boldsymbol{r}}_1,\hat{\boldsymbol{r}}_2)\Big], \tag{7.18}$$

for $n_1\ell_1 \neq n_2\ell_2$, and

$$\chi^{LS\Pi}_{n_1\ell_1 n_2\ell_2}(\hat{\boldsymbol{r}}_1,\hat{\boldsymbol{r}}_2) = \frac{1}{r_1}u_{n_1\ell_1}(r_1)\frac{1}{r_2}u_{n_2\ell_2}(r_2)\mathcal{Y}_{\ell_1\ell_2 LM_L}(\hat{\boldsymbol{r}}_1,\hat{\boldsymbol{r}}_2), \tag{7.19}$$

for $n_1\ell_1 = n_2\ell_2$.

Here we have only considered the space part of the wavefunction, which is symmetric when $S = 0$ and antisymmetric when $S = 1$ (where S is the total spin of the two electrons) in accordance with the Pauli exclusion principle. The coupled angular functions are defined as

$$\mathcal{Y}_{\ell_1\ell_2 LM_L}(\hat{\boldsymbol{r}}_1,\hat{\boldsymbol{r}}_2) = \sum_{m_1 m_2}(\ell_1\, m_1, \ell_2 m_2; L\, M_L)\, Y_{\ell_1 m_1}(\hat{\boldsymbol{r}}_1)\, Y_{\ell_2 m_2}(\hat{\boldsymbol{r}}_2), \tag{7.20}$$

where L is the total angular momentum and M_L is its z component. Furthermore, $(\ell_1\, m_1, \ell_2 m_2; L\, M_L)$ are Clebsch–Gordan coefficients (see Rose [7.24a]), $Y_{\ell m}(\hat{\boldsymbol{r}})$ are spherical harmonics, and Π is the parity defined by $(-1)^{\ell_1+\ell_2}$. We note that L, S, and Π are conserved in the collision.

One of the most important decisions that has to be made in the R-matrix method is the choice of radial orbital basis functions $u_{n\ell}(r)$. The range of energies over which the method is accurate and its convergence properties depend critically on this choice. In the Queen's University programs, the $u_{n\ell}(r)$ are divided into two classes called bound and continuum orbitalss. The bound orbitals denoted by $u^{\mathrm{b}}_{n\ell}(r)$ correspond to 1s, 2s, 2p, ... target eigenstates and possible pseudostates retained in the calculation. These orbitals are orthogonal over the internal region and the choice of boundary radius a ensures that they have decayed to be essentially zero by the boundary. The continuum orbitals denoted by $u^{\mathrm{c}}_{n\ell}(r)$ describe the radial motion of the scattered electron. They are chosen to satisfy the differential equation

$$\left(\frac{\mathrm{d}^2}{\mathrm{d}r^2} - \frac{\ell(\ell+1)}{r^2} + V(r) + k^2_{n\ell}\right)u^{\mathrm{c}}_{n\ell}(r) = \sum_{n'}\lambda_{nn'\ell}u^{\mathrm{b}}_{n'\ell}(r) \tag{7.21}$$

subject to the boundary conditions

$$u^{\rm c}_{n\ell}(0) = 0\,, \tag{7.22a}$$

$$\frac{a}{u^{\rm c}_{n\ell}}\frac{{\rm d}u^{\rm c}_{n\ell}}{{\rm d}r}_{|r=a} = b\,, \tag{7.22b}$$

where the Lagrange multipliers $\lambda_{nn'\ell}$ are chosen so that

$$\int_0^a u^{\rm c}_{n\ell}(r)u^{\rm b}_{n'\ell}(r){\rm d}r = 0 \qquad \text{for all } n, n', \ell\,. \tag{7.23}$$

For any choice of potential $V(r)$, (7.21) defines a set of continuum orbitals which, when taken together with the bound orbitals, is complete over the internal region. However, the choice of $V(r)$ is important in order to obtain good convergence properties. In practice it is important to ensure that the continuum orbitals have the correct behaviour near the nucleus. This means that $V(r)$ must behave as $2Z/r$ near the origin (where Z is the nuclear charge). In practice the potential $V(r)$ is usually chosen to be the static potential of the hydrogen-like ion. Finally, b is an arbitrary constant. Provided enough terms are retained in the R-matrix expansion, the results are independent of b and for convenience b is usually chosen to be zero. Of course, in any practical calculation only a relatively small number of continuum orbitals can be included in the expansion. Since these orbitals satisfy homogeneous boundary conditions (7.22) at $r = a$, it is necessary to apply a correction to the R-matrix expansion first proposed by Buttle [7.20] to ensure rapid convergence of the cross section as the number of such orbitals retained in the calculation is increased. The details of this correction will be discussed in the next section.

Once the energy-independent basis functions, $\chi^{LS\Pi}_{n_1\ell_1n_2\ell_2}(\boldsymbol{r}_1,\boldsymbol{r}_2)$, are chosen, expansion (7.5) in the internal region can be written explicitly as

$$\Psi^{LS\Pi}_k(\boldsymbol{r}_1,\boldsymbol{r}_2) = \sum_{\substack{n_1\ell_1\\ [b,c]}}\sum_{\substack{n_2\ell_2\\ [b]}} \chi_{n_1\ell_1n_2\ell_2}(\boldsymbol{r}_1,\boldsymbol{r}_2)\,\beta_{n_1\ell_1n_2\ell_2k}\,, \tag{7.24}$$

where the summation $n_1\ell_1$ is over the bound and continuum orbital basis and the summation $n_2\ell_2$ is over only the bound orbital basis where all combinations of ℓ_1 and ℓ_2 consistent with the particular values of L and Π being considered are retained. These ranges are denoted by [b,c] and [b] in expansion (7.24).

The approximation described by (7.24) is usually called the close-coupling approximation. We see that the expansion is incomplete since one electron is always restrained to be in a bound orbital. However, this expansion has been widely used in calculations, and if the bound orbital basis also includes a few suitably chosen pseudostates then this method is capable of yielding accurate results. However, recent work [7.21] has extended this expansion, writing it as

$$\Psi_k^{LS\Pi}(\boldsymbol{r}_1,\boldsymbol{r}_2) = \sum_{\substack{n_1\ell_1 \\ [\mathrm{b,c}]}} \sum_{\substack{n_2\ell_2 \\ [\mathrm{b,c}]}} \chi^{LS\Pi}_{n_1\ell_1 n_2\ell_2}(\boldsymbol{r}_1,\boldsymbol{r}_2)\,\beta_{n_1\ell_1 n_2\ell_2 k}\,, \tag{7.25}$$

where the summations over $n_1\ell_1$ and $n_2\ell_2$ now both go over the bound and continuum orbital basis. This expansion is complete over the internal region and is capable of yielding highly accurate results at low and intermediate energies.

We now return to expansion (7.24). The coefficients $\beta_{n_1\ell_1 n_2\ell_2 k}$ are obtained by diagonalizing $H + L$ in (7.6) where H is the two-electron Hamiltonian which can be written in atomic units as

$$H = -\frac{1}{2}\nabla_1^2 - \frac{1}{2}\nabla_2^2 - \frac{Z}{r_1} - \frac{Z}{r_2} + \frac{1}{r_{12}}\,. \tag{7.26}$$

In this particular case the channel functions $\bar{\phi}_{n_2\ell_1\ell_2}$ (where for notational simplicity we now omit the $LS\Pi$ superscript), which are formed by coupling the wavefunction for the target electron with the angular functions of the scattered electron, are given by

$$\begin{aligned}\bar{\phi}_{n_2\ell_1\ell_2} = \frac{1}{\sqrt{2}}\Big\{ & r_2^{-1}u_{n_2\ell_2}(r_2)\mathcal{Y}_{\ell_1\ell_2 LM_L}(\hat{\boldsymbol{r}}_1,\hat{\boldsymbol{r}}_2) \\ & + (-1)^{L+S+\ell_1+\ell_2} r_1^{-1}u_{n_2\ell_2}(r_1)\mathcal{Y}_{\ell_2\ell_1 LM_L}(\hat{\boldsymbol{r}}_1,\hat{\boldsymbol{r}}_2)\Big\}\,,\end{aligned} \tag{7.27}$$

and the Bloch operator defined schematically in (7.3) then becomes

$$\begin{aligned} L = \frac{1}{4}\sum_{n_2\ell_1\ell_2} & \Big|\Big\{ r_2^{-1}u_{n_2\ell_2}(r_2)\mathcal{Y}_{\ell_1\ell_2 LM_L}(\hat{\boldsymbol{r}}_1,\hat{\boldsymbol{r}}_2)\delta(r_1-a)\left(\frac{\mathrm{d}}{\mathrm{d}r_1}-\frac{b}{r_1}\right) \\ & + (-1)^{L+S+\ell_1+\ell_2} r_1^{-1}u_{n_2\ell_2}(r_1)\mathcal{Y}_{\ell_2\ell_1 LM_L}(\hat{\boldsymbol{r}}_1,\hat{\boldsymbol{r}}_2)\delta(r_2-a)\left(\frac{\mathrm{d}}{\mathrm{d}r_2}-\frac{b}{r_2}\right)\Big\}\Big\rangle \\ & \Big\langle r_2^{-1}u_{n_2\ell_2}(r_2)\mathcal{Y}_{\ell_1\ell_2 LM_L}(\hat{\boldsymbol{r}}_1,\hat{\boldsymbol{r}}_2) \\ & + (-1)^{L+S+\ell_1+\ell_2} r_1^{-1}u_{n_2\ell_2}(r_1)\mathcal{Y}_{\ell_2\ell_1 LM_L}(\hat{\boldsymbol{r}}_1,\hat{\boldsymbol{r}}_2)\Big|\,. \end{aligned} \tag{7.28}$$

The **R** matrix, reduced radial functions, and reduced width amplitudes are as given in (7.9), (7.10), and (7.11), respectively. We find in this particular case that the reduced width amplitudes ω_{ik} can be expressed as

$$\begin{aligned}\omega_{n_2\ell_1\ell_2 k} &= r\langle\bar{\phi}_{n_2\ell_1\ell_2}|\Psi_k\rangle\big|_{r=a} \\ &= r\langle\bar{\phi}_{n_2\ell_1\ell_2}|\sum_{\substack{n_1'\ell_1' \\ n_2'\ell_2'}} \beta_{n_1'\ell_1' n_2'\ell_2' k}\chi_{n_1'\ell_1' n_2'\ell_2'}\rangle|_{r=a}\,.\end{aligned} \tag{7.29}$$

Substituting for $\bar{\phi}_{n_2\ell_1\ell_2}$ and $\chi_{n_1'\ell_1' n_2'\ell_2'}$ from (7.27), (7.18), and (7.19), respectively, and integrating over the spatial coordinates of the bound electron and the angular coordinates of the scattered electron we obtain

$$\omega_{n_2\ell_1\ell_2k} = \frac{r}{2}\sum_{\substack{n'_1\ell'_1\\ n'_2\ell'_2}} \beta_{n'_1\ell'_1n'_2\ell'_2k}\Big[2r^{-1}u_{n'_1\ell'_1}(r)\delta_{n_2n'_2}\delta_{\ell_1\ell'_1}\delta_{\ell_2\ell'_2}$$
$$+ (-1)^{L+S+\ell_1+\ell_2}2r^{-1}u_{n_2\ell_2}(r)\delta_{n'_1n_2}\delta_{\ell_2\ell'_1}\delta_{\ell_1\ell'_2}\Big]_{|r=a} . \quad (7.30)$$

We assume that the summation over $n_2\ell_2$ includes only bound orbitals which decay to zero by the R-matrix boundary. Hence

$$\omega_{n_2\ell_1\ell_2k} = \sum_{n_1}\beta_{n_1\ell_1n_2\ell_2k}u_{n_1\ell_1}(a) . \quad (7.31)$$

We now turn to the solution of the problem in the outer region. In the specific problem of electron scattering from hydrogen-like ions, the total wavefunction, Ψ_E, can be expanded as

$$\Psi_E = \sum_{n_2\ell_2\ell_1} r_2^{-1}u_{n_2\ell_2}(r_2)\mathcal{Y}_{\ell_1\ell_2LM_L}(\hat{\boldsymbol{r}}_1,\hat{\boldsymbol{r}}_2)r_1^{-1}F_{n_2\ell_2\ell_1(r_1)} , \quad (7.32)$$

where $u_{n_2\ell_2}(r)$ is the radial function for the residual hydrogen-like ion or atom, $\mathcal{Y}_{\ell_1\ell_2LM_L}(\hat{\boldsymbol{r}}_1,\hat{\boldsymbol{r}}_2)$ is the usual angular function for the two electrons in the system, and $F_{n_2\ell_2\ell_1}(r)$ is the radial function of the scattered electron when the angular momentum of this electron is ℓ_1 and the residual hydrogen-like ion or atom is in the $n_2\ell_2$ state. If we substitute (7.32) into the Schrödinger equation and project onto the functions $r_2^{-1}u_{n_2\ell_2}(r_2)\mathcal{Y}_{\ell_1\ell_2LM_L}(\hat{\boldsymbol{r}}_1,\hat{\boldsymbol{r}}_2)$ we obtain the usual set of n coupled differential equations for the reduced radial functions $F_i(r)$ [see (7.13)] where for clarity we have replaced the combination of subscripts $n_1\ell_1\ell_2$ by the one subscript i. In general, the potential matrix elements V_{ij} can be expanded in inverse powers of the radius r of the scattered electron as

$$V_{ij} = \sum_{\lambda=0}^{\lambda_{\max}} a_{ij}^{\lambda}r^{-\lambda-1} ; \qquad r \geq a . \quad (7.33)$$

In this case a_{ij}^{λ} simplifies to

$$a_{ij}^{\lambda} = f_\lambda(\ell_1\ell_2\ell'_1\ell'_2;L)R_{ij}^{\lambda} , \quad (7.34)$$

where $f_\lambda(\ell_1\ell_2\ell'_1l'_2;L)$ is the angular factor described by Percival and Seaton [7.22] which we will discuss in Sect. 7.4. Furthermore,

$$R_{ij}^{\lambda} = \int_0^a u_{n\ell}(r)r^{\lambda}u_{n'\ell'}(r)\mathrm{d}r \quad (7.35)$$

where $u_{n\ell}(r)$ and $u_{n'\ell'}(r)$ refer to the atomic orbital in channel i and channel j, respectively. The evaluation of the cross section then follows as discussed in Sect. 7.2.

7.4 Computational Solution of the Electron–Hydrogen-like Ion-Collision Problem

7.4.1 The Inner Region

In this section we consider in more detail the solution of the problem in the inner region for the particular case of electron scattering from atomic hydrogen where the total orbital angular momentum and the total spin angular momentum of the two-electron system are zero. We are interested in evaluating the Hamiltonian matrix elements

$$\langle \chi_{n_1\ell_1 n_2\ell_2} | H | \chi_{n_1'\ell_1' n_2'\ell_2'} \rangle . \tag{7.36}$$

Let us consider the case when $n_1\ell_1 \neq n_2\ell_2$. Then, expanding $\frac{1}{r_{12}}$ in terms of Legendre polynomials as $\frac{1}{r_{12}} = \sum_\lambda \frac{r_<^\lambda}{r_>^{\lambda+1}} P_\lambda(\cos\theta_{12})$, (7.36) becomes

$$\begin{aligned}
&\delta_{\ell_1\ell_1'}\delta_{\ell_2\ell_2'}\left[I^{\ell_1}_{n_1n_1'}\delta_{n_2n_2'} + I^{\ell_2}_{n_2n_2'}\delta_{n_1n_1'}\right] \\
&+(-1)^{L+S+\ell_1+\ell_2}\delta_{\ell_1\ell_2'}\delta_{\ell_2\ell_1'}\left[I^{\ell_1}_{n_1n_2'}\delta_{n_2n_1'} + I^{\ell_2}_{n_2n_1}\delta_{n_1n_2'}\right] \\
&+\sum_\lambda f_\lambda(\ell_1\ell_2\ell_1'\ell_2';L)R_\lambda(n_1\ell_1n_2\ell_2n_1'\ell_1'n_2'\ell_2') \\
&+(-1)^{L+S+\ell_1+\ell_2}\sum_\lambda f_\lambda(\ell_1\ell_2\ell_2'\ell_1';L)R_\lambda(n_1\ell_1n_2\ell_2n_2'\ell_2'n_1'\ell_1'),
\end{aligned} \tag{7.37}$$

where

$$R_\lambda(n_1\ell_1n_2\ell_2n_1'\ell_1'n_2'\ell_2') = \int_0^a\int_0^a u_{n_1\ell_1}(r_1)u_{n_2\ell_2}(r_2)\frac{r_<^\lambda}{r_>^{\lambda+1}}u_{n_1'\ell_1'}(r_1)u_{n_2'\ell_2'}(r_2)\mathrm{d}r_1\mathrm{d}r_2 \tag{7.38}$$

and

$$f_\lambda(\ell_1\ell_2\ell_1'\ell_2';L) = \langle \mathcal{Y}_{\ell_1\ell_2LM_L}(\hat{r}_1,\hat{r}_2)|P_\lambda(\cos\theta_{12})|\mathcal{Y}_{\ell_1'\ell_2'LM_L}(\hat{r}_1,\hat{r}_2)\rangle \tag{7.39}$$

and

$$I^\ell_{n_1n_2} = -\frac{1}{2}\int_0^a u_{n_1\ell}(r)\left[\frac{\mathrm{d}^2}{\mathrm{d}r^2} - \frac{\ell(\ell+1)}{r^2} + \frac{2Z}{r}\right]u_{n_2\ell}(r)\mathrm{d}r . \tag{7.40}$$

7.4.1.1 Calculation of Orbitals

The orbitals $u_{n\ell}(r)$ are divided into two types – bound and continuum. The bound orbitals are simply the usual hydrogenic orbitals while the continuum orbitals are taken as the solution of (7.21) satisfying the stated boundary conditions (7.22). This eigenvalue problem is solved using the package

BASFUN [7.23]. An initial eigenvalue is found using the subroutine FINDER and the differential equation is then integrated using de Vogelare's method. The eigenvalue and eigenfunction are subsequently found using Newton's iteration method and orthogonality to the bound orbitals ensured by the use of Lagrange undetermined multipliers.

7.4.1.2 The Angular Factors $f_\lambda(\ell_1, \ell_2, \ell_1', \ell_2'; L)$

The angular factors $f_\lambda(\ell_1, \ell_2, \ell_1', \ell_2'; L)$ [7.22] are evaluated by expanding

$$P_\lambda(\cos\theta_{12}) = \frac{4\pi}{2\lambda+1}\sum_{m_\lambda} Y^*_{\lambda m_\lambda}(\hat{r}_1)Y_{\lambda m_\lambda}(\hat{r}_2) \tag{7.41}$$

to give

$$\begin{aligned} f_\lambda(\ell_1, \ell_2, \ell_1', \ell_2'; L) &= (-1)^{\ell_1+\ell_1'+L}\left[(2\ell_1+1)(2\ell_2+1)(2\ell_1'+1)(2\ell_2'+1)\right]^{1/2} \\ &\quad \times (2\lambda+1)^{-1}(\ell_1 0 \ell_1' 0|\lambda 0)(\ell_2 0 \ell_2' 0|\lambda 0)W(\ell_1\ell_2\ell_1'\ell_2'; L\lambda)\,, \end{aligned} \tag{7.42}$$

where $(\ell_1 m_1 \ell_2 m_2|LM)$ are Clebsch–Gordan and $W(\ell_1\ell_2\ell_1'\ell_2'; L\lambda)$ are Racah coefficients (see Rose [7.24a]). In obtaining (7.42) we make use of the formula [7.24b]

$$\begin{aligned} &\int Y^*_{\ell_1 m_1}(\hat{r})Y_{\ell_2 m_2}(\hat{r})Y_{\ell m}(\hat{r})\mathrm{d}\hat{r} \\ &\qquad = \left[\frac{(2\ell+1)(2\ell_2+1)}{(2\ell_1+1)4\pi}\right]^{1/2}(\ell m \ell_2 m_2|\ell_1 m_1)(\ell 0 \ell_2 0|\ell_1 0) \end{aligned} \tag{7.43}$$

and the reduction formula for Clebsch–Gordan coefficients [7.24c]. Subroutines for evaluating the Clebsch–Gordan coefficients and the Racah coefficients are given in the sample code.

7.4.1.3 Setting up and Diagonalizing the Hamiltonian Matrix in the Inner Region

To complete the calculation of the Hamiltonian matrix we need to evaluate the radial integrals $I^{\ell}_{n_1 n_2}$ and $R_\lambda(n_1\ell_1 n_2\ell_2 n_1'\ell_1' n_2'\ell_2')$. The integrals $I^{\ell}_{n_1 n_2}$ are evaluated using the subroutine ONEELE and stored in main memory to be used in the subroutine HMAT, which sets up the Hamiltonian matrix. The integrals $R_\lambda(n_1\ell_1 n_2\ell_2 n_1'\ell_1' n_2'\ell_2')$ are calculated as required in the subroutine HMAT by calling the subroutine RKINT. The upper right-hand triangle of the Hamiltonian matrix is then stored in a one-dimensional array ready for diagonalization using the Householder method (see, for example, STG3 of the RMATRX packages [7.25]).

7.4.1.4 Evaluation of the Surface Amplitudes

Diagonalization of the Hamiltonian matrix yields a set of eigenvalues E_k and eigenvectors a_k where the components of a_k are the coefficients $\beta_{n_1\ell_1 n_2\ell_2 k}$. The surface amplitudes ω_{ik} are given by

$$\omega_{n_2\ell_1\ell_2 k} = \sum_{n_1} \beta_{n_1\ell_1 n_2\ell_2 k} u_{n_1\ell_1}(a) . \tag{7.44}$$

7.4.1.5 Buttle Correction

We are now in a position to calculate the **R** matrix at any energy E in the range of interest using

$$R_{ij} = \frac{1}{2a} \sum_k \frac{\omega_{ik}\omega_{jk}}{E_k - E} . \tag{7.45}$$

However, we note that in practice the sum over k has to be truncated. In order to take account of the higher-lying contributions to the **R** matrix we include a correction to the diagonal elements of the **R** matrix, which was introduced by Buttle [7.20]. We have

$$R^c_{ii} = \left[\frac{a}{y_i} \frac{\mathrm{d}y_i}{\mathrm{d}r} - b \right]^{-1}_{r=a} - \frac{1}{a} \sum_{j=1}^{N} \frac{\left[u_{ij}(a)\right]^2}{k^2_{ij} - k^2_i} , \tag{7.46}$$

where k^2_i is the channel energy and k^2_{ij} is the energy eigenvalue associated with (7.21), and u_{ij} are the corresponding continuum orbitals satisfying the orthogonality conditions (7.23); y_i is the solution of the differential equation (7.21) at energy k^2_i. Evaluation of R^c_{ii} at each energy is time consuming and we can adopt one of two approximations to simplify this. For small channel energies ($-0.5 < k^2_i < 2.0$ Rydbergs) the Buttle correction is accurately represented by a quadratic with respect to the channel energy k^2_i. In this case we evaluate the Buttle correction explicitly at a few energies and obtain a quadratic expression for the correction in the form $a_0 + a_1 k^2_i + a_2 k^4_i$. An alternative procedure given by Seaton [7.26], is to consider (7.21) with $\ell = 0$ and $V(r) = 0$. In this case the correction, with $b = 0$, is given by

$$B(N_B, u) = \frac{\tan(k)}{K} - 2 \sum_{n=0}^{N_B} (u_n - u)^{-1} , \tag{7.47}$$

where $K = ak_i$, $u = K^2$, $u_n = K^2_n$, and $K_n = a(n + 1/2)\pi$. N_B is chosen so that the eigenvalue $k^2_{N_B} \geq k^2_N$, i.e., $k^2_{N_B}$ is at least as great as the highest eigenvalue associated with (7.21).

7.4.2 Outer Region Problem

In the outer region ($r > a$) where exchange effects between the scattered electron and the residual atomic electron can be ignored, we are required to solve the set of radial equations

$$\left(\frac{\mathrm{d}^2}{\mathrm{d}r^2} - \frac{\ell_i(\ell_i+1)}{r^2} + \frac{2(Z-N)}{r} + k_i^2\right)F_i(r)$$
$$= 2\sum_{j=1}^{n}\sum_{\lambda=0}^{\lambda_{\max}} a_{ij}^{\lambda} r^{-\lambda-1} F_j(r); \qquad i = 1, n;\ r \geq a. \tag{7.48}$$

The **K** matrix is defined by the asymptotic form of the solution of these equations which is given in (7.15). To relate the $n \times n$ dimensional **R** matrix to the $n_a \times n_a$ dimensional **K** matrix (where n_a is the number of open channels, i.e., the number of channels for which $k_i^2 > 0$) we introduce $n + n_a$ linearly independent solutions v_{ij} of (7.48) satisfying the asymptotic boundary conditions

$$v_{ij} \underset{r\to\infty}{\sim} k_i^{-1/2} \sin\theta_i \delta_{ij} + O(r^{-1}); \quad j = 1, n_a; \quad i = 1, n; \tag{7.49a}$$

$$v_{ij} \underset{r\to\infty}{\sim} k_i^{-1/2} \cos\theta_i \delta_{ij} + O(r^{-1}); \quad j = n_a + 1, 2n_a; \quad i = 1, n; \tag{7.49b}$$

$$v_{ij} \underset{r\to\infty}{\sim} \exp(-\phi_i)\delta_{ij-n_a} + O(r^{-1}\exp(-\phi_{j-n_a}));$$
$$j = 2n_a + 1, n + n_a;\ i = 1, n\,. \tag{7.49c}$$

In (7.49), we have defined

$$\phi_i = |k_i|r - \frac{Z-N}{k_i}\ln(2|k_i|r)\,. \tag{7.50}$$

A number of techniques such as de Vogelare's, Fox-Goodwin, variable phase and various propagation methods exist to solve the set of equations (7.48). These continue the integration to sufficiently large radial distances that the coupling potentials may be neglected and the solutions matched to the appropriate asymptotic form. The radial distances over which we need to integrate these equations may, however, become quite large. To overcome this problem, analytic solutions have been proposed by Burke and Schey [7.27] and Gailitis [7.28] so that the matching procedure may be applied at smaller radii. A further reduction in the boundary radius is achieved using the `CFASYM` program of Noble and Nesbet [7.29] in which the asymptotic expansions suggested by Burke and Schey and by Gailitis are analytically continued and calculated in the form of a continued fraction using an algorithm devised by Nesbet. Assuming a solution for v_{ij} we have

$$F_{ij}(r) = \sum_{l=1}^{n+n_a} v_{il}(r)x_{lj}\,; \qquad i = 1, n; \quad j = 1, n_a\,. \tag{7.51}$$

It follows that

$$x_{ij} = \delta_{ij}; \quad i = 1, n_a \,, \tag{7.52}$$

and

$$K_{ij} = x_{i+n_a j}\,; \quad i, j = 1, n_a \,. \tag{7.53}$$

Substituting (7.51) into (7.8) gives

$$\sum_{l=1}^{n+n_a} \left\{ v_{il}(a) - \sum_{m=1}^{n} R_{im} \left(a \frac{\mathrm{d}v_{ml}}{\mathrm{d}r} - b v_{ml} \right)_{|r=a} \right\} x_{lj} = 0; \quad i = 1, n \,. \tag{7.54}$$

This system must be solved for each $j = 1, n_a$. Equations (7.53) and (7.54) enable us to evaluate the x_{ij} and hence the K_{ij}.

The **S** matrix is determined by taking linear combinations of the solutions in (7.15) such that the new asymptotic solutions satisfy the boundary conditions

$$z_{ij} \underset{r\to\infty}{\sim} k_i^{-1/2} [\exp(-\mathrm{i}\theta_i)\delta_{ij} - \exp(\mathrm{i}\theta_i) S_{ij}]\,; \quad i = 1, n_a\,; \quad j = 1, n_a\,; \tag{7.55a}$$

$$z_{ij} \underset{r\to\infty}{\sim} O(r^{-2})\,; \qquad i = n_a + 1, n; \; j = 1, n_a \,. \tag{7.55b}$$

The $n_a \times n_a$ dimensional **S** matrix is then related to the **K** matrix by the matrix equation

$$\mathbf{S} = [\mathbf{1} + \mathrm{i}\mathbf{K}]\,[\mathbf{1} - \mathrm{i}\mathbf{K}]^{-1} \,, \tag{7.56}$$

and the **T** matrix is given in terms of the **S** and **K** matrices by

$$\mathbf{T} = \mathbf{S} - \mathbf{1} = 2\mathrm{i}\mathbf{K}\,[\mathbf{1} - \mathrm{i}\mathbf{K}]^{-1} \,. \tag{7.57}$$

In the particular situation where all the channels are closed we define n linearly independent solutions v_{ij} which satisfy the boundary conditions

$$v_{ij} \underset{r\to\infty}{\sim} \exp(-|k_i|r)\delta_{ij}; \qquad i, j = 1, n \,. \tag{7.58}$$

The required solution can then be expanded as

$$F_i = \sum_{j=1}^{n} v_{ij} x_j; \qquad i = 1, n; \quad a \le r \le \infty \,. \tag{7.59}$$

Substituting into (7.8) gives

$$\sum_{j=1}^{n} \left\{ v_{ij}(a) - \sum_{k=1}^{n} R_{ik} \left(a \frac{\mathrm{d}v_{kj}}{\mathrm{d}r} - b v_{kj} \right)_{|r=a} \right\} x_j = 0; \quad i = 1, n \,. \tag{7.60}$$

This can be written as

$$\sum_{j=1}^{n} B_{ij} x_j = 0; \quad i = 1, n. \tag{7.61}$$

These equations only have nontrivial solutions at the negative energy eigenvalues corresponding to bound states of the electron–atom or electron–ion system. The condition for solution is

$$\det \{B\} = 0. \tag{7.62}$$

We refer the reader to Burke and Seaton [7.30] and Seaton [7.31] for the solution of this equation. We recall that

$$\Psi_E = \sum_k A_{Ek} \psi_k . \tag{7.63}$$

The expansion coefficients A_{Ek} can be evaluated from (7.7) giving

$$A_{Ek} = \frac{1}{2a(E_k - E)} \omega_k^T \cdot \mathbf{R}^{-1} \cdot F(a) \tag{7.64}$$

(see also Burke and Robb [7.32]) and, since the ψ_k are known, the bound state wavefunction Ψ_E can be obtained. This wavefunction is used in the calculation of photodetachment and photoionization cross sections [7.7].

7.5 Computer Program

The computer program supplied with this chapter will calculate cross sections and collision strengths for electron hydrogen-atom scattering. It is written in FORTRAN77. The program package comes in six modules.

1. `HMAT2E` – this module sets up the Hamiltonian matrix in the inner region.
2. `DIAG2E` – this module diagonalises the Hamiltonian matrix.
3. `SURA2E` – this module evaluates the surface amplitudes on the **R**-matrix boundary;
4. `RKMAT2E` – this module controls the energy loop for the evaluation of the **R** matrix on the boundary of the internal and external regions. It initiates the call to the asymptotic package `VPM` [7.33], through the interface subroutine `VFACE` and finally feeds the **K** matrix to the module `XSEC` to evaluate the cross sections and collision strengths.
5. `VPM` – this module solves the external region problem given the **R** matrix on the boundary at energy E, returning the required **K** matrix. We refer the reader to the article [7.33] where more detailed information may be found.
6. `XSEC` – given the **K** matrix at energy E, this module returns the cross section or collision strength as required.

The program is best run as two separate packages. The first, `PACK1`, consisting of the energy-independent modules (1), (2), and (3) and the second,

PACK2, consisting of the energy-dependent modules (4), (5), and (6). However, the construction of each package is such that each module can be easily separated out and either run independently or replaced. This will facilitate the use of, for example, a different asymptotic module (5) or, when the physical problem is very large, it allows modules (1) and (2) to be run separately.

7.5.1 Description of Input Data for the Program Package PACK1

In this package the input data are read from unit IREAD which is set in subroutine READS (currently to IREAD=5). Integers are read in I5 format and real numbers in F14.7 format.

Record 1. LTOT,ISTOT,L1MAX,L2MAX,N1MAX,N2MAX,NPTY,NZ

LTOT — Total orbital angular momentum of the two-electron system.
ISTOT — Total spin angular momentum of the two-electron system.
L1MAX — Maximum orbital angular momentum of the scattered electron.
L2MAX — Maximum orbital angular momentum of the target electron.
N1MAX — Maximum value of n_1 retained in expansion (7.24).
N2MAX — Maximum value of n_2 retained in expansion (7.24).
NPTY — Parity of the two-electron system; set to 0 if $(-1)^{\ell_1+\ell_2} = 1$ and to 1 if $(-1)^{\ell_1+\ell_2} = -1$.
NZ — Nuclear charge.

Record 2. RA,BSTO

RA — The boundary radius. This should be chosen to be large enough to envelope the target state of interest. The value of the bound orbitals should have decayed to less than 0.001 on the boundary.
BSTO — This is the constant b referred to in (7.22).

Record 3. NBUG1,NBUG2,NBUG3,NBUG4,NBUG5,NBUG6,NBUG7,NBUG8

NBUG1 — Set to 1 for debug printout of radial orbitals.
NBUG2 — Set to 1 for debug printout of potential $V(r)$ [see (7.21)] at each mesh point.
NBUG3 — Set to 1 for debug printout from subroutine BASFUN.
NBUG4 — Set to 1 for debug printout from FINDER.
NBUG5 — Set to 1 for debug printout of one-electron integrals.
NBUG6 — Set to 1 for debug printout of Hamiltonian matrix.
NBUG7 — Set to 1 for debug printout of radial part of bound orbitals.
NBUG8 — Set to 1 for debug printout of angular coefficients.

Record 4. IWRITE,ITAPE1,ITAPE2,ITAPE3,ITAPE4,ITAPE5

IWRITE — Lineprinter output unit number.
ITAPE1 — Output unit number for data required for module SURA2E.
ITAPE2 — Output unit number for data required for module DIAG2E.
ITAPE3 — Output unit number for basic data from DIAG2E.
ITAPE4 — Output unit number for direct-access file to store the eigenvalues and eigenvectors from DIAG2E.

ITAPE5 Output unit number for surface amplitudes from SURA2E.

Record 5. NBOUND,LBOUND,IMESH

NBOUND Maximum principal quantum number of the bound orbitals in expansion (7.24).

LBOUND Maximum orbital angular momentum of the bound orbitals in expansion (7.24).

For electron–hydrogen-atom scattering, where we expand the basis $\psi_k^{LS\Pi}$ as in (7.24), NBOUND=N2MAX and LBOUND=L2MAX. However, when we use expansion (7.25) this is not the case. NBOUND and LBOUND are also used to define the hydrogenic target states retained in the outer region, for example, if NBOUND=3 and LBOUND=1 then we retain the following states: 1s, 2s, 3s, 2p and 3p.

IMESH If the mesh for the radial integrals $R_\lambda(n_1\ell_1n_2\ell_2n'_1\ell'_1n'_2\ell'_2)$ and R_{ij}^λ is to be determined automatically then IMESH is set to 0. If we wish to read in a set of parameters defining this mesh then IMESH is set to 1.

Record 6. NTERM(N,L)
Record 7. IRAD(N,L,I)
Record 8. COEF(N,L,I)
Record 9. ALPHA(N,L)

These records are used to input the bound orbital information. Each bound orbital is defined in the following way:

$$\sum_{\mathtt{I=1}}^{\mathtt{NTERM(N,L)}} r^{\mathtt{IRAD(N,L,I)}} * \mathtt{COEF(N,L,I)} * e^{-\mathtt{ALPHA(N,L)}r} . \qquad (7.65)$$

The arrays NTERM, ALPHA, COEF, and IRAD are read from the input unit using the following coding:

```
      DO 20 L=0,LRANG1
        DO 10 N=L+1,NRANG1
          READ(IREAD,1000) NTERM(N,L)
          NT=NTERM(N,L)
          READ(IREAD,1000) (IRAD(N,L,I),I=1,NT)
          READ(IREAD,1001) (COEF(N,L,I),I=1,NT)
          READ(IREAD,1001) ALPHA(N,L)
   10   CONTINUE
  20 CONTINUE
```

Records 6–9 are repeated for each bound orbital.

Record 10. NIX
Record 11. IHX(I),I=1,NIX

Record 12. IRX(I),I=1,NIX
Record 13. HINT

Records 10–13 are only required if IMESH=1 and we wish to define our own mesh for the radial integrations. The integration is carried out between $r = 0$ and $r =$ RA. This distance is divided into NIX subranges each of which has a step length described as a multiple of a basic step length HINT. There are IRX(I) integration steps between the origin and the end of the Ith subrange.

NIX	Number of subranges in the interval $0 > r >$ RA.
IHX()	Array storing the multiple of basic step length which defines integration step in the Ith subrange.
IRX()	Array storing number of integration steps from the origin to the end of the Ith subrange.
HINT	Basic step length.

Record 14. NPOT
Record 15. IPOT(I),I=1,NPOT
Record 16. CPOT(I),I=1,NPOT
Record 17. XPOT(I),I=1,NPOT

The potential *V(r)* in (7.21) is read in using records 14–17. This is then calculated at each of the radial mesh points and at the half mesh points. The potential is defined in the following way:

$$\sum_{\mathtt{I=1}}^{\mathtt{NPOT}} \mathtt{CPOT(I)} * r^{\mathtt{IPOT(I)}} * e^{-\mathtt{XPOT(I)}\, r} . \tag{7.66}$$

NPOT	Number of terms in expansion (7.66).
IPOT()	Array storing the powers to which r is raised in expansion (7.66).
CPOT()	Array storing the coefficients for each term in expansion (7.66).
XPOT()	Array storing the exponential factor of r in expansion (7.66).

Finally, we note that the dimensions of important arrays are set up using the parameters IDIM1, IDIM2, ... and KDIM1, KDIM2, ...; these are defined as follows:

IDIM1	Maximum order of the Hamiltonian matrix.
IDIM2	Maximum L1MAX, L2MAX value.
IDIM3	Maximum N1MAX, N2MAX value.
IDIM4	Maximum number of terms allowed in expansion (7.66).
IDIM5	Maximum number of subranges for radial integration mesh.
IDIM6	Maximum number of points in the radial integration mesh.
IDIM7	Maximum number of bound orbitals allowed for a given angular momentum.
IDIM8	Maximum number of channels allowed.
KDIM1	Maximum number of half mesh points in the radial integration mesh – defined as IDIM6*2.

KDIM2 IDIM7+1.

KDIM3 Maximum λ value for the evaluation of the angular coefficients. This is given by IDIM2*2.

7.5.2 Description of Input Data for PACK2

We will now discuss the input data for PACK2, which is the second computer package consisting of the modules (4), (5), and (6), i.e., RKMAT2E, the asymptotic package VPM, and XSEC. The input data are read from unit IREAD=5 (set in the main program). All integer data are read in the format 12I5, whereas the format 5F14.7 is used for real numbers. All data are read in the main program unit.

Record 1. ITAPE1,ITAPE2,ITAPE3,IWRITE

ITAPE1 Unit number for the file containing Buttle correction coefficients and AIJ coefficients from module HMAT2E.

ITAPE2 Unit number for the file containing the surface amplitudes from module SURA2E.

ITAPE3 Unit number for output of **K** matrices from module RKMAT2E.

IWRITE Unit number for line printer output.

Record 2. LTOT,ISTOT

LTOT Total orbital angular momentum of the two-electron system.

ISTOT Total spin angular momentum of the two-electron system.

Record 3. NSTS,NCHAN

NSTS Number of residual atomic states.

NCHAN Total number of channels.

Record 4. ESTAT(I),I=1,NSTS

ETHR() Array containing the threshold energies of the residual atomic states. These are required in Rydbergs and should be given with respect to the ground state of the atom.

Record 5. ICHST(I),I=1,NSTS

ICHST() Array containing the number of channels coupled to the residual atomic states.

Record 6. LTST(I),I=1,NSTS

LTST() Array containing the orbital angular momenta of the residual atomic states.

Record 7. ISTST(I),I=1,NSTS

ISTST() Array containing the spin angular momenta of the residual atomic states, expressed as $2S+1$.

Record 8. `IRMAT,IKMAT,IDELTA,ICOLL,ICROSS`

`IRMAT` If set to 1 the Buttle corrected **R** matrix will be printed out at each energy.

`IKMAT` If set to 1 the **K** matrix will be printed out at each energy.

`IDELTA` If set to 1 and there is only one open channel at the energy in question the phase shift will be printed out.

`ICOLL` If set to 1 the collision-strength matrix will be printed out.

`ICROSS` If set to 1 the cross-section matrix will be printed out.

Record 9. `EKMIN,EKMAX,EKINCR,ENABS`

These parameters define the energy range for which the **K** matrices will be calculated.

`EKMIN` Minimum scattered electron energy.

`EKMAX` Maximum scattered electron energy.

`EKINCR` Energy step.

These parameters are all given in Rydbergs with respect to the ground state of the atom.

`ENABS` The absolute energy of the ground state of the atom in atomic units.

Important dimensions in `PACK2`, with the exception of the asymptotic module `VPM`, are as given in `PACK1` with the addition of the following dimensions used in the interface subroutine to the asymptotic package:

`NVDIM1` Maximum number of channels allowed in `VPM`.

`NVDIM2` Maximum number of terms in the asymptotic expansion (see `VPM` write-up [7.33]).

`NVDIM3` As `NVDIM1`.

`NVDIM4` See `VPM` write-up [7.33].

`NVDIM5` The maximum order for the `AIJ` coefficients.

`NVDIM6` See `VPM` write-up [7.33].

These parameters are referred to as `NDIM1`, ..., `NDIM6` throughout the `VPM` module. Although they are related to dimensions in other modules in `PACK1` and `PACK2`, it is not essential that the dimensions are the same. However, they must be consistent; for example, `NVDIM1` and `IDIM8` refer to the maximum number of coupled channels allowed. They do not have to be identical, but we can only study a problem where the maximum number of channels is less than or equal to `MIN(NVDIM1,IDIM8)`.

7.5.3 Sample Input and Output

The input file (read from unit `IREAD=5`) for `PACK1` for electron scattering from atomic hydrogen when the total orbital angular momentum of the two-electron system is 0 and the spin angular momentum of the two-electron system is 0 is given below. We consider the case where the 1s, 2s, and 2p states are included in the $n_2\ell_2$ expansion in (7.24).

```
 0    0    1    1   28    2    0    1
 25.0000000        0.0000000
 0    0    0    0    0    0    0    0
 6    1    2    3    4    7
 2    1    0
 1
 1
  2.0000000
  1.0000000
 1    2
  0.7071068       -0.3535534
  0.5000000
 1
 2
  0.2041241
  0.5000000
 1
-1
  2.0000000
  0.0000000
```

Note that we have set the number of continuum orbitals retained in (7.24) to 28 and have set a boundary radius of 25.0 a_0 (where a_0 is the Bohr radius).

The input data required for PACK2 to evaluate the phase shift (where appropriate) and the cross-section matrix in the energy region 0.05 Ryd to 2.0 Ryd in steps of 0.05 Ryd is

```
 1    2    3    6
 0    0
 3    3
  0.0000000       0.7500000       0.7500000
 1    1    1
 0    0    1
 2    2    2
 1    1    1    1    1
  0.1000000       2.5000000       0.1000000       -0.5000000
```

This input data will also ensure that the Buttle corrected **R** matrix, the **K** matrix and the collision-strength matrix are printed out at each energy.

In Table 7.1 we give the results for the elastic $\ell = 0$ phase shift for scattered electron energies in the range 0.1–0.7 Rydbergs, and Tables 7.2 and 7.3 present the Buttle-corrected **R** matrix, **K** matrix, and the cross-section matrix $\sigma_{L_iS_i \to L_jS_j}$ (in units of πa_0^2) at an incident energy of 2.0 Rydbergs for the $^1S^e$ and $^1P^o$ symmetries, respectively.

Table 7.1. $^1S^e$ elastic phase shift δ for electron–hydrogen scattering, as obtained with the 1s, 2s, 2p R-matrix method.

k^2 / Ryd	0.1	0.2	0.3	0.4	0.5	0.6
δ / rad	1.5432	1.2001	1.0089	0.8879	0.8111	0.7714

Table 7.2. Buttle-corrected **R** matrix, **K** matrix, and cross sections for electron–hydrogen scattering at an incident energy of 2.0 Ryd, as obtained in the 1s, 2s, 2p R-matrix method for the $^1S^e$ symmetry.

	R matrix			**K** matrix	
0.066647	0.012919	−0.039190	0.210207	1.066139	0.596310
0.012919	0.026569	0.044911	1.066192	−2.286896	−2.374396
−0.039190	0.044911	0.007872	0.596358	−2.374401	−1.412339
	cross-section matrix				
0.090825	0.015097	0.022673			
0.024157	0.371293	0.146140			
0.012093	0.048714	0.066534			

Table 7.3. Buttle-corrected **R** matrix, **K** matrix, and cross sections for electron–hydrogen scattering at an incident energy of 2.0 Ryd, as obtained in the 1s, 2s, 2p R-matrix method for the $^1P^o$ symmetry.

	R matrix		
−0.023560	−0.004137	−0.031679	−0.005475
−0.004137	−0.030305	−0.048473	−0.024166
−0.031679	−0.048473	0.201048	0.039807
−0.005475	−0.024166	0.039807	0.003557
	K matrix		
0.034530	−0.402809	0.179552	−0.254452
−0.402904	1.120690	1.489583	0.185563
0.179650	1.489564	−3.965188	1.461632
−0.254533	0.185576	1.461639	−0.061634
	cross-section matrix		
0.002602	0.031427	0.027388	
0.050301	1.188167	0.256707	
0.014616	0.085570	0.764097	

7.5.4 Notes on Running the Codes

The codes have been compiled and run on a number of machines. In some cases specific compiler options are necessary. For example, on UNIX-based workstations such as the HP Apollo series, the codes were compiled using the

−K option, whereas the −Wf "–a static" option was used when compiling under UNICOS using CF77.

7.6 Summary

In this chapter we have introduced the R-matrix method as applied to electron–atom and electron–ion excitation, and in particular to the specific problem of electron–hydrogen-like ion scattering with excitation. The computer codes given apply only to the two-electron problem. For computer codes for the general N-electron problem we refer the reader to the R-matrix codes of Berrington *et al.* [7.25]. As discussed in the introduction to this chapter, the R-matrix method has been developed as an *ab initio* approach to a broad range of atomic and molecular processes; again, we refer the reader to the suggested articles for more details of these processes.

7.7 Suggested Problems

1. Study the $^1S^e$ cross section near the resonance at 9.6 eV in the approximation where the 1s, 2s, and 2p states are retained in the $n_2\ell_2$ expansion in (7.24).
2. Calculate the elastic-scattering phase shift in the case where the hydrogen atom is confined to the 1s state. Choose the total orbital angular momentum of the two-electron system as 0 and the total spin angular momentum of the two-electron system as 0. This corresponds to the static exchange approximation considered by John [7.34]. It will be possible to reduce the R-matrix boundary radius since the sphere no longer has to envelope the $n = 2$ orbitals. As a consequence of this, the number of continuum orbitals in (7.24) can also be reduced. Investigate.
3. Extend Problem 2 to allow for the use of expansion (7.25). Restrict both electrons to be in s orbitals by setting `L1MAX=L2MAX=0`. Allow `N2MAX` to equal `N1MAX`. This corresponds to the spherically symmetric model studied by Poet [7.35,36] and by Burke and Mitchell [7.37].
4. Extend Problem 3 so that both electrons can be in s, p, or d orbitals. This corresponds to approximation (iii) discussed by Burke *et al.* [7.21]. (Note: the dimension setup parameters `IDIM1`, `IDIM2`, ... may need to be increased.)
5. Amend the original test calculation to include the $n = 3$ states of hydrogen. Remember to increase the R-matrix boundary radius and also the number of continuum orbitals in (7.24).
6. Consider the use of the codes to study electron scattering from other one-electron targets such as He^+. The module `VPM` will need to be replaced in this case, since it assumes a neutral target.

Acknowledgments

Some of the subroutines used in the accompanying computer codes have been adapted from similar subroutines in the `RMATRX` suite of computer programs (Berrington *et al.* [7.25]) and we would like to acknowledge the work of these authors. We are grateful to Drs. K.L. Bell and N.S. Scott for providing us with a recent version of `VPM` (Croskery *et al.* [7.33]).

References

7.1 E.P. Wigner, Phys. Rev. **70** (1946) 15
7.2 E.P. Wigner and L. Eisenbud, Phys. Rev. **72** (1947) 29
7.3 A.M. Lane and R.G. Thomas, Rev. Mod. Phys. **30** (1958) 257
7.4 G. Breit, Handbuch der Physik **41** Part 1 (1959) 1
7.5 P.G. Burke, A. Hibbert and W.D. Robb, J. Phys. B **4** (1971) 153
7.6 D.C.S. Allison, P.G. Burke and W.D. Robb, J. Phys. B **5** (1972) 55
7.7 P.G. Burke and K.T. Taylor, J. Phys. B **8** (1975) 2620
7.8 M. LeDourneuf, VoKyLan and P.G. Burke, Comm. At. Mol. Phys. **7** (1977) 1
7.9 P.G. Burke, I. Mackey and I. Shimamura, J. Phys. B **10** (1977) 2497
7.10 J.C. Light and R.B. Walker, J. Chem. Phys. **65** (1976) 4222
7.11 K.L. Bell, P.G. Burke and A.E. Kingston, J. Phys. B **10** (1977) 3117
7.12 B.I. Schneider, M. LeDourneuf and P.G. Burke, J. Phys. B **12** (1979) L375
7.13 J. Gerratt, Phys. Rev. A **30** (1984) 1643
7.14 J. Tennyson, C.J. Noble and P.G. Burke, Int. J. Quant. Chem. **29** (1986) 1033
7.15 K. Higgins, P.G. Burke and H.R.J. Walters, J. Phys. B **23** (1990) 1345
7.16 J. Tennyson, J. Phys. B **19** (1986) 4255
7.17 K. Bartschat and P.G. Burke, J. Phys. B **20** (1987) 3191
7.18 C. Bloch, Nucl. Phys. **4** (1957) 503
7.19 N.F. Mott and H.S.W. Massey, *The Theory of Atomic Collisions*, 3rd edition (Oxford University Press, London and New York, 1965)
7.20 P.J.A. Buttle, Phys. Rev. **160** (1967) 719
7.21 P.G. Burke, C.J. Noble and M.P. Scott, Proc. Roy. Soc. A **410** (1987) 289
7.22 I.C. Percival and M.J. Seaton, Proc. Camb. Phil. Soc. **53** (1957) 654
7.23 W.D. Robb, Comp. Phys. Commun. **1** (1970) 457
7.24 M.E. Rose, *Elementary Theory of Angular Momentum* (Wiley, New York, 1957) (a) chapter 3; (b) Eq. (4.34), p. 62; (c) §23, p. 110
7.25 K.A. Berrington, P.G. Burke, M. LeDourneuf, W.D. Robb, K.T. Taylor and VoKyLan, Comp. Phys. Commun. **14** (1978) 367
7.26 M.J. Seaton, J. Phys. B **20** (1987) L69
7.27 P.G. Burke and H.M. Schey, Phys. Rev. **126** (1962) 147
7.28 M. Gailitis, Zh. Eksp. Theor. Fiz. **47** (1964) 160 [Engl. transl.: Sov. Phys.–JETP **20** (1965) 107]
7.29 C.J. Noble and R.K. Nesbet, Comput. Phys. Commun. **33** (1984) 399
7.30 P.G. Burke and M.J. Seaton, J. Phys. B **17** (1984) L683
7.31 M.J. Seaton, J. Phys. B **18** (1985) 2111
7.32 P.G. Burke and W.D. Robb, Adv. At. Molec. Phys. **11** (1975) 143
7.33 J.P. Croskery, N.S. Scott, K.L. Bell and K.A. Berrington, Comput. Phys. Commun. **27** (1982) 385
7.34 T.L. John, Proc. Phys. Soc. **76** (1960) 532
7.35 R. Poet, J. Phys. B **11** (1978) 3081
7.36 R. Poet, J. Phys. B **13** (1980) 2995
7.37 P.G. Burke and J.F.B. Mitchell, J. Phys. B **6** (1973) 320

8. Momentum-Space Convergent-Close-Coupling Method for a Model e–H Scattering Problem

Igor Bray[1] and Andris Stelbovics[2]

[1] Electronic Structure of Materials Centre, The Flinders University of South Australia, G.P.O. Box 2100, Adelaide 5001, Australia
[2] Centre for Atomic, Molecular and Surface Physics, School of Mathematical and Physical Sciences, Murdoch University, Perth 6150, Australia

Abstract

The convergent-close-coupling (CCC) method is illustrated for the Temkin–Poet model of electron–hydrogen scattering. This model treats only states of zero orbital angular momentum, but has played an important role in the development and testing of methods of solution for general electron–atom scattering systems. This chapter describes the features of the model and discusses the application of the CCC approach to its numerical solution. All scattering processes, elastic, inelastic excitation, and ionization, are found from a single unified calculation. The computer program associated with this chapter is the much reduced version of the general CCC code, but contains the essential techniques for solving large sets of coupled equations.

8.1 Introduction

The aim of this chapter is to introduce the reader to an approach for solving the coupled equations, which often arise in scattering theory, that is based on solving coupled integral equations in momentum space. The strength of the technique is that the linear integral equations have as their driving terms the Born approximations to the scattering amplitudes, thereby enabling higher-order perturbative approximations to be derived as iterative refinements in addition to the full integral-equation solution. The integral-equation system is readily solved by efficient algorithms that permit the solution of relatively large sets of close-coupling equations.

In the past decade, enormous advances have been made in the theoretical modeling of electron–hydrogen scattering. It is the one problem in electron impact scattering theory we have been able to solve essentially without approximation and for which error bounds can be placed on scattering amplitudes over a wide range of energies. The reason for the recent progress has been the rapid development of the computing power available for the computational work. The close-coupling approach utilizes the fact that the complete set of eigenstates of the hydrogen atom (discrete plus continuum) forms a basis for expanding the three-body electron–hydrogen wavefunction. An expansion of the wavefunction using the target states leads naturally to the desired asymptotic states of the system consisting of a plane wave for the incident electron and a hydrogen atom in one of the discrete states. Expanding the Schrödinger equation by means of these target states converts it to an (infinite) system of coupled equations, which is formidable to solve directly because of the infinite sum over the discrete states and the integration over the continuum. However, it is physically reasonable to argue that at least at low incident electron energies, where the excited-state channels are closed, their contribution to the elastic scattering amplitudes should be small and this indeed is found to be the case. Thus early calculations truncated the target space to a few low-lying excited states (hence the name "close-coupling"), as these small-scale models were the only ones that were numerically tractable.

Subsequent studies gradually increased the target space subset by including, first, more target states and then square-integrable functions. The latter were not eigenstates but were chosen to span, efficiently, the remainder of the target space. These functions are now commonly called pseudo-states. A major effect of including these states, which approximate the higher discrete as well as continuum contributions, was to reduce the elastic cross section because of the additional loss of flux to these new channels. A new problem arose with the introduction of the pseudo-states. Although they improved the agreement of theory with experimental data for many energies, the singlet scattering amplitudes developed unphysical resonances above the ionization threshold. These resonances are referred to as pseudo-resonances and their position and number vary with the number of the pseudo-state functions included in the expansions. In this next generation of models and calculations, ansatzes were used to average over the pseudo-resonances in order to produce smooth cross sections for elastic and inelastic scattering.

A parallel development occurred in the study of approximations to the spectrum of the hydrogen atom, which has had a profound influence on shaping the present-day calculations. It is a standard result of mathematics that one can choose a set of Laguerre functions in such a manner as to form a basis for the Hilbert space of square-integrable functions. Yamani and Reinhardt [8.1] demonstrated that the hydrogen continuous spectrum (whose functions are not square integrable) could be expanded in such a basis in analytical form. They were also able to show that when a finite set of these functions was used, the resultant pseudo-states obtained from a diagonalization of the hydrogen-atom Hamiltonian in the subset, were just the Fourier expansions of the continuum restricted to the subspace. For the first time it was possible to understand the nature of the slow convergence of calculations that employ pseudo-states; they are a consequence of the slow convergence of the Fourier-series expansions of continuum states. A further important observation flowing from their analysis is that the completeness sum over the target wavefunctions calculated in a subspace of the Laguerre basis forms a Gaussian quadrature approximation for the completeness relation of the exact discrete plus continuum states. This association of a quadrature rule with the discretization of the spectrum strongly suggests that a proper implementation of target-state expansions should utilize the complete set of approximate target states in order to obtain the full benefits of convergence accruing to the implicit quadrature rule. The present generation of calculations, which incorporate these ideas and investigate convergence by systematically increasing the Laguerre subspace of the target, are referred to as the convergent-close-coupling (CCC) method [8.2].

Rather than presenting the full complexity of the method applied to the full problem, we consider in this chapter the simpler problem, whereby the spectrum of the hydrogen target is restricted to S symmetry target states. Within this manifold the S states form a complete set and hence the ideas of the CCC method can be fully tested. It is also common but not essential to restrict the calculations to the $L = 0$ partial wave of the full three-body scattering wavefunction. This model was first discussed by Temkin [8.3] and later by Poet [8.4] who used the specialized nature of the model to derive a highly accurate numerical solution which has evolved into a standard against which to check alternative methods. In this chapter we present the elements of the model, followed by a description of the Laguerre basis and then the method of solution of the CCC equations. Details of the use of the computer program are outlined through a number of examples.

8.2 Theory

The electron–hydrogen Schrödinger equation, for total energy E of the scattering system, is written as

$$(H - E)\,|\Psi^S\rangle = 0\,, \tag{8.1}$$

where the Hamiltonian operator is

$$H = K_0 + v_0 + K_1 + v_1 + v_{01}\,, \tag{8.2}$$

and the incident and target electrons are labeled 0 and 1, respectively. The proton is assumed to be static and infinitely heavy, so $K_{0,1}$ refer to the electron kinetic energies, $v_{0,1}$ the electron–proton and v_{01} the electron–electron potentials. The three-body wavefunction $\Psi^S(\boldsymbol{r}_0, \boldsymbol{r}_1)$ is symmetric (singlet channel $S = 0$) or antisymmetric (triplet channel $S = 1$) as required by the Pauli exclusion principle. If the space of the target hydrogen atom is restricted to S states and further if only those collisions which result in zero total angular momentum are considered we have the Temkin–Poet model. In coordinate space the Schrödinger equation becomes

$$\left(\frac{1}{2}\frac{\partial^2}{\partial {r_0}^2}+\frac{1}{2}\frac{\partial^2}{\partial {r_1}^2}+\frac{1}{r_0}+\frac{1}{r_1}-\frac{1}{r_>}+E\right)\Psi^S(r_0,r_1)=0, \tag{8.3}$$

where $r_> \equiv \max(r_0, r_1)$. Poet noted that if the domain was restricted to the region $r_0 > r_1$ the equation reduced to

$$\left(\frac{1}{2}\frac{\partial^2}{\partial {r_0}^2}+\frac{1}{2}\frac{\partial^2}{\partial {r_1}^2}+\frac{1}{r_1}+E\right)\Psi^S(r_0,r_1)=0, \qquad r_0 > r_1. \tag{8.4}$$

This is of a separable form and has solutions comprising a plane wave in the coordinate r_0 and a Coulomb function in r_1. Of course the *physical solution* is formed by taking a superposition of these separable solutions which individually do not satisfy the physical boundary conditions. Poet showed that the imposition of these boundary conditions along the boundary $r_0 = r_1$ leads to an integral equation that is much simpler to solve than the original differential equation. The interested reader is referred to the original paper [8.4] for details of his novel approach.

8.2.1 Close-Coupling Equations

The close-coupling method rewrites the Schrödinger equation (8.1) by making an expansion of the three-body wavefunction in terms of a complete set of target states of the hydrogen atom:

$$(K_1 + v_1)|\phi_i\rangle = \epsilon_i|\phi_i\rangle. \tag{8.5}$$

The completeness of the states may be used to form an explicitly symmetrized expansion for the full wavefunction,

$$\Psi^S(\boldsymbol{r}_0,\boldsymbol{r}_1)=\frac{1}{2}\sum\!\!\!\!\!\!\int_i \left(\phi_i(\boldsymbol{r}_1)F_i^S(\boldsymbol{r}_0)+(-1)^S\phi_i(\boldsymbol{r}_0)F_i^S(\boldsymbol{r}_1)\right), \tag{8.6}$$

where the generalized summation indicates a summation over the discrete states and an integration over the continuum ones. By applying this expansion to (8.1) and folding on the left with target states, then using the orthogonality properties, the two-variable equation is replaced by a single variable system of coupled equations:

$$\sum\!\!\!\!\!\!\int_i (K_0\delta_{ij}+V_{ji}^S)F_i^S=(E-\epsilon_j)F_j^S, \tag{8.7}$$

where

$$V_{ji}^S=\langle\phi_j|V^S|\phi_i\rangle=\langle\phi_j|v_0+v_{01}+(-1)^S(H-E)P_r|\phi_i\rangle. \tag{8.8}$$

Here P_r is the space-exchange operator whose effect is to interchange r_0 and r_1. In this form the equations are still intractable because of the generalized summation over states. But they do have the advantage that they lend themselves to approximation schemes which allow them to be solved to high accuracy. One method we favor is to choose a finite subset of a countably infinite set of square-integrable (L^2) functions, which form a basis for the hydrogen-atom Hilbert space.

8.2.2 Target-State Manifold

We choose the functions

$$\xi_i(r) = \left(\frac{\lambda(i-1)!}{(1+i)!}\right)^{1/2} (\lambda r) \exp(-\lambda r/2) L^2_{i-1}(\lambda r), \qquad i = 1, 2, \ldots, \tag{8.9}$$

which are L^2 and for $\lambda > 0$ form a basis for the target states with S symmetry. The $L^2_i(\lambda r)$ are associated Laguerre polynomials of order 2. Now, if we take the first N functions from the set (8.9) and diagonalize the hydrogen-atom Hamiltonian in this subspace we obtain a set of approximate eigenstates. Formally we can describe this process as one in which the Hamiltonian $H_1 = K_1 + v_1$ is replaced by $I^N H_1 I^N$ where I^N is the unit projection operator onto the subspace:

$$I^N = \sum_{i=1}^{N} |\xi_i\rangle\langle\xi_i|. \tag{8.10}$$

The eigenstates are defined by

$$I^N H_1 I^N |\phi^N_n\rangle = \epsilon^N_n |\phi^N_n\rangle, \quad \langle\phi^N_n|\phi^N_{n'}\rangle = \delta_{nn'}, \quad n = 1, 2, \ldots, N. \tag{8.11}$$

The pattern of eigenenergies is such that the lowest lying states are essentially exact with a reasonable choice for λ, whereas higher negative-energy and positive-energy states are a superposition of the true eigenstates, with greatest weighting from the nearby energies. One very useful effect of projecting onto a discrete subspace is to replace the integration over the continuum states by a summation. Indeed, if one pursues the nature of this further one can establish that the induced summation is equivalent to a Gaussian quadrature approximation to the generalized summation over the exact target states.

8.2.3 Ensuring the Uniqueness of Solution

The close-coupling equations (8.7) do not have a unique solution and this may therefore lead to numerical instabilities in their numerical solution. The reason for this is due to the form adopted for the symmetrized expansion in (8.6). It is best illustrated by taking an example from triplet (S=1) scattering.

If F_i^1 is one solution, then so must be $F_i^1 + \alpha_i \phi_i$ for arbitrary constant α_i. The reason that the symmetrization procedure does not ensure uniqueness is because such an expansion is too general. The Pauli principle may be used once more to eliminate the non-uniqueness problem. This matter is discussed in detail elsewhere [8.2]. It may be shown that there are N^2 linearly independent spurious solutions for an N function subspace. Their elimination is deceptively simple; as a consequence of the Pauli principle, one finds that the additional condition

$$\langle \phi_i | F_j^S \rangle = (-1)^S \langle \phi_j | F_i^S \rangle \tag{8.12}$$

is necessary. Once this is implemented the solution may be demonstrated to be unique. The argument is tortuous so we merely state the result here. It is sufficient to replace the potential V^S by

$$V^S(\theta) = v_0 + v_{01} - E\theta I_0 + (-1)^S \big(H - E(1-\theta)\big) P_r. \tag{8.13}$$

Here I_0 is the unit operator for the incident electron and θ is an arbitrary non-zero scalar. Since the solution does not depend on θ, a test of the numerical accuracy of a solution is obtained by repeating the calculation for several values of θ and ascertaining the variation in the numerical solution.

8.2.4 Momentum-Space Convergent-Close-Coupling

From now on, we assume that the target states and energies are those associated with the N-basis approximation defined in (8.11). In order to economize on notation the explicit dependence on N will be omitted from the equations where no confusion arises. A consequence of using the N basis is that the generalized sum in the close-coupling equations (8.7) is replaced by a finite sum. This means that our scattering wavefunction will have an implicit dependence on N. However, as we have already observed, the completeness relation over the subspace target states is a Gaussian quadrature approximation to the completeness relation over the exact target states. Hence one should expect convergence of the scattering wavefunction as N is increased using the Laguerre basis (8.9). This is the convergent-close-coupling (CCC) approach to calculation of electron–atom scattering. For ever increasing N we take the first N functions from the Laguerre basis and solve the resultant close-coupling equations until satisfactory convergence in the quantities of interest is obtained.

All of the information about the scattering processes is contained in scattering amplitudes which may be obtained from the scattering wavefunction $\Psi^S(r_0, r_1)$ matched to outgoing spherical-wave boundary conditions. There is a more direct way we can find the scattering amplitudes through use of a Lippmann–Schwinger equation for the $\mathbf{K}$ matrix and the closely associated $\mathbf{T}$ matrix. The momentum-space matrix elements of these operators,

$$\langle k'n'|K^S(E)|nk\rangle \equiv \langle k'|K^S_{n'n}(E)|k\rangle, \tag{8.14}$$

are directly related to the scattering amplitudes when the momenta k, k' are put on the energy shell in each open channel. We denote these on-shell momenta by k_n, $k_{n'}$. They satisfy the relation $\epsilon_n + \frac{1}{2}{k_n}^2 = \epsilon_{n'} + \frac{1}{2}{k_{n'}}^2 = E$. The **K** matrix operator is defined by

$$K^S(E) = V^S(E) + V^S(E)\mathcal{P}G_0(E)K^S(E), \tag{8.15}$$

where $\mathcal{P}G_0$ is a free Green's function with standing-wave boundary conditions. In momentum space this equation becomes a Fredholm integral equation of the second kind:

$$\begin{aligned}\langle k'n' |K^{SN} \mid nk_n\rangle &= \langle k'n' \mid V^{SN}(\theta) \mid nk_n\rangle \\ &+ \sum_{m=1}^{N} \mathcal{P} \int_0^\infty \mathrm{d}k k^2 \frac{\langle k'n' \mid V^{SN}(\theta) \mid mk\rangle}{E - \epsilon_m - \frac{1}{2}k^2} \langle km \mid K^{SN} \mid nk_n\rangle.\end{aligned} \tag{8.16}$$

Note that we deliberately allow k' to be off the energy shell, as is k. This is necessary when solving this equation for the **K** matrix.

The integrand in (8.16) is singular when the momentum in the integration takes on the on-shell momentum for each open channel, but this singularity can be treated by numerical methods (see below). We note that the equation for **K** uses only real arithmetic and hence provides a considerable practical advantage over the corresponding **T** matrix Lippmann–Schwinger equation, which differs from that of the **K** matrix through the replacement of the standing-wave Green's function by one with outgoing wave boundary conditions. Unfortunately this has the effect of making the integral equation complex and hence complex arithmetic must be used to solve the system numerically, thereby requiring twice as much array storage. The on-shell **T** matrix elements, on the other hand, are obtained from the **K** matrix ones through the Heitler equation:

$$\begin{aligned}\langle k_{n'}n' \mid K^{SN} \mid nk_n\rangle = &\sum_{m=1}^{N_{\text{open}}} \langle k_{n'}n' \mid T^{SN} \mid mk_m\rangle \\ &\times \left(\delta_{mn} + \mathrm{i}\pi k_m \langle k_m m \mid K^{SN} \mid nk_n\rangle\right).\end{aligned} \tag{8.17}$$

The sum in the above equation only extends over the open channels.

8.2.5 Cross Sections

We define the spin-dependent cross section for excitation from an initial state n to a final state n' in atomic units (a_0^2) as

$$\sigma^{SN}_{n'n} = (2S+1)\pi^3 \frac{k_{n'}}{k_n} \left|\langle k'n'|T^{SN}|kn\rangle\right|^2, \tag{8.18}$$

with the spin-averaged cross section being given by the sum of the $S = 0$ and $S = 1$ cross sections. The total cross section for excitation from initial state n is obtained by either summing over individual cross sections or by the use of the optical theorem,

$$\sigma^{SN}_{\mathrm{Tot}\ n} = \sum_{n'=1}^{N_{\mathrm{open}}} \sigma^{SN}_{n'n} = -\pi^2(2S+1)\mathrm{Im}(\langle k'n'|T^{SN}|kn\rangle)/k. \tag{8.19}$$

We define the total ionization cross section by summing that subset of excitations that lead to final states in which both the scattered and excited electron target have positive energies, i.e.,

$$\sigma^{SN}_{\mathrm{Ion}\ n} = \sigma^{SN}_{\mathrm{Tot}\ n} - \sum_{n':\epsilon_{n'}<0}^{N_{\mathrm{open}}} \sigma^{SN}_{n'n} = \sum_{n':\epsilon_{n'}>0}^{N_{\mathrm{open}}} \sigma^{SN}_{n'n}. \tag{8.20}$$

When only a few states are used we may take more care to ensure that there is no contribution to the estimated ionization cross section from the discrete subspace by projecting the total non-breakup cross section ($\epsilon_{n'} < 0$) onto the true discrete subspace, formed using exact hydrogen eigenstates.

It is important to make it clear what we mean by convergence with increasing N. Suppose we require a cross section to a specified accuracy ε. Then we say we have convergence at $N = N_0$ if for $N > N_0$ we have $|\sigma^{SN}_{n'n} - \sigma^{SN_0}_{n'n}| < \varepsilon$. In practice we use the converse procedure, where by presenting a set of results for various N we are able to give an estimate of ε and hence of the accuracy of our calculations. The rate of convergence, the values of N_0 and ε all depend on the transition of interest, total spin, and the incident energy. Typically, the least detailed and largest cross sections are the easiest to obtain accurately. On the other hand, individual **T** matrix elements are the slowest to converge, particularly when small.

8.3 Numerical Solution

The required computation may be split into three parts. First we need to diagonalize the target Hamiltonian using our chosen Laguerre basis. We then need to calculate the fully off-shell **V** matrix elements for the various channel combinations. And finally, we form and solve a set of linear equations yielding half-on-shell **K** matrices. These are then used to form the on-shell **K** matrices, which are in turn used to generate the on-shell **T** matrix elements and cross sections.

8.3.1 Diagonalizing the Target Hamiltonian

We wish to obtain states $\langle r|n\rangle \equiv \phi_n(r)/r$ which are linear combinations of our basis functions $\xi_i(r)$ (see (8.9-11)) and satisfy

$$\langle n|H_1|n'\rangle = \delta_{nn'}\epsilon_n. \tag{8.21}$$

In other words, we wish to find the eigenvalues and eigenvectors of H_1 within the Hilbert subspace spanned by the N orthonormal basis functions. To do this we write

$$0 = \langle \xi_n|\epsilon_{n'} - H_1|n'\rangle = \sum_{m=1}^{N} \left(\delta_{nm}\epsilon_{n'} - \langle \xi_n|H_1|\xi_m\rangle\right)c_{mn'}, \tag{8.22}$$

where

$$|n'\rangle = \sum_{m=1}^{N} c_{mn'}|\xi_m\rangle. \tag{8.23}$$

This forms a standard eigenvalue problem that we solve using the LINPACK routine RS. The coefficients $c_{mn'}$ depend on the subspace dimension N. The major difficulty in this process is calculating the matrix elements of the target Hamiltonian H_1 using the specified basis. We leave it to the interested reader as an exercise to evaluate analytical expressions for these matrix elements.

8.3.2 Calculating V Matrix Elements

We wish to calculate matrix elements of the form

$$\begin{aligned}
&\langle k'n'|\frac{1}{r_>} - \frac{1}{r_0} - EI_0^N\theta + (-1)^S(H_0 + H_1 + \frac{1}{r_>} - E(1-\theta))P_r|nk\rangle \\
&=\langle n'k'|\frac{1}{r_>} - \frac{1}{r_0} + (-1)^S\frac{1}{r_>}P_r|nk\rangle - \delta_{n'n}E\theta\sum_{m=1}^{N}\langle k'|m\rangle\langle m|k\rangle \\
&\quad + (-1)^S\Bigg[\left(\frac{k^2}{2} + \frac{k'^2}{2} - E(1-\theta)\right)\langle k'|n\rangle\langle n'|k\rangle \\
&\quad - \langle k'|\frac{1}{r}|n\rangle\langle n'|k\rangle - \langle k'|n\rangle\langle n'|\frac{1}{r}|k\rangle\Bigg].
\end{aligned} \tag{8.24}$$

Only the first part of (8.24) is not trivial, so we will write it out explicitly:

$$\langle k'n'|\frac{1}{r_>}-\frac{1}{r_0}|nk\rangle = \frac{2}{\pi kk'}\int_0^\infty \mathrm{d}r_0 \sin(k'r_0)\sin(kr_0)$$
$$\times \left[\left(\int_0^{r_0}\frac{\mathrm{d}r_1}{r_0}+\int_{r_0}^\infty\frac{\mathrm{d}r_1}{r_1}\right)\phi_{n'}(r_1)\phi_n(r_1)-\frac{\delta_{nn'}}{r_0}\right] \quad (8.25)$$

and

$$\langle k'n'|\frac{1}{r_>}P_r|nk\rangle = \frac{2}{\pi kk'}\int_0^\infty \mathrm{d}r_0 \sin(k'r_0)\phi_n(r_0)$$
$$\times \left(\int_0^{r_0}\frac{\mathrm{d}r_1}{r_0}+\int_{r_0}^\infty\frac{\mathrm{d}r_1}{r_1}\right)\phi_{n'}(r_1)\sin(kr_1). \quad (8.26)$$

Note that square integrability of our states $\phi_n(r)$ ensures the existence of all of the integrals. We evaluate the **V** matrix elements for all channel (n, n') combinations and various values of k and k'.

Another important point to consider is the consequence of choosing a non-zero θ in (8.24). Such a choice introduces the $\langle k'|I^N|k\rangle$ term which tends to $\delta(k-k')$ with increasing N. Therefore, the integration over k' becomes increasingly more difficult (more quadrature points are necessary) as N is increased. This is unfortunate, as the size of the calculations grows more rapidly than if just N is increased. Fortunately in the case of the Temkin–Poet model, the on-shell results remain stable even for the $\theta = 0$ case. These issues will be discussed in more detail later on.

8.3.3 Solution of the Coupled Equations

We now discuss the numerical techniques used in solving the coupled integral equations. Here we restrict ourselves to the simple Temkin–Poet model of electron–hydrogen scattering but the method of solution is general. For example, the same techniques may be used for calculating scattering of electrons from general multi-electron targets. To date they have been applied to one-electron atoms and ions [8.5], as well as helium [8.6]. The only differences are that the size of the matrices is larger due to the inclusion of target states of general angular momenta and that we have to solve the coupled equations for many partial waves.

Another important point is that the momentum-space method of solution of the coupled equations will yield very similar results to the intermediate-energy **R** matrix (IERM) method [8.7,8.8] or variational approaches [8.9] involving coupled integro-differential equations, assuming a similar level of sophistication in describing the target states. Because convergent amplitudes are obtained only by solving large sets of coupled equations, a key issue is that of solving them with maximum efficiency. The momentum-space method discussed here can be implemented on desk-top workstations, making it accessible to a large audience. This quest for efficiency has been paramount in our design of the codes.

8.3.3.1 Formation of the Linear Equations

Our first step is to discretize the integral (8.16) by taking a quadrature rule for the integration from zero to infinity. Choosing an efficient quadrature grid is very important as this leads to smaller matrices. However, this process is complicated by the fact that we have a principal-value-type singularity whose position varies with the channel energy ϵ_n. Only open channels have the singularity. For closed channels the denominator remains negative for all values of k.

We deal with this problem by breaking the integration range into a number of parts with the boundaries being chosen dependent on the position of the singularity, if any. Generally, we have N_{k_1} Gaussian points in the interval $(0,k_1)$, N_{k_2} points in the interval (k_1, k_2), and N_{k_3} points in the interval (k_2, ∞). For those channels that contain the singularity we take an additional even number (N_{k_4}) of Gaussian points in an interval of width w positioned symmetrically about the singularity. This is an efficient way of dealing with this type of singularity. In order to complete the quadrature rule to infinity we suppose that the integrand falls off as k^{-p} for $k > k_2$.

Upon the replacement of the integral in channel m by a quadrature rule using N_m points with weights w_{jm} $(j = 1, \ldots, N_m)$, we may write (8.16) as

$$\begin{aligned}\langle k'n'|K^{SN}|nk\rangle &= \langle k'n'|V^{SN}(\theta)|nk\rangle \\ &+ \sum_{m=1}^{N}\sum_{j=1}^{N_m} \frac{w_{jm}k_{jm}^2\langle k'n'|V^{SN}(\theta)|mk_{jm}\rangle\langle k_{jm}m|K^{SN}|nk\rangle}{E-\varepsilon_m-k_{jm}^2/2},\end{aligned} \tag{8.27}$$

or in matrix notation as

$$\begin{aligned}K^{SN}(k_{j'n'}, n) =& V^{SN}(k_{j'n'}, n, \theta) \\ &+ \sum_{jm} W_{jm} V^{SN}(k_{j'n'}, k_{jm}, \theta) K^{SN}(k_{jm}, n),\end{aligned} \tag{8.28}$$

where we included the energy denominator in the new generalized weights W_{jm}. In order to solve (8.28) we allow the $k_{j'n'}$ to run over the same range as the k_{jm} and form a set of linear equations of the form $AX = B$, where the square matrix $\mathbf{A}$ is given by $A = I - WV^{SN}(\theta)$, and $B = V^{SN}(n, \theta)$ and $X = K^{SN}(n)$ are vectors. I is the identity matrix and W is the diagonal matrix with entries W_{jm}. For compactness of notation, we omitted the entrance channel momentum since it is fixed to the on-shell value.

Now let us consider the optimal choice of quadrature rule in (8.27) above. We mentioned that the introduction of non-zero θ in the $\mathbf{V}$ matrix elements leads to a δ-function-like term. The quadrature rule must be able to integrate accurately over such terms. In particular, we need to make sure that we are able to integrate with respect to k' over terms like $\sum_m \langle m|k'\rangle$. The quality of the channel-dependent quadrature rule may be tested independently of doing the full calculation by checking that $\int \mathrm{d}k' \langle m|k'\rangle\langle k'|m\rangle = 1$ to sufficient precision for each m.

8.3.3.2 Solution of Linear Equations

It is the maximum size of the matrix **A**, for a given set of computational resources, that determines the number of coupled equations we may solve. The matrix **A** must fit into the core memory as the time required to solve the linear equations of a $n \times n$ system is proportional to n^3, and so takes a considerable part of the computational time. To minimize storage requirements we take advantage of the fact that $V^{SN}(k_{j'n'}, k_{jn}, \theta)$ is symmetric upon interchange of $j'n'$ and jn and that for electron-impact excitation we may have at most two possible total spins S (S=0,1 in the case considered here). As a result we are able to use the same matrix, of dimension $n \times (n+1)$, for storing the **V** matrix elements for both spins. The extra dimension is so that there is room for both the diagonals of the S=0,1 cases.

Having stored the **V** matrix elements efficiently, we rewrite the linear equation $AX = B$ in such a way so that the matrix **A** is symmetric. This is achieved by writing (8.28) as

$$(IW^{-1} - V^{SN}(\theta))WK^{SN}(n) = V^{SN}(n, \theta). \tag{8.29}$$

Thus we are able to form **A** using the same array that is used for storage of the **V** matrix elements. Using the `LAPACK` routine `SSYSV`, we may solve the linear equations separately for the two spin channels using only the space provided for the matrix **A**. After solution of the linear equations for each total spin, the corresponding **V** matrix elements are destroyed, since they are no longer necessary. Note that we do need to save the half-on-shell **V** matrix elements prior to the solution of the linear equations, which take very little space.

In summary, we are able to solve the coupled equations using primarily only the space necessary for storing the fully-off-shell **V** matrix elements for the two spin cases. This efficient usage of memory has enabled the CCC method to solve large sets of close-coupling equations using relatively inexpensive desk-top computational resources (see, for example, Bray [8.10]).

8.4 Computer Program

The program associated with this chapter is a much reduced version of the CCC program used to calculate electron or positron scattering on hydrogen-like and helium-like atoms and ions. For the sake of simplicity and compactness, the reduced version applies only to the Temkin–Poet model. In other words, the target states are the S states of hydrogen, and only the $\ell = 0$ partial-wave of the projectile is treated.

The file `README` contains the information on the associated source codes and on how to make the run-file `ccc`.

8.4.1 Input

There is a single input file `ccc.in` with four lines containing the following input parameters:

1. `ENERGY, NZE, NATOP, NPS, ALPHA`

`ENERGY` Incident energy (eV) on the ground state. It may range from approximately 0.1 eV to 1000 eV.

`NZE` -1 for electron scattering
1 for positron scattering.

`NATOP` Normally the number of states to be coupled. For convenience it may be used as a switch to make the size of the Laguerre basis determine the number of states to be coupled. A value of 0 means all `NPS` states will be coupled. A value of $-n$ means that all open plus $-n+1$ closed states will be coupled.

`NPS` $= N$, the size of the Laguerre basis that generates `NPS` states.

`ALPHA` Exponential fall-off factor of the Laguerre basis, $\alpha = \lambda/2$.

2. `NUNIT, NNBTOP, NENT, IFIRST, NOLD, THETA`

`NUNIT` 1 for the cross section units to be in a_0^2;
2 for πa_0^2;
3 for cm^2.

`NNBTOP` Used to form overlaps between the first `NNBTOP` exact hydrogen discrete S states and those obtained by diagonalization.

`NENT` Number of entrance channels.

`IFIRST` 1 if direct and exchange is to be calculated;
0 for the direct case only.

`NOLD` 1 for the case of using the Laguerre basis states;
0 for using the exact discrete states.

`THETA` Parameter that, when non-zero, ensures uniqueness of the solution of the Lippmann–Schwinger equation.

3. `NQM, QCUT, RMAX, SLOWE`

`NQM` -1 for using the full set of k integration points;
1 for a unitarized Born approximation.

`QCUT` Determines the spacing in the coordinate space mesh. For $q <$ `QCUT`, $\sin(qr)$ will be reliably integrated.
`RMAX` Largest value in the coordinate mesh.
`SLOWE` If non-zero, then `ALPHA` is iterated until one of the pseudo-states has `SLOWE` as the energy.

4. `(NK(I),SK(I),I=1,4)`

These are the k-grid parameters that are used to distribute the k integration points in each channel.
`NK(1)` Number of points in the interval (0,`SK(1)`);
`NK(2)` Number of points in the interval (`SK(1)`,`SK(2)`);
`NK(3)` Number of points in the interval (`SK(2)`,∞);
`SK(3)` Power p used to define the distribution of points in the last interval.
`NK(4)` Number of extra points distributed about the on-shell point, if any, and lying inside one of the previous intervals. This number must be even.
`SK(4)` Half the width w of the interval containing the singularity, if any. See the output for the actual distribution of points.

In addition to the above input parameters, there are a set of parameters in the file `par.f` which control the sizes of available arrays. Informative messages are returned if the array sizes are exceeded. The major requirements of memory are the array `VMAT` in which the **V** matrix is stored, and the array `CHIL` in which the plane waves ($\sin(kr)$) are stored. The array `VMAT` requires space for (`KMAX * NCHAN`)^2 real numbers, where `KMAX` is the maximum number of k-grid points in any channel, and `NCHAN` is the maximum number of channels (or states). The array `CHIL` requires space for `KMAX * NCHAN * MAXR` real numbers, where `MAXR` is the maximum number of r-integration points.

8.4.2 Output

There is standard output which monitors the progress of the program as well as the creation of the files **totalcs**, **singlet.**n**1**, and **triplet.**n**1**. The file **totalcs** presents the total cross sections. The **singlet.**n**1** and **triplet.**n**1** files contain the half-on-shell **K** and **V** matrix elements for the excitation of the initial channel to channel n. These are useful for making plots to show uniqueness, or lack thereof, in the solution of the Lippmann–Schwinger equations.

8.4.2.1 Standard Output

The standard output displays the set of states used in the multi-channel expansion. All of the states arising out of the diagonalization of the target Hamiltonian are listed. Those to be included in the multi-channel expansion have a "+" in the final column. The negative energy states are projected onto the true discrete subspace, using the earlier calculated hydrogen bound states. This projection is used when forming ionization cross sections. The results, given in the `totalcs` file, differ significantly from (8.20) only when very few states are used.

After presenting the target state information, the distribution of the k integration points is given in each channel m. A "$*$" appearing in an interval indicates that the singularity is in this interval. The tests consist of checking that $\int_0^\infty \mathrm{d}k k^2 |\langle m|k\rangle\langle k|m\rangle| = 1$. Each channel-dependent integration grid is tested with all $m = 1, \ldots, N$. The variable `NBAD` indicates how many tests did not yield unity to better than three significant figures. This is only a rough indication of the quality of the k quadratures. In the final analysis, a number of calculations must be run with varying distribution and number of k points to ensure accuracy of the final results.

After the k-grids have been set up and displayed, the $\mathbf{V}$ matrix elements are calculated, followed by the formation and solution of the coupled Lippmann–Schwinger equations. The final on-shell $\mathbf{T}$, $\mathbf{S}$, $\mathbf{K}$, and $\mathbf{V}$ matrix elements are then printed out. Eigenphases and their sums are also given.

8.4.2.2 The file `totalcs`

The `totalcs` file contains the integrated cross sections. The following abbreviations are used. `TNBCS` is the total non-breakup cross section and is evaluated by summing the integrated cross sections for negative energy states. `TICS` is the total ionization cross section which is initially evaluated by simply summing the cross sections for positive energy states. Finally, it is estimated by subtracting from the total cross section (`TCS`) the cross sections from states with negative energies multiplied by the corresponding projection. The spin asymmetry is also given.

8.4.2.3 The files `singlet.`*n*`1` and `triplet.`*n*`1`

These are plot files containing the half-off-shell $K(k)$ and $V(k,\theta)$ matrix elements for the $1 \to n$ transition. These are the first two vectors in (8.28). Suppose we have two calculations which only vary in their k-grid parameters. Then we would expect that the two corresponding half off-shell $K(k)$ and $K(k')$, and $V(k,\theta)$ and $V(k',\theta)$ vectors should interpolate smoothly. Furthermore, one would expect that a particular $V(k,\theta)$ generates a unique $K(k)$, whereas another $V(k,\theta')$ yields $K'(k)$. For the $\theta = 0$ case, we find that two different k-grids generate distinct $K(k)$ and $K'(k')$, but $K(k_f) = K'(k_f)$ if k_f is the on-shell point. However, in the case of $\theta \neq 0$ two distinct $V(k,\theta)$ and $V(k,\theta')$ yield identical vectors $K(k)$ and $K'(k)$.

These ideas may be illustrated by taking the static-exchange model of electron–hydrogen scattering. In this model, only the single physical $(1\mathrm{s})^2\mathrm{S}$ state is included in the close-coupling expansion. The question of uniqueness may then be studied in the triplet channel only. Looking at the $\mathbf{V}$ matrix elements (8.24) reveals that the θ term disappears for $S = 0$, but remains for

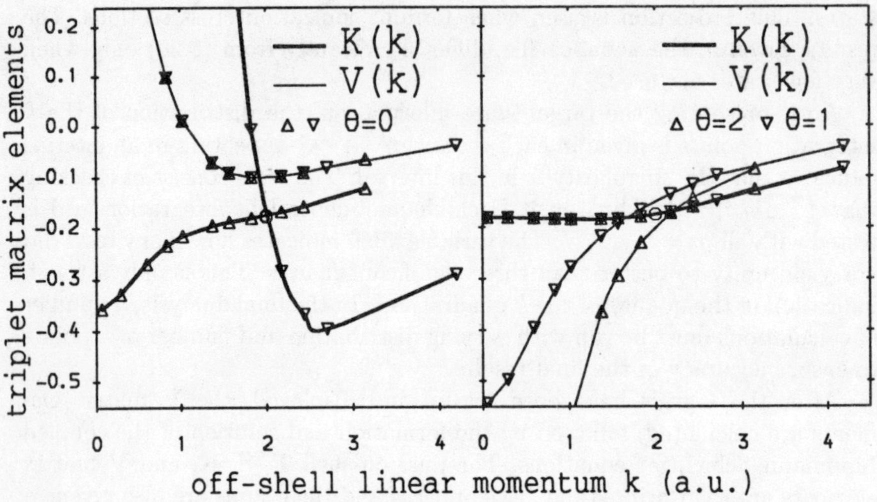

Fig. 8.1. Half-off-shell triplet **V** and **K** matrix elements arising in the static-exchange model of e–H scattering are shown for the "old" CC equations ($\theta = 0$), and the "new" form with $\theta = 1, 2$. The two integration k-grids (upright and inverted triangles) differ only in the last few points. The incident energy of 54.42 eV corresponds to the on-shell value of $k = 2$ atomic units (denoted by ○). The lines connecting the points are provided as a visual guide.

the triplet ($S = 1$) channel. Figure 8.1, from [8.11], gives this example. In the left-hand picture, we have two k-grids that differ only in their last two points. Yet the difference between the two resulting half-off-shell **K** matrix elements is enormous. However, the two curves do intersect at the on-shell point. In the right-hand picture, we keep the same k-grids but perform calculations for different non-zero values of θ. We see that though the **V** vectors are quite different they yield the same **K** vectors. This demonstrates how any non-zero θ ensures uniqueness of the solution of our coupled equations.

In practice, for large-scale calculations, the use of a non-zero θ is crucial for obtaining numerically stable results. In the case of the simple model considered here, we find that both cases lead to reliable on-shell results. For reasons outlined above, the $\theta = 0$ case often requires fewer k points to solve for the on-shell matrix elements of the Temkin–Poet model – considerably fewer in the case of a relatively large basis size N (20 or so).

8.4.2.4 Sample Input and Output

A sample input file `book.in`, given together with the source code, has the following input parameters:

```
16.0,-1,0,5,0.5    ENERGY, NZE, NATOP, NPS, ALPHA
2,5,1,1,1,0.5      NUNIT, NNBTOP, NENT, IFIRST, NOLD, THETA
```

```
-1,8.0,100.0,2.0  NQM, QCUT, RMAX, SLOWE
14,0.9,22,3.0,4,4.0,10,0.2 (NK(I),SK(I),I=1,4) K grid params
```

which yields, in part, the following output:

```
JS f i  real(T)    imag(T)     real(S)   imag(S)
0+ 1 1-1.213E-01-1.061E-01  2.770E-01 8.266E-01
0+ 2 1-7.162E-02 2.502E-02  1.323E-01 3.786E-01
0+ 3 1-4.349E-02 1.976E-02  9.364E-02 2.060E-01
0+ 4 1 5.527E-02 2.798E-02  7.556E-02-1.493E-01
N=  240 eigenphase sum: -1.401
0- 1 1-7.067E-02-2.749E-01 -8.733E-01 4.815E-01
0- 2 1-8.361E-03-1.035E-02 -5.472E-02 4.419E-02
0- 3 1-2.077E-04-4.301E-03 -2.038E-02 9.839E-04
0- 4 1-1.423E-03 2.710E-03  7.317E-03 3.844E-03
N=  240 eigenphase sum:  1.856
```

Here the singlet case is denoted by + and the triplet case by −. The S wave is denoted by $J = 0$.

8.4.3 Typical Usage

Typically the program is used to find the singlet and triplet 1s→ ns cross sections for the Temkin–Poet model at a particular energy. The rate of convergence with basis size N depends on the energy E, spin S, and principal quantum number n. If we are interested in the case $n \leq 2$ we may start with $N = 10$ and $\alpha = 0.5$. This yields a good representation of both the discrete and the continuum subspaces.

A more general usage of the program is to run it several times over a range of energies to see if the resultant cross sections are smooth or if pseudo-resonance behavior is evident. Pseudo-resonances are typically associated with pseudo-thresholds. To make this study particularly convenient, therefore, we have incorporated an iterative process which varies `ALPHA` to ensure that one of the target-state energies comes out to be `SLOWE`. This results in a fixed pseudo-threshold, irrespective of N, thereby allowing for a detailed study of cross sections in its vicinity. Such a study is shown in Fig. 8.2, also presented in [8.11]. Here we have set `SLOWE` = 2 eV for all three values of $N = 5, 10, 30$, resulting in a common pseudo-threshold at 15.6 eV. In the energy range considered, there is only one pseudo-threshold for the $N = 5$ case, three in the case of $N = 10$, and fourteen in the case of $N = 30$. The energy of the projectile determines which of the pseudo-states may be genuinely excited. The pseudo-thresholds below the projectile energy generate open channels, whereas the other channels are closed. In the case of $N = 5$ and $N = 10$, we see that associated with each pseudo-threshold is a pseudo-resonance in the cross section, which substantially diminishes in magnitude

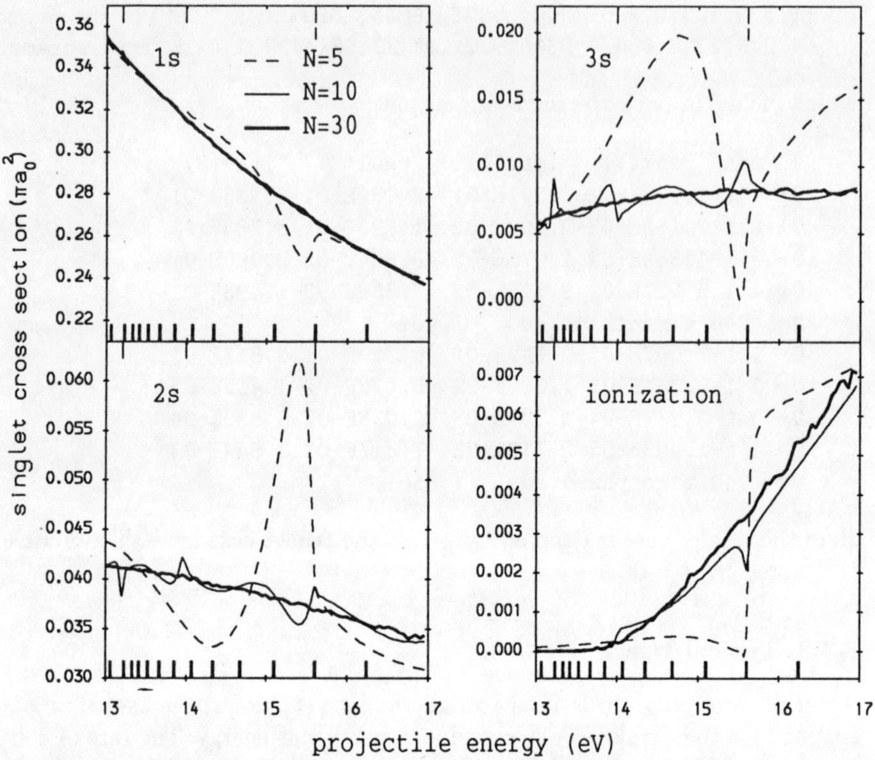

Fig. 8.2. The 1s, 2s, 3s, and ionization singlet cross sections in the Temkin–Poet model are shown in a small energy range for three basis sizes. The large tick marks on the horizontal axes indicate energy thresholds corresponding to the indicated basis size. For a given projectile energy E, states corresponding to thresholds less than E are open, those greater than E are closed. The exponential fall-off factor λ has been varied so that all three basis sizes have a state with energy 2 eV, leading to a common pseudo-threshold at 15.6 eV.

and width with increasing N. The significance of the above results is that pseudo-resonances are a manifestation of an insufficiently large N, though accurate answers may still be obtained if a little care is exercised when using small N to ensure that there are no nearby pseudo-thresholds.

8.5 Summary

We have given a brief outline of the simplified model of electron–hydrogen scattering that treats only states with zero orbital angular momentum. The strength of this model is that treatment of exchange and the continuum must be accurate in order to obtain reliable results. One of the very first applications of the CCC method was to this model. We were able to demonstrate conclusively that the continuum may be treated systematically by the use of square-integrable states. The program, associated with this chapter, allows the reader to verify this for themselves. The program presented here is limited to this restricted model. However, all of the successes of the CCC method in full scattering problems, including ionization, are due to the success of the method for this simple model. Applications of the method to calculations of realistic problems are discussed in several of the references.

8.6 Suggested Problems

The program is a substantially reduced version of the general code for one-electron targets, for the sake of simplicity. In order to make it applicable to real problems a number of extensions are necessary. The following list provides a series of suggestions for extending the code to the general case. The numbered items are graded in order of increasing complexity.

1. Copy the file `book.in` to `ccc.in`, run the `ccc` program, and check the output against the file `book.out`. Note that the input `ALPHA` has been iterated to obtain a pseudo-state with energy 2 eV. Check that the resultant `totalcs` file contains the 16 eV singlet cross sections given in Fig. 8.2.
2. Reproduce the plotted data for `THETA` $= 0$ of Fig. 8.1 by using, in turn, the two given input examples `kgrid1.in` and `kgrid2.in`. The required data are created in the corresponding `triplet.11` files. Edit the input files to vary `THETA` and reproduce data presented on the right side of Fig. 8.1. Choose other values of `THETA` and check for stability of the **K** matrix elements and variation of the **V** matrix elements.
3. We have discussed convergence of the on-shell **K** matrix elements with increasing basis size N. Check for convergence of the half-off-shell **K** matrix elements. This is discussed in [8.11]. Begin with the `book.in` file and increase `NPS`. For large `NPS` you may need to take more care with the k-grids and/or change `THETA` to zero.
4. The unitarized Born approximation (UBA) for the **T** matrix is often used as a high-energy approximation in electron scattering. It is obtained from the Heitler equation (8.17) by replacing the **K** matrix terms by the on-shell potentials. Work out an explicit form for the UBA **T** matrix.
5. If one iterates the **K** matrix equation (8.15) one obtains the series $K = V + VG_0V + VG_0VG_0V + \dots$ Show that the second-order approximation to the **K** matrix requires only on-shell and half-on-shell potentials. You should find it relatively easy to develop an algorithm from (8.28) to do this numerically.
6. What new feature is needed to extend the second-Born calculation to arbitrary order? If you calculate the higher-order approximations you are not guaranteed to converge to the same answer you get by solving

the set of linear equations (8.29). You may find that convergence is absent even in the high-energy limit. The interested reader is referred to [8.12] for further details.

7. The solution of the Lippmann–Schwinger equation discussed here assumed a plane-wave formalism. A distorted-wave formalism may be used instead. Derive a distorted-wave version of the close-coupling equations, see [8.5]. In this case the driving term of the linear equations becomes more dominant, thereby reducing the difficulty of solution of the coupled equations.
8. The implementation of the distorted-wave formalism also allows for extending the program code to ionic targets. Note that a major change is required in the construction of the channel Green's functions in the kernels. Can you explain why? See [8.5] for a full discussion of this point.
9. For incorporation of target states with arbitrary angular momentum the Laguerre function set (8.9) must be be modified. What would be a suitable modification? See [8.2] for details. This generality has been left in the existing code; see routine `MAKEPS`.
10. Calculation of the higher ℓ projectile angular momenta must be added to describe the collision generally. Assuming LS coupling holds, write down the form of the general partial-wave three-body wavefunction expanded in terms of a complete set of target states coupled to general projectile states.
11. Incorporation of non-zero angular momentum states leads to more complicated $\mathbf{V}$ matrix elements. Using your answer to problem 10, derive an expression for the general momentum-space matrix element. For the case of hydrogenic targets this may be found in [8.5], for the case of helium in [8.6].

References

8.1 H.A. Yamani and W.P. Reinhardt, Phys. Rev. A **11** (1975) 1144
8.2 I. Bray and A.T. Stelbovics, Phys. Rev. A **46** (1992) 6995
8.3 A. Temkin, Phys. Rev. A **126** (1962) 130
8.4 R. Poet, J. Phys. B **11** (1978) 3081
8.5 I. Bray and A.T. Stelbovics, Adv. At. Mol. Opt. Phys. **35** (1995) 209
8.6 D.V. Fursa and I. Bray, Phys. Rev. A **502** (1995) 1279
8.7 M.P. Scott, T.T. Scholz, H.R.J. Walters and P.G. Burke, J. Phys. B **22** (1989) 3055
8.8 T.T. Scholz, H.R.J. Walters, P.G. Burke and M.P. Scott, J. Phys. B **24** (1991) 2097
8.9 Y.D. Wang and J. Callaway, Phys. Rev. A **50** (1994) 2327
8.10 I. Bray, Phys. Rev. A **49** (1994) 1066
8.11 I. Bray and A.T. Stelbovics, Comp. Phys. Comm. A **85** (1995) 1
8.12 A.T. Stelbovics, Phys. Rev. A **41** (1990) 2536

9. The Calculation of Spherical Bessel and Coulomb Functions

A.R. Barnett

Department of Physics and Astronomy, University of Manchester, Manchester, M13 9PL, England[1]

Abstract

An account is given of the Steed algorithm for calculating Coulomb functions and, as a special case, both spherical Bessel and Riccati–Bessel functions. These functions are needed for boundary-condition matching in scattering problems in atomic and nuclear physics. Central to the technique is the evaluation of continued fractions and for this calculation the forward method of Lentz (modified by Thompson) is recommended. The FORTRAN programs `SBESJY`, `RICBES`, and `COUL90` are presented and described, and some test cases are given. In each program the algorithm returns both the regular and irregular functions as well as their derivatives. The programs are written for real arguments and real orders; references guide the reader to more general codes.

9.1 Introduction

Coulomb wave functions arise in many problems of physical interest when charged particles scatter from each other. Such scattering is characterized by a relative angular momentum $L\hbar$ (with L a non-negative integer) and by a Sommerfeld parameter $\eta = Z\alpha/\beta$ which gives the strength of the Coulomb interaction. The product of the particle charges is Ze^2, the fine-structure constant is $\alpha = e^2/\hbar c(4\pi\epsilon_0)$, and the relative velocity of the particles is βc. With charges of opposite sign, as is frequently the case in atomic physics, η is negative; whereas in nuclear physics problems generally the charges have the same sign and η is positive. Positron scattering from a nucleus will also have $\eta > 0$. For the scattering of a neutral particle (for example, a neutron) or by a neutral target the parameter η is zero, resulting in Riccati–Bessel functions in which only the angular momentum effects appear. These functions differ

[1] Work completed at Physics Department, University of Auckland, Auckland, New Zealand

in a minor way from spherical Bessel functions and both can be properly regarded as special cases of the Coulomb functions.

This chapter is mostly concerned with the atomic physics area and thus primarily with the cases where η is not positive. In atomic units the relative energy E, in Rydberg units, is given by $E = E(\mathrm{eV})/13.605\ (\mathrm{eV})$. Then $E = k^2$ and $\eta = -Z/k$. Typical energies are $0.5-200$ eV, with corresponding k values from $0.2\ a_0^{-1}$ to $3.8\ a_0^{-1}$, and η values between -5.2 and -0.26. For matching radii of $r = 10\ a_0$ to $r = 200\ a_0$ the dimensionless variable $x = kr$ lies in the range $1-1000$. The angular momentum quantum number L will vary from 0 to perhaps 50, whereas Bessel functions of orders up to 150 may be required. As an example, the parameters for the calculations in Chap. 4 of this book with energy 54.4 eV are $k = 2.00$, $\eta = -0.50$ (0.50) for electron (positron) scattering from hydrogen ions, and $\eta = 0$ for either particle scattering from neutral hydrogen.

The second-order differential equation satisfied by the Coulomb functions is

$$w''(x) + \left[1 - \frac{2\eta}{x} - \frac{L(L+1)}{x^2}\right] w(x) = 0 \qquad (9.1)$$

(where the primes indicate differentiation with respect to x) and it has two linearly independent solutions. They are chosen to be the regular solution $F_L(\eta, x)$ which is zero at $x = 0$, and the irregular solution $G_L(\eta, x)$ which is infinite at $x = 0$. See also references [9.1–3,9,10] for alternative forms of (9.1).

The 'turning point' for the Lth partial wave occurs at

$$x_L = \eta + (\eta^2 + L^2 + L)^{1/2}, \qquad (9.2)$$

which is where the bracket [...] is zero in (9.1); it is also the first point of inflexion for the functions. For negative η, say $\eta = -h$, the $L = 0$ turning point is at the origin, $x_0 = 0$, and for other L values x_L is always less than L; it lies between $L - h$ (for small h) and $\frac{1}{2}L(L/h)$ (for small L). It transpires that the computational task is easier when $x > x_L$, so that functions with negative η are obtained more straightforwardly than those with positive η.

For values of x which are greater than x_L the functions take on an oscillatory character, although the 'period' slowly changes. For $\eta > 0$ the regular-function magnitude is greater than unity, and it slowly decreases towards unity as x grows larger. When $\eta < 0$ the magnitude of $F_L(\eta, x)$ is less than unity and it increases steadily for larger x. Examples of the functions are shown in Fig. 9.1. In the asymptotic region as $x \to \infty$ they become circular with $F \to \sin\theta_L$ and $G \to \cos\theta_L$, where θ_L is the asymptotic phase given by

$$\theta_L = x - \eta \log(2x) - \tfrac{1}{2}L\pi + \arg\Gamma(L + 1 + \mathrm{i}\eta), \qquad (9.3)$$

which for $\eta = 0$ becomes linear in x: $\theta_L = x - \frac{1}{2}L\pi$. An additional phase of $\frac{1}{2}\pi$ will be required for spherical Bessel functions because of a different definition, which is given below.

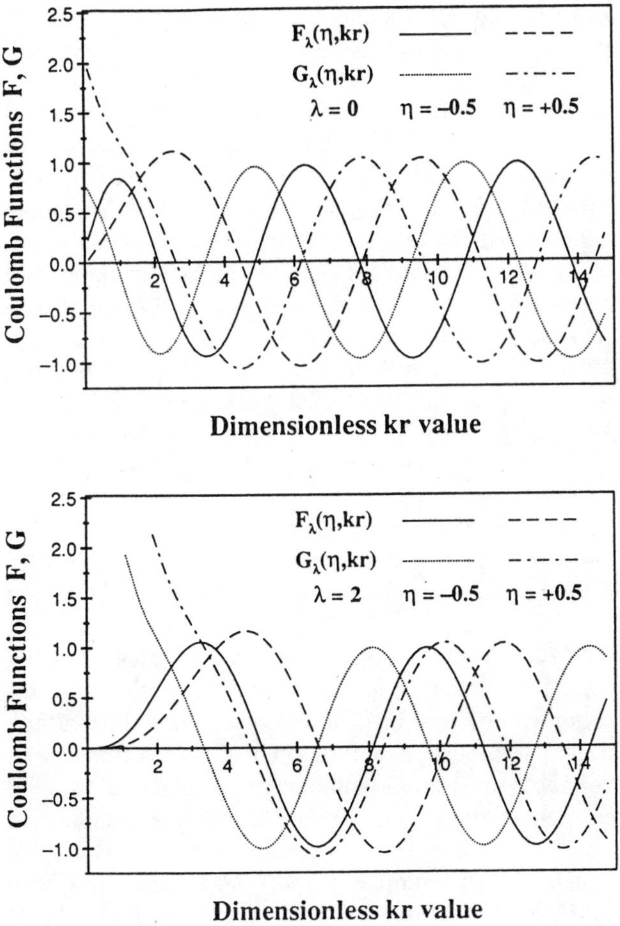

Fig. 9.1. Examples of Coulomb functions

This chapter deals with solutions to the Coulomb scattering problem in which the particles have a positive relative energy. Curtis [9.1] published a detailed discussion for $\eta < 0$ in 1964 for $L = 0, 1, 2$ for functions closely related to F and G for use in electron scattering calculations. More recent work for electron scattering is that of Seaton [9.2] and of Bell and Scott [9.3]: Seaton dealing with positive, negative and zero energies; Bell and Scott treating negative energies. The work of Curtis for positive energy was verified numerically by Barnett [9.4] in 1974. Bardin *et al.* [9.5] published a comprehensive suite of programs in 1972 containing codes for all real η and x. All of these methods involve some appropriate series expansion, either in powers of x, or asymptotically in powers of x^{-1}. Both F and G are calculated separately, as are their x derivatives. A new approach to the calculation of Coulomb functions was developed by Steed in the late 1960s and was published by Barnett *et al.* [9.6] in 1974. The functions and their derivatives are calculated *together* in an interdependent way, and two continued fractions are used (see Sect. 9.6). Virtually all previous work, especially that of Bardin *et al.* [9.5], is verified and in some cases extended in this chapter. A detailed description of this algorithm and a careful comparison with other approaches was given by Barnett [9.7] in 1982. The original program [9.6] `RCWFN` was superseded by a more comprehensive version [9.4] called `COULFG`, which includes the calculation of both Bessel and spherical Bessel functions (of the first and second kind, i.e., $j_n(x), y_n(x)$) as well as the Coulomb functions. The version presented here in Sect. 9.7, `COUL90`, has been further improved in one important respect which, however, does not affect the algorithm, the range of the calculations or the results already given in [9.4] and the earlier work. Comments on the methods of `COULFG` have been made by Nesbet [9.8] with suggestions for improvements for Coulomb function calculations when $x < x_{L=0}$, and especially for spherical Bessel functions.

Solutions for negative energy (including the bound states) are given by Whittaker functions and, for both $\eta < 0$ and $\eta > 0$, can be obtained by using the program `COULN` of Noble and Thompson [9.9] which is well suited to electron scattering.

The algorithm of `COULFG` is not restricted to integer L values [9.4,10], and it can hence be used for scattering solutions to relativistic problems for which the Klein–Gordon equation or the Dirac equation are appropriate. Here an equivalent non-integer L is required; for small L values this can be imaginary. Extensions of the concepts underlying the calculations to *complex arguments* are presented in the program `COULCC`, which was published in 1985 by Thompson and Barnett [9.11]. The range of each of the three variables has been extended into the complex plane. The program has been incorporated into the IMSL SFUN library. A description is given in [9.12] of the various approaches required in the different parameter regions, with references to earlier and more restricted work. The code `COULCC` also computes Bessel functions with complex arguments and order; however, for real order, `BESSCC` for modified Bessel functions is a more efficient code [9.11] which incorporates similar principles. Further specialization is found in the *Numerical Recipes* [9.14] routines `BESSJY` and `BESSIK` which provide J_ν, Y_ν, and I_ν, K_ν for real arguments and order, using a restricted version of `BESSCC`.

9.2 Spherical Bessel Functions

Spherical Bessel functions, and their close relatives the Riccati–Bessel functions, are required frequently in atomic physics calculations and in many other branches of physics; an example is the treatment of Mie scattering in optics. In Chap. 4 of this book they emerge as the asymptotic wavefunctions for the problem of charged particles scattering from neutral atoms or molecules. Similarly, they are required for the description of the scattering of neutrons from nuclei. Despite their importance, however, they tend to be the poor relations of the numerical analysis world. They are covered in the encyclopedic *Handbook* of Abramowitz and Stegun [9.13], but they were not mentioned until the second edition of *Numerical Recipes* [9.14] where programs based on the author's work were given in Sect. 6.7. Few, if any, large-scale libraries (such as NAG, IMSL, etc.) have suitable subroutines. The reason is simply that all orders are expressible as sums of polynomials in x^{-1} multiplied by $\sin x$ and $(-\cos x)$, and recurrence relations connect consecutive orders. The computational task appears trivial. Nevertheless there are difficulties, shared with the evaluation of more general Bessel functions and Coulomb functions, which stem directly from the relevant differential equation. A useful discussion, intended for the physicist exploring numerical analysis, of some of these difficulties in the case of cylindrical Bessel functions is found in Sect. 4.1 of Koonin [9.15], which also contains the details of a fixed-accuracy program to evaluate them. Similarly, *Numerical Recipes* by Press *et al.* [9.14] contains an excellent coverage of the counter-intuitive aspects of the evaluation of special functions, difference and differential equations, and numerous programs. Two programs have been published for the calculation of spherical Bessel functions, by Gillman and Fiebig [9.16] and by Lentz [9.17]. They are discussed in Sect. 9.5.

For *spherical* Bessel Functions, with solutions $j_n(x)$ and $y_n(x)$, the differential equation is:

$$w'' + \frac{2w'}{x} + \left[1 - \frac{n(n+1)}{x^2}\right] w = 0 . \tag{9.4}$$

In this equation (see [9.13], Eq. 10.1.1) x is a real variable while n is an integer which is positive, negative or zero and which will be identified with the non-negative angular momentum. Many of the properties of j_n and y_n follow directly from their identification as Coulomb functions with $\eta = 0$, as is done below through (9.11). The results are

$$j_n(x) = x^{-1} F_n(0,x) , \qquad y_n(x) = -x^{-1} G_n(0,x) . \qquad (9.5\text{a, b})$$

From this and the properties of F and G we immediately deduce that j and y will display an oscillatory nature for x values larger than $\sqrt{(n^2+n)}$, their magnitude will decrease as x^{-1} for these larger x values, and that y will diverge towards $-\infty$ as x approaches zero. The first two orders of the spherical Bessel functions are:

$$j_0(x) = x^{-1}\sin x , \qquad y_0(x) = -x^{-1}\cos x , \qquad (9.6\text{a, b})$$

$$j_1(x) = x^{-2}\sin x - x^{-1}\cos x , \quad y_1(x) = -x^{-2}\cos x - x^{-1}\sin x , \quad (9.6\text{c, d})$$

and recurrence relations (Sect. 9.3) link sets of three consecutive orders. The derivatives $j_0'(x)$ and $y_0'(x)$ follow from (9.6).

The solutions j_n and y_n are called the *regular solution* and the *irregular solution*, respectively. This notation describes their behavior near the origin ($x \to 0$) where the irregular solution diverges to $-\infty$ as $-(2n-1)!!x^{-(n+1)}$, whereas the regular one goes to zero as $x^n/(2n+1)!!$ (with $j_0(0) = 1$). The irregular solution, $y_n(x)$, is the spherical Neumann function and is occasionally given the symbol $n_n(x)$. The following linear combinations are the spherical Hankel functions,

$$h_n^{(1)}(x) = j_n(x) + iy_n(x)\,, \tag{9.7a}$$

$$h_n^{(2)}(x) = j_n(x) - iy_n(x)\,. \tag{9.7b}$$

As x becomes large the equations of [9.13] (9.2.1, 9.2.17, and those following 10.1.26) show that

$$j_n(x) \to x^{-1}\cos(x - \tfrac{1}{2}n\pi - \tfrac{1}{2}\pi) \;=\; x^{-1}\sin(\theta_n)\,, \tag{9.8a, b}$$

$$y_n(x) \to x^{-1}\sin(x - \tfrac{1}{2}n\pi - \tfrac{1}{2}\pi) \;=\; -x^{-1}\cos(\theta_n)\,. \tag{9.8c, d}$$

This difference in phase definition (the additional $\frac{1}{2}\pi$ interchanges the role of sine and cosine in (9.8)) is carried over to the equivalents of the Hankel functions for Coulomb scattering. For this Coulomb case the expression for the outgoing wave is chosen to be

$$H_L^+(\eta, x) = G_L(\eta, x) + \mathrm{i}F_L(\eta, x) \to \exp(\mathrm{i}\theta_L) \qquad \text{as } x \to \infty \tag{9.9}$$

and similarly for the incoming wave H^-. For $\eta = 0$ the relations are

$$h^{(1)}(x) = -\frac{\mathrm{i}}{\mathrm{x}}\,H^+(0, x)\,, \qquad h^{(2)}(x) = \frac{\mathrm{i}}{\mathrm{x}}\,H^-(0, x)\,, \tag{9.10a, b}$$

with the $\frac{1}{2}\pi$ phase difference remaining. The $H^\pm$ scattering functions must not be confused with the Hankel functions for cylindrical Bessels, $H_n^{(1)}$ and $H_n^{(2)}$ (see [9.13], Eqs. 9.1.3, 9.1.4, 9.1.6), which are analogous to (9.7).

To obtain the Riccati–Bessel functions we transform (9.4), by removing the first-derivative term, into

$$w'' + \left[1 - \frac{n(n+1)}{x^2}\right] w = 0\,, \tag{9.11}$$

whose solutions ([9.13], Sect. 10.3) are $xj_n(x)$ and $xy_n(x)$. No special symbol has been given to them in the mathematical literature although, for complex argument z, a consistent notation is used [9.17–19] for Mie scattering:

$$\psi_n(z) \;=\; zj_n(z)\,, \tag{9.12a}$$

$$\chi_n(z) \;=\; -zy_n(z)\,, \tag{9.12b}$$

$$\zeta_n(z) \;=\; \psi_n(z) + \mathrm{i}\chi_n(z)\,. \tag{9.12c}$$

It should be noted that this definition carries the opposite sign on the irregular function from both [9.10] and the program RICBES. The Riccati–Bessel properties are briefly treated in [9.10, Sect. 10.3]. For the first two values of n we have

$$xj_n(x) = \sin x\,, \qquad xy_0(x) = -\cos x\,, \qquad (9.13\text{a, b})$$
$$xj_1(x) = x^{-1}\sin x - \cos x\,, \quad xy_1(x) = -x^{-1}\cos x - \sin x\,. \qquad (9.13\text{c, d})$$

On comparison of (9.11) with (9.1) it is evident that the Riccati–Bessel functions are the same as the Coulomb functions with $\eta = 0$, whereas (9.8) and (9.9) show that the sign of the irregular function is reversed. The Riccati–Bessel functions thus behave like $\sin\theta_n$ and $(-\cos\theta_n)$ as x becomes large, where the asymptotic phase, θ_n, is given by (9.3). Alternatively, this can be written as $\cos(\theta_n - \frac{1}{2}\pi)$, $\sin(\theta_n - \frac{1}{2}\pi)$; see (9.8).

9.3 Recurrence Relations for Spherical Bessel Functions

Each of the four spherical Bessel functions j_n, y_n, $h_n^{(1)}$, and $h_n^{(2)}$ obeys recurrence relations ([9.13], Eqs. 10.1.19–22) which connect the functions of order $(n-1)$, n, and $(n+1)$. If g_n is used to represent any of these four functions the relations are:

$$g_{n-1} - \frac{2n+1}{x} g_n + g_{n+1} = 0 \qquad (9.14)$$

and

$$ng_{n-1} - \frac{2n+1}{x} g_n' - (n+1)g_{n+1} = 0\,. \qquad (9.15)$$

These can be rewritten in a form which is suitable for *downward recurrence*, connecting two successive orders and a derivative,

$$g_{n-1} = S_{n+1}g_n + g_n'\,, \qquad (9.16)$$
$$g_{n-1}' = S_{n-1}g_{n-1} - g_n\,, \qquad (9.17)$$

in which $S_n = n/x$. The equivalent expressions for *upward recurrence* are equations (9.17) and (9.16) rearranged:

$$g_{n+1} = S_n g_n - g_n' \qquad (9.18)$$

and

$$g_{n+1}' = g_n - S_{n+2}g_{n+1}\,. \qquad (9.19)$$

The recurrence relations (9.14)–(9.17) are an alternative way of expressing the differential equation (9.4) as difference equations.

It is well known (see, for example, [9.14], Sect. 5.4; [9.15], Sect. 4.1; [9.8], Sect. 3) that a recurrence relation is numerically unstable in the direction in which the function is decreasing. Successive values are computed as small differences between nearly equal terms and all accuracy is soon lost. This occurs for the regular function $j_L(x)$ once $L > x$ for any fixed value of x: $j_L(x)$ *decreases monotonically* as a function of L, so that upward recurrence in n of $j_n(x)$ is unstable. Conversely, since the irregular function $y_L(x)$ *increases* as L increases, once $L > x$, upward recurrence is stable. Thus for $L > x$ we must use *downward recurrence in* n to calculate the values of $j_n(x)$ ($n = L, L-1, ..., 3, 2, 1, 0$); similarly *upward recurrence* must be used for $y_n(x)$. In the region where $L < x$ the functions have an oscillatory character and recurrence in both directions is stable.

Hence, to compute both $j_n(x)$ and $j'_n(x)$ for all orders from 0 to L by using (9.16) and (9.17) we need $j_L(x)$ and $j'_L(x)$, or their ratio, for the maximum $n = L$ and need to find the smaller n values by downward recurrence, normalizing at $j_0(x)$ with (9.6). On the other hand, starting with $y_0(x)$ and $y'_0(x)$ from (9.6), the upward recurrence equations (9.18) and (9.19) will yield stable values for the irregular functions and their derivatives for each value of the order.

From (9.18), which is satisfied by $j_n(x)$, we see that the logarithmic derivative is given by

$$\frac{j'_n}{j_n} = S_n - \frac{j_{n+1}}{j_n}, \tag{9.20}$$

and from (9.14) it is easy to derive a continued fraction for the ratio of successive orders:

$$\frac{j_{n+1}}{j_n} = \cfrac{1}{(2n+3)/x - \cfrac{1}{(2n+5)/x - \cfrac{1}{(2n+7)/x - ...}}}. \tag{9.21}$$

The coefficients in the denominators are $S_k + S_{k+1} \equiv T_k$ for k starting at $n+1$. The equation is implicit in [9.10], Eqs. (9.1.73) and (10.1.1); it was derived (for Coulomb functions) as Eq. (2.19) of [9.7] and by a different method in [9.6]. (It was quoted erroneously in Eqs. (37), (38), and (39) of [9.23] – in each case the first term should be dropped.) A most important point is that (9.21) *applies only to the regular solution.* It is not apparent from the derivation that the formula does not hold for the irregular functions y_n or h_n even though they, too, satisfy all the recurrence relations. The explanation of this remarkable result, and an indication of its generality, appears in [9.24] and also in [9.20,22]: only the minimal (i.e., *regular*) solution has the property (9.21). Also, combining (9.20) and (9.21) we find that

$$f \equiv \frac{j'_n}{j_n} = S_n - \cfrac{1}{T_{n+1} - \cfrac{1}{T_{n+2} - \cfrac{1}{T_{n+3} - ...}}}. \tag{9.22}$$

The numerical evaluation of the continued fraction f uses the fact that eventually the denominator $T_k = (2k+1)/x$ will become large enough so that the value of f can be found by terminating (9.22) at the kth step while retaining a chosen accuracy.

The problem of calculating the spherical Bessel functions becomes that of computing the continued fraction f to sufficient accuracy. Equation (9.22) will be referred to as CF1, whereas CF2 will refer to a second continued fraction: $p + iq = H'/H$ which is discussed in Sect. 9.6. It may be noted that the reciprocal of (9.21) is also used (for example, by Lentz [9.17]) and that it is a more complicated expression.

9.4 Evaluation of the Continued Fraction

Continued fractions are mentioned by Abramowitz and Stegun [9.13] in their Sect. 3.10 and are covered in *Numerical Recipes* [9.14] in Chap. 5.2. There is an intimate relation between recurrence relations of the type (9.14) and continued fractions: a full discussion and literature guide appears in Chap. 4 of van der Laan and Temme [9.20], particularly Sect. 4.8, and much valuable and detailed information appears in Wynn [9.21] and in Wimp [9.22]. The forward evaluation suggested in [9.13,14] is not to be recommended: a backward recurrence algorithm originally due to Miller (see [9.14], Sects. 5.4 and 6.4; [9.20], Sect. 3.3; or [9.15], Chap. 4 and 5) is superior and is very widely used in the improved form given by Gautschi [9.24].

Steed's method [9.6] for continued fractions is a stable forward recurrence and it has further advantages (discussed in [9.23], Sect. 4). It was adopted in the author's earlier programs [9.4,6,10,23]. This method involves a *summation* when updating f_{n-1} to become f_n and was hence subject to rare numerical cancellation errors when a certain denominator approached zero ([9.11]. Eq. (11); [9.8,25,26]). Thus Steed's method for continued fractions does not compute the result everywhere to uniform accuracy. In cases where there are no zeros involved (e.g., in calculating CF2 rather than CF1, see end of Sect. 9.6) there can be no objection in principle to using it (as does BESSCC and part of the COULCC code). The same method, although unnamed, has been used by Gautschi and Slavik [9.27] and is also described and recommended in the classic Gautschi paper [9.24]. Subsequent authors have not been alert to the manifest advantages of Steed's method. Thompson and Barnett [9.11] in 1987 released the code BESSCC, which treats complex-argument Bessel functions and evaluates the complex continued fraction involved (CF2) by Steed's method.

The method of choice for continued fractions, however, is not one of the above but the method of Lentz [9.17] together with the 'zero shifts' of both numerator and denominator discussed by Jaaskelainen and Ruuskanen [9.25]. That paper, however, proposed an elaborate change to the algorithm, whereas

Thompson shows in Appendix III of [9.12] how to achieve these shifts with minimum change to the Lentz method. The explicit form of the Lentz–Thompson (L–T) algorithm is given below.

L–T algorithm for the forward evaluation of continued fractions

Given the n th convergent of a CF, i.e., the sum to n terms of

$$f_n = b_0 + \cfrac{a_1}{b_1 + \cfrac{a_2}{b_2 + \ldots \cfrac{a_{n-1}}{b_{n-1} + \cfrac{a_n}{b_n}}}} \tag{9.23}$$

evaluate $f = \lim_{n\to\infty}(f_n)$ to an accuracy acc with the algorithm:

```
f0:=b0;     if  (f0 = 0)  f0=small
C0:=f0, D0:=0
for  n=1,limit do
    Cn:=bn+an / Cn-1;    if  (Cn=0)  Cn = small
    Dn:=bn+an × Dn-1;    if  (Dn=0)  Dn = small          (9.24)
    Dn:=1/Dn
    Δn:=Cn × Dn;    fn:=fn-1Δn
    if  (| Δn-1|< acc ) exit
end
```

Notes:

1. If at any stage we represent f_n as A_n/B_n(before cancellation), then the two quotients which are used in the L–T algorithm are $C_n = A_n/A_{n-1}$ and $D_n = B_{n-1}/B_n$.
2. The parameter `small` should be some non-zero number less than typical values of the quantity `acc` $\cdot|\mathrm{b_n}|$; for example, 10^{50} is typical for double precision calculations in which a sensible choice of `acc` is 10^{-14}.
3. The zero tests are to a working accuracy, say `tol`, which can be chosen to avoid divide-by-zero error checks. It could be taken as equal to `small`. Thus algebraic conditions such as '`if (Dn = 0)`' are to be read as computation conditions: '`if (abs(Dn) < tol)`'.
4. The constant `limit` is an integer designed to abort the loop if necessary. (Our programs use `limit = 20 000`. For small L values the algorithm would require this number of iterations for an x value of about 20 000. Graphs, estimates and explanations for this behavior are given in [9.7].)
5. All the methods above, and others ([9.12], Sect. 3.3), apply for *complex* values of a_k, b_k.

Lentz's method was developed in 1975 to deal with complex arguments in calculations of Mie scattering. Lentz required the Riccati–Bessel logarithmic derivative $[zj_n(z)]'/[zj_n(z)]$ for a range of values of n and $z = x + iy$. More extreme parameter values were treated by Wiscombe [9.18] in 1980. It must be realized that when z lies away from the real axis, the natural directions for recurrence in order may change, and considerably more care needs to be exercised in the computations. Examples of difficulties are given in Lentz [9.17,26] and in [9.12], Sect. 3.1; the effect complicates the coding [9.11] of COULCC by forcing the monitoring of the moduli of the functions during recurrence.

The Lentz–Thompson method given above is quite general and it seems to retain the advantages of Steed's method and to remove the (minor) problems. In Lentz's papers [9.17,26] he advocates a slightly different error-correcting procedure. The choice is probably a matter of taste. However, it should be restated that nothing is particularly critical in the selection of the number *small* (see 2. above.) In addition, the Lentz–Thompson method can also be recommended for complex arguments (where these numerical cancellation problems are even less likely). It is not at all clear why [9.26] discards it for this use.

In the BESSCC paper [9.11] Thompson gave a listing of a complex-argument spherical Bessel code SBESJH which uses the new L–T continued fraction algorithm. (The range of usefulness of the program in fact is greater than that assumed by Thompson.) The code SBESJY in the next section also adopts this technique.

9.5 The Programs SBESJY and RICBES

The spherical Bessel program SBESJY, for real argument, returns the four functions $j_L(x)$, $j'_L(x)$, $y_L(x)$, and $y'_L(x)$ for all orders from 0 to LMAX. It is an extension of the program with the same name which was published by Barnett [9.23] in 1981, and although the principles are identical, the realization is rather different. FORTRAN 77 coding is used, the arrays are indexed from 0 to MAXL, the L–T algorithm replaces the Steed algorithm for the calculation of f, i.e., CF1 (which, as programmed in [9.23], may fail for isolated values of x, e.g., for $x = \sqrt{15}$), the cosine and sine functions are used to find the $L = 0$ solution rather than the square-root function, the results for $x = 0$ are given, and all four functions are computed.

The closely related program RICBES calculates the Riccati–Bessel functions of (9.12), $\psi_L(x)$, $-\chi_L(x)$ and their x derivatives $\psi'_L(x)$, $-\chi'_L(x)$. These functions are equally $F_L(0,x)$, $-G_L(0,x)$, $F'_L(0,x)$, and $-G'_L(0,x)$ and so can be obtained as a special case of COULFG or COUL90 (see Sect. 9.7).

In the program SBESJY both the two recurrences and the continued fraction are evaluated as Bessel functions. First $\mathtt{CF1} = f$ is calculated from (9.22) with $n = \mathtt{LMAX}$, $a_k = -1$ and $b_k = T_k$. This yields a value $j'_{\mathtt{LMAX}} = f \cdot j_{\mathtt{LMAX}}$ when the unnormalized $j_{\mathtt{LMAX}}$ is taken as unity. Then, if $\mathtt{LMAX} > 0$, a DO loop implements the downward recurrence (9.16) and (9.17), until unnormalized values for j_0 and j'_0 are found. Relative values of j_L and j'_L for each L from this procedure are stored ready for normalization, which is accomplished by (9.6). The values of y_0, y'_0 are also obtained from (9.6) and the recurrence relations (9.18) and (9.19) are used to find the remaining L values from 1 to `LMAX`. It is clear that if a program without derivatives were required then (9.16) and (9.17) could be replaced by (9.15), with (9.21) used instead of (9.22, `CF1`). `SBESJY` was tested against `SBESJ` [9.23], `COULFG` [9.4], the values in [9.13], and the two programs described below, `DPHRIC` [9.16] and [9.26]. The range of parameters was $x = 0.01$ to $x = 1000.0$ and $L = 0$ to $L = 1000$; in general, all results agreed to better than a relative accuracy of 10^{-12} when the `acc` parameter is set to 10^{-15}. The first two programs are unusual in *not* using the functions $\cos(x)$ and $\sin(x)$ in the calculation. Those which do may suffer from truncation error when a large value of x is reduced to the range $\pm\frac{1}{2}\pi$ by subtracting $N\pi$, where N is a suitable integer. This reduction will be compiler dependent and can be programmed, to some extent, by methods described in Cody and Waite [9.33], Chap. 8, p. 136. The value π is written as `C1 + C2`, where `C1 = 3217/1024 = 3.1416 01562 50000` is exactly machine representable and `C2 = -8.9089 10206 76153 73566 E-6`.

For the program `RICBES` both the two recurrences and the continued fraction are now evaluated as Coulomb functions. First $\mathtt{CF1} = f$ is calculated from (9.36) with $\eta = 0$ (and hence $R_k^2 = 1$ and $T_k = (2k+1)/x$) and with $n = \mathtt{LMAX}$. It differs from (9.22) only by $1/x$. This yields a value $F'_{\mathtt{LMAX}} = f \cdot F_{\mathtt{LMAX}}$ when the unnormalized value of $F_{\mathtt{LMAX}}$ is taken as unity. Then, if $\mathtt{LMAX} > 0$, a `DO` loop implements the downward recurrence (9.26) and (9.27) until unnormalized values for F_0 and F'_0, i.e., ψ_0 and ψ'_0, are found. Relative values of all the F_L and F'_L are stored ready for normalization, which is accomplished by (9.13). The values of $\chi_0 = -G_0$ and $\chi'_0 = -G'_0$ are also obtained from (9.13) and the upward recurrences (9.29) and (9.30) are used to find the remaining L values.

A recent program to compute $j_L(x)$ and $y_L(x)$ (called $n_L(x)$) is that of Gillman and Fiebig [9.16]. Constructive criticisms of its style have been given by Welch [9.28] who has presented rewritten versions in FORTRAN 77 and FORTRAN 90. The method of Gillman and Fiebig is to calculate modified functions $u_L(x)$ and $v_L(x)$ in place of $j_L(x)$ and $y_L(x)$ from which factors of $x^L/(2L+1)!!$ and $-(2L-1)!!/x^{L+1}$ have been extracted. These scaled functions tend to unity as $x \to 0$ (see Sect. 9.2). A version of Miller's method is used to obtain the relative values of u_L for all the L values considered (0–1000), with x in the range 0.01–100.0. The values are normalized by the Wronskian relation for the u and v functions.

The reason for choosing $u_L(x)$ and $v_L(x)$ is that computational overflow and underflow conditions are removed. When representable real numbers were restricted to $10^{\pm 38}$ or $10^{\pm 70}$ this was indeed a problem. For modern FORTRAN 77 compilers, however, it is commonplace to offer $10^{\pm 308}$ in double precision variables (as does the Lahey F77L compiler, whose version 4.0

was used in this work) which is large enough to remove the need for special programming in normal physical applications.

A second point is that the u, v recurrences are *assumed* to be stable in the same directions as are the j, y recurrences. This is not an obvious fact (see Lentz [9.17,26]) but the stability was proved rigorously in 1980 by O'Brien [9.29] for real arguments. He claimed, further, that no Miller's method is necessary, replacing it by two evaluations of 'a rapidly convergent series'. (No numerical details were given and O'Brien's results have not yet appeared as a program.)

The third point is that tests for overall accuracy in [9.16] only address numerical consistency, and not the method. This is recognized by Gillman and Fiebig who say that 'it is of some value'. It is rather puzzling to a user who sees tests aimed at a relative accuracy of 2.10^{-8} apparently produce results to better than 10^{-15}, which is about machine accuracy. One reason may be the tests measure the *rms* Wronskian deviation from unity, which reduces by the square root of the number of L values, and here by a factor of 30. But the *same Wronskian* is used to normalize the u_L values, so the tests are not independent. There is no need for this choice of normalization; the exact value of $u_0(x)$ is $\sin(x)/x$ and should be chosen instead. However, the choice of this correct normalization gives exactly the same result to machine precision. The correct answer involves a more subtle point. By choosing to compute the rms deviation *over all L values* Gillman and Fiebig use the fact that the downwards recurrence of Miller's method is most at error *at the high L values* which is worst when x/L is largest. An independent calculation [9.26] shows that $j_{1000}(x)$ is in error by a fraction 10^{-3} for $x = 100$, which falls to 10^{-7} for $x = 0.5$. The healing step in L is small: even for $L = 999$ the errors improve to 10^{-5} and 10^{-15}. The last few L values should not be used, and it would seem that the choice of 1000 for the starting L is too large and their estimate of an error $\epsilon \simeq (x/2L)^6$ is optimistic for L itself. Making the correct choice of a starting L is the hard part of Miller's method. Of course these comments relate only to the regular solution $j_n(x)$; the $y_n(x)$ values depend only on their (accurate) starting values, $y_0(x)$ and $y_0'(x)$, since the upward recurrence is stable. The value of $j_{1000}(100)$ is $5.32338\ 16172 \times 10^{-872}$, while $j_{1000}(0.5) = 6.06344\ 55462 \times 10^{-3172}$, and these are clearly of mathematical rather than of physical interest.

The approach of [9.16] in factorizing out the small-x behavior, when reinforced by theory [9.29] works well for spherical Bessel functions. The underflows and overflows in reapplying the normalization to recover j_L and y_L can easily be trapped by extracting, say, powers of $10^{\pm 200}$ when appropriate, as is done by Lentz [9.26]. There is nothing to prevent the CF1 calculation in SBESJY replacing Miller's method to remove the inaccuracy near the maximum L value.

The program CBESSEL of Lentz [9.26] calculates a single value of $j_n(z)$, for complex z. His algorithm for the forward evaluation of continued fractions creates an infinite product which is terminated when a given accuracy is reached. It is a flexible and elegant concept, applicable to a wide range of complex arguments, although just what the limits are is not discussed by Lentz. Despite the different appearances it is the same algorithm as is described in Sect. 9.4. As given, the Lentz program only results in $j_n(z)$ and not $y_n(z)$ or the derivatives, and intermediate n values are not calculated. Thompson's program SBESJH given in [9.11] includes all these features. These two codes will be compared in detail in a subsequent publication.

9.6 Recurrence Relations for Coulomb Functions

Each of the four Coulomb functions F_n, G_n, H_n^+ and H_n^- (see (9.19)) obeys recurrence relations ([9.13], Eqs. 14.2.3, 14.2.1 and 14.2.2) which connect the functions of order $(n-1)$, n, and $(n+1)$, thus

$$R_n w_{n-1} - T_n w_n + R_{n+1} w_{n+1} = 0\,, \tag{9.25}$$

$$w_{n-1} = \frac{S_n w_n + w_n'}{R_n}\,, \tag{9.26}$$

$$w_{n-1}' = S_n w_{n-1} - R_n w_n\,. \tag{9.27}$$

The coefficients are:

$$R_k = \sqrt{1+\frac{\eta^2}{k^2}}\,, \tag{9.28a}$$

$$S_k = \frac{k}{x} + \frac{\eta}{k}\,, \tag{9.28b}$$

$$T_k = S_k + S_{k+1} = (2k+1)\left[x^{-1} + \frac{\eta}{k^2+k}\right]. \tag{9.28c}$$

Equations (9.26) and (9.27), connecting two successive orders and a derivative, are in a form suitable for *downward recurrence* in n. The equivalent expressions for *upward recurrence* in n are rearrangements of (9.27) and (9.26),

$$w_{n+1} = \frac{S_{n+1} w_n - w_n'}{R_{n+1}}\,, \tag{9.29}$$

and

$$w_{n+1}' = R_{n+1} w_n - S_{n+1} w_{n+1}\,. \tag{9.30}$$

The recurrence relations (9.25), (9.26), and (9.27) are an alternative way of expressing the differential equation (9.4) as a difference equation. As Fröberg [9.30] shows, not even two of these recurrence relations are independent.

The equation analogous to the spherical Bessel (9.15) loses its simplicity and becomes

$$S_{n+1}R_n w_{n-1} - T_n w_n' - S_n R_{n+1} w_{n+1} = 0\,, \tag{9.31}$$

and this is not particularly useful.

The boundary between solutions of oscillating and monotonic character occurs at the turning point x_L of (9.2) when L is fixed; alternatively, for fixed x it is found at L_{TP}. This is also given by (9.2) and equivalently by

$$x^2(R_L^2 - S_L^2) = 1 \tag{9.32}$$

or

$$L_{TP} = (x^2 - 2\eta x + \tfrac{1}{4})^{1/2} - \tfrac{1}{2}\,. \tag{9.33}$$

Since $F_L(x)$ decreases to zero as L increases beyond L_{TP}, the stable direction (the function must not decrease) is *downward recurrence in* n to calculate the values of $F_n(x)$ $(n = L, L-1, ..., 3, 2, 1, 0)$. Similarly *upward recurrence* must be used for $G_n(x)$. In the region where $L < L_{TP}$ all the functions have an oscillatory character and recurrence in both directions is stable.

Hence to compute both $F_n(x)$ and $F_n'(x)$ for all orders from 0 to L it appears that we need both functions for the maximum order L and to use (9.26) and (9.27) to find lower n values, as was the case for the spherical Bessels in Sect. 9.3. Now, however, there is no easy way to normalize at $n = 0$. Methods of finding $F_0(\eta, x)$ and $F_0'(\eta, x)$ directly are given in the programs of Bardin *et al* [9.5]. They demand detailed numerical analysis and a knowledge of most of the properties of Coulomb functions near the origin, in order to deal with the full range of x and η. Similarly intricate study is required for the separate evaluation of $G_0(\eta, x)$ and $G_0'(\eta, x)$, as is also shown by Strecock and Gregory [9.31], so that the upward recurrence equations, (9.29) and (9.30), will yield the irregular functions and their derivatives for each value of the order. In all, ten separate subroutines for the different methods are required. Bardin *et al.* [9.5] check their independent evaluation of the F and G functions, and their derivatives, by computing the value of the Wronskian. This is unity for Coulomb functions:

$$F_L'(\eta, x)G_L(\eta, x) - F_L(\eta, x)G_L'(\eta, x) = 1\,, \tag{9.34}$$

although the test is not foolproof (as was shown in [9.6], Sect. 5).

Steed's algorithm for calculating Coulomb functions (and hence Bessel functions, spherical Bessel and Riccati–Bessel functions, Airy functions, etc. [9.4,10,23]) is based on a different approach, which has the significant merit that *no detailed information about the function behavior at the origin is required* [9.7]. It also has the remarkable property that an *individual* L value (which need not be an integer) can be found, without computing a range of L values. No other method in the literature has this property. Program

KLEIN [9.10] illustrates this feature, which is useful for relativistic calculations in which the effective L values for each angular momentum channel are not integer-spaced. The algorithm consists in combining the ratio $F'_L/F_L \equiv f \equiv$ CF1 with the ratio $H'_L/H_L \equiv p + iq \equiv$ CF2 and the Wronskian (9.31) to solve first for $F_L(\eta, x)$ and then for G_L, F'_L and G'_L all at the same time. Naturally the Wronskian relation may then not be used also as an independent check of the solution. The details are explained in several references [9.4,10,23,27]. In practice, the lowest L is chosen and CF2 is evaluated, almost invariably, for $L = 0$. The value f is obtained from CF1 evaluated for the highest L required, followed by a downwards recurrence. The functions G_L and G'_L are found by upward recurrence from G_0 and G'_0.

In order to obtain CF1 we proceed as in Sect. 9.3. From (9.28), which is satisfied by $F_n(x)$, we see that the logarithmic derivative is given by

$$\frac{F'_n}{F_n} = S_{n+1} - \frac{F_{n+1}}{F_n}, \tag{9.35}$$

and from (9.25) a continued fraction for the ratio of successive orders can be derived [9.6,7] which is analogous to (9.22). Combining it with (9.35) we find

$$\text{CF1}: f \equiv \frac{F'_n}{F_n} = S_{n+1} - \cfrac{R^2_{n+1}}{T_{n+1} - \cfrac{R^2_{n+2}}{T_{n+2} - \dots \cfrac{R^2_k}{T_k - \dots}}} . \tag{9.36}$$

Eventually T_k will become large enough so that the value of f can be determined by terminating the evaluation at the k th step, as in (9.24). To recover the specialized equation (9.22), set $\eta = 0$, i.e., $R^2_k = 1$ and $T_k = (2k + 1)/x$, and take the derivatives of (9.5), which turns S_{n+1} into S_n.

A major feature of Steed's algorithm is the conversion of two (slow) asymptotically convergent series, for $H_L(\eta, x)$ and its derivative, into a ratio which can be expressed as a rapidly convergent continued fraction. This second continued fraction, CF2, is derived in [9.6,7] and reads:

$$\text{CF2}: p + iq \equiv \mathrm{i}(1 - \eta/x) + \mathrm{i}x^{-1} \cfrac{ab}{2(x - \eta + \mathrm{i}) + \cfrac{(a+1)(b+1)}{2(x - \eta + 2\mathrm{i}) + \dots}} \tag{9.37}$$

where $a = \mathrm{i}\eta - L$ and $b = \mathrm{i}\eta + L + 1$. It may also be computed by the L–T algorithm. When x becomes smaller than x_L of (9.2), p loses accuracy relative to q. The reasons are explained in [9.7,10] and are discussed by Nesbet [9.8] who suggests an alternative algorithm to minimize the difficulty. In practice it is only of minor concern when $\eta < 0$ because of (9.2), and because the minimum L needed is zero.

9.7 The Program COUL90

The new version of the COULFG program, COUL90, follows the logical flow of its predecessor [9.4] and differs, first, in the choice of the L–T algorithm for evaluating CF1 and CF2, and, second, in adopting FORTRAN 77 conventions. Assume that the minimum L required is XM (not necessarily an integer) and the maximum is L+XM. (Except in rare circumstances [9.32] it is computationally desirable to set XM = 0.0. A serious loss of accuracy is possible when XM is chosen to be large enough that the required x value is well below the turning point, x_{XM}.) CF1 is calculated for L+XM and then downward recurrence using (9.26) and (9.27) carries the relative values of $F_n(\eta, x)$ and $F_n'(\eta, x)$ from $n = L + \mathrm{XM} - 1$ to $n = \mathrm{XM}$. The fraction CF2 is calculated at angular momentum XM and all the lowest-order functions are found, with the help of (9.31). Equations (9.29) and (9.30) are used to obtain all higher L values, $\mathrm{XM} + 1,\ \mathrm{XM} + 2, \ldots, \mathrm{XM} + L$. The subroutine is called as

```
CALL COUL90 (X,ETA, XLMIN,LRANGE, F,G,FP,GP, KFN,IFAIL)
```

where X, ETA, XLMIN are double-precision variables; LRANGE is the integer number of additional L values required; F, G, FP, and GP are double-precision arrays dimensioned at least (0:LRANGE); KFN selects the particular function (0 for Coulomb, 1 for spherical Bessel, and 2 for cylindrical Bessel); and IFAIL is an integer which is set to zero after a successful computation (and should be checked by the user). Generally XLMIN, the minimum L value, will lie between 0.0 and 1.0 (most usually 0.0) and LRANGE will just be the maximum L value. To compute the oscillating Airy function, for example, $\mathrm{XLMIN} = -\frac{1}{6}$ and a suitable [9.23] normalizing constant is used. It is an easy matter to vary the calculation (as was done in the COULFG program) to obtain the various Bessel functions, and the KFN parameter in the argument list allows for this. In the earlier code, provision was made for a mode choice, whereby storage could be saved by not storing the derivatives. This option has been removed, since most matching in scattering problems is formulated in terms of the logarithmic derivatives, and earlier core memory restrictions are no longer a factor.

The code produces results which are virtually the same as the test cases for COULFG [9.4] to which reference should be made for the details. Selected test data and the corresponding output using a simplified driver program (COUL90_S) are given in Sect. 9.8; more test cases that should be run with the more complicated driver (COUL90_T) are provided on the disk.

9.8 Test Data

Below are test data which can be run using a simplified driver program COUL90_S. This program reads the values of ETA, X, XLMIN, LRANGE, and KFN (one per line) listed in Sect. 9.8.1 and after calculating the results prints them for $L = 0,\ 50, 100, \ldots$. The output is shown in Sect. 9.8.2. A complete test run using the full driver program COUL90_T can be performed using the more complicated input data provided on disk.

9.8.1 Sample Input Data

```
--ETA-----..X.......--XM------ LRANGE KFN( 0=COUL ,1=SBES, 2=CYLBES)
 -0.5        20.0       0.0        50   0
 -0.5       200.0       0.0        50   0
 -0.5       200.0      50.0         0   0  test of XM offset
  0.0        20.0       0.0        50   0
  0.0       200.0       0.0        50   0
  0.5        20.0       0.0        50   0
  0.5       200.0       0.0        50   0
 -5.2         1.0       0.0        50   0
 -5.2        30.0       0.0        50   0
 -5.2      1000.0       0.0        50   0
  0.0         1.0       0.0       100   1  Sph.Bessels
  0.0       100.0       0.0       100   1  Sph.Bessels
  0.0         1.0       0.0       100   2  Cyl.Bessels
  0.0       100.0       0.0        50   2  Cyl.Bessels
  0.0        -1.0       0.0         0   0  error condition
```

9.8.2 Sample Output Data

```
  TEST OF THE CONTINUED-FRACTION COULOMB & BESSEL PROGRAM - COUL90
            WHEN LAMBDA IS AN INTEGER (L-VALUE)

          F IS REGULAR AT THE ORIGIN ( X = 0 ) WHILE
          G IS IRREGULAR ( => -INFINITY AT X = 0 )

   L       F(ETA,X,L)       G(ETA,X,L)     D/DX (F)         D/DX (G)


 ETA =    -0.500      X =     20.000    XLMIN =    0.00     KFN =  0 (Coulomb)
    0   -1.0237230D-01  -9.8257145D-01  -1.0068693D+00   1.0431459D-01
   50    2.0210828D-15   1.0720740D+14   4.7241181D-15  -2.4419563D+14
 ETA =    -0.500      X =    200.000    XLMIN =    0.00     KFN =  0 (Coulomb)
    0    8.1994516D-01  -5.7026249D-01  -5.7168125D-01  -8.2199599D-01
   50   -9.2809974D-02  -1.0110000D+00  -9.8083865D-01   9.0206999D-02
```

```
ETA =    -0.500      X =     200.000    XLMIN =  50.00    KFN =  0 (Coulomb)
  50  -9.2809974D-02   -1.0110000D+00   -9.8083865D-01    9.0206999D-02
ETA =     0.000      X =      20.000    XLMIN =   0.00    KFN =  0 (Coulomb)
        The 3 sets of results are COUL90(KFN), SBESJY & (1/X) RICBES
   0   9.1294525D-01    4.0808206D-01    4.0808206D-01   -9.1294525D-01
   0   4.5647263D-02   -2.0404103D-02    1.8121740D-02    4.6667468D-02
   0   4.5647263D-02   -2.0404103D-02    2.0404103D-02    4.5647263D-02
  50   1.1300162D-15    1.9085083D+14    2.6533576D-15   -4.3681189D+14
  50   5.6500808D-17   -9.5425417D+12    1.2984284D-16    2.2317722D+13
   0   8.4147098D-01   -5.4030231D-01   -3.0116868D-01    1.3817733D+00
   0   8.4147098D-01   -5.4030231D-01    5.4030231D-01    8.4147098D-01
 100   7.4447277-190   -6.6830795+186    7.4443610-188    6.7495744+188
 100   7.4447277-190   -6.6830795+186    7.4443610-188    6.7495744+188
 100   7.4447277-190   -6.6830795+186    7.5188083-188    6.6827436+188
    j,y Bessels      X =     100.000    XLMIN =   0.00    KFN =  1 (SphBess)
        The 3 sets of results are COUL90(KFN), SBESJY & (1/X) RICBES
   0  -5.0636564D-03   -8.6231887D-03    8.6738253D-03   -4.9774245D-03
   0  -5.0636564D-03   -8.6231887D-03    8.6738253D-03   -4.9774245D-03
   0  -5.0636564D-03   -8.6231887D-03    8.6231887D-03   -5.0636564D-03
 100   1.0880477D-02   -2.2983850D-02    2.2873004D-03    4.3590946D-03
 100   1.0880477D-02   -2.2983850D-02    2.2873004D-03    4.3590946D-03
 100   1.0880477D-02   -2.2983850D-02    2.3961052D-03    4.1292561D-03
    J,Y Bessels      X =       1.000    XLMIN =   0.00    KFN =  2 (CylBess)
   0   7.6519769D-01    8.8256964D-02   -4.4005059D-01    7.8121282D-01
 100   8.4318288-189   -3.7752878+185    8.4314114-187    3.7750971+187
    J,Y Bessels      X =     100.000    XLMIN =   0.00    KFN =  2 (CylBess)
   0   1.9985850D-02   -7.7244313D-02    7.7145352D-02    2.0372312D-02
  50  -3.8698340D-02    7.6505264D-02   -6.6001492D-02   -3.4025651D-02
 COUL90 ERROR! IFAIL   -1
 SBESJY ERROR! IFAIL   -1
 RICBES ERROR! IFAIL   -1
ETA =     0.000      X =      -1.000    XLMIN =   0.00    KFN =  0 (Coulomb)
```

9.9 Summary

A detailed discussion regarding the numerical aspects in the calculation of Coulomb, Bessel, and Riccati-Bessel functions was presented, with emphasis on the negative η case appropriate for electron scattering. Some comparison with other methods and computer codes was given. The subroutines COUL90, SBESJY, and RICBES are used in several programs described in other chapters of this book.

9.10 Suggested Problems

1. Compute the values of $j_{100}(1)$, $y_{100}(1)$, $j_1(100)$, and $y_1(100)$. Also find their derivatives (with respect to x) and evaluate the Wronskian relation. Use the subroutines `COUL90` and `SBESJY`. What accuracy (compared to machine accuracy) do you expect? [Note that the Wronskian relation for spherical Bessel functions is

$$j_L(x)\, y'_L(x) - y_L(x)\, j'_L(x) = 1/x^2 \tag{9.38}$$

(see [9.13], Eq. 10.1.6). The first four values are `7.44472E-190`, `-6.68308E+186`, `1.08808E-2`, and `-2.29839E-2`.]
2. Verify selected entries from the tables on pages 465 and 466 of Abramowitz and Stegun [9.13] for $j_n(x)$ and $y_n(x)$. The values of n are 0,1,2, ...,10,20, ...,50,100 while x takes values 1,2,5,10,50,100. Observe the size of the function for $x \ll n$ and $x \gtrsim n$, the turning point.
3. Use Newton's method to search for the first and second zeros of $j_0(x)$, and hence obtain estimates of π. You will need the derivative as well.
4. Tables in Abramowitz and Stegun [9.13] give zeros of $j_n(x)$ and $y_n(x)$ for $n = 10$. The first zero of $j_{10}(x)$ is at $x_1 = 15.033469$, and the third zero is at $x_3 = 22.662721$. The first zero of $y_{10}(x)$ is at $x_2 = 12.659840$.
   (a) Verify these by direct evaluation.
   (b) Show that Newton's method finds these results from nearby starting values.
   (c) Calculate values of $j_0(x_1)$ and $j_1(x_1)$ from (9.6), and use the recurrence relations (9.18) and (9.19) to find the value of $j_{10}(x_1)$. Compare it with zero.
   (d) Repeat the computations for $j_{10}(x_3)$ which should also be zero, but is greatly in error.
   (e) Finally repeat the computations for $y_{10}(x_2)$, but this time starting with the values of $y_0(x_2)$ and $y_1(x_2)$ from (9.6). The answer, which should also be zero, is now much closer than when the regular spherical Bessels were used. If you can explain the reasons for this behavior (all of which are discussed in the chapter), then you will understand much of the subtleties of these special functions.
5. Evaluate the sum rule for regular spherical Bessel functions,

$$\sum_{n=0}^{\infty} (2n+1)\, j_n^2(x) = 1\,, \tag{9.39}$$

(Abramowitz and Stegun [9.13], Eq. 10.1.50) for $x = 0.1$, 0.875, and 8.75. In each case determine the n values required to produce convergence to within 10^{-6} and 10^{-15}, respectively.
6. Evaluate the sum rule for regular cylindrical Bessel functions,

$$J_0^2(x) + 2\sum_{k=1}^{\infty} J_k^2(x) = 1\,, \tag{9.40}$$

(Abramowitz and Stegun [9.13], Eq. 9.1.76) for $x = 0.1,\ 0.875$, and 8.75. In each case determine the k values required to produce convergence to within 10^{-6} and 10^{-15}, respectively. Use COUL90 with KFN = 2 for the values of $J_k(x)$. This relation is sometimes used for normalization, as is a similar result,

$$J_0(x) + 2\sum_{k=1}^{\infty} J_{2k}(x) = 1\,, \tag{9.41}$$

quoted in [9.14], Eq. (5.5.16).

7. For the functions plotted in Fig. 9.1 for x values close to $x_F^{(0)} = x_G^{(0)} = 6.6$, observe that both $F_2(+0.5, x_F^{(0)}) \approx 0.0$ and $G_2(-0.5, x_G^{(0)}) \approx 0.0$. Find more precise values for these zeros by using Newton's method in the form

$$x_F^{(n+1)} = x_F^{(n)} - F_2(+0.5, x_F^{(n)})/F_2'(+0.5, x_F^{(n)}) \tag{9.42a}$$
$$x_G^{(n+1)} = x_G^{(n)} - G_2(-0.5, x_G^{(n)})/G_2'(-0.5, x_G^{(n)}) \tag{9.42b}$$

and obtaining all the functions from calls to COUL90. These zeros can be brought into coincidence by varying the η value between -0.5 and $+0.5$. What η value is needed to achieve this coincidence ?

Acknowledgments

This work was completed while the author was on sabbatical leave at the University of Auckland. He is most appreciative of the help and welcome that he received from the Physics Department there. The comments of Dr. I.J. Thompson on a draft version are also appreciated.

References

9.1 A.R. Curtis, *Coulomb Wave Functions* Vol.11 *Royal Soc. Math. Tables* (Cambridge University Press, London, 1964) 9
9.2 M.J. Seaton, Comp. Phys. Comm. **25** (1982) 87; ibid **32** (1984) 115
9.3 K.L. Bell and N.S. Scott, Comp. Phys. Comm. **20** (1980) 447
9.4 A.R. Barnett, Comp. Phys. Comm. **27** (1982) 147 (program COULFG)
9.5 C. Bardin, Y. Dandeu, L. Gauthier, J. Guillermin, T. Lena, J.-M. Pernet, H.H. Wolter and T. Tamura, Comp. Phys. Comp. **3** (1972) 73
9.6 A.R. Barnett, D.H. Feng, J.W. Steed and L.J.B. Goldfarb, Comp. Phys. Comp. **8** (1974) 377 (program RCWFN)
9.7 A.R. Barnett, J. Comp. Phys. **46** (1982) 171
9.8 R.K. Nesbet, Comp. Phys. Comm. **32** (1984) 341

9.9 C.J. Noble and I.J. Thompson, Comp. Phys. Comm. **33** (1984) 413 (program `COULN`)
9.10 A.R. Barnett, Comp. Phys. Comm. **24** (1981) 141 (program `KLEIN`)
9.11 I.J. Thompson and A R. Barnett, Comp. Phys. Comm. **36** (1985) 363 (program `COULCC`); ibid **47** (1987) 245 (program `BESSCC`)
9.12 I.J. Thompson and A.R. Barnett, J. Comp. Phys. **64** (1986) 490
9.13 H.A. Antosiewicz (Ch. 10) and M. Abramowitz (Ch.14), *Handbook of Mathematical Functions*, eds. M. Abramowitz and I. Stegun (Nat. Bur. Stds., New York, 1964)
9.14 W.H. Press, B.P. Flannery, S.A. Teukolsky, and W.T. Vetterling *Numerical Recipes: The Art of Scientific Computing*, 2nd edition (Cambridge U. Press, New York, 1992)
9.15 S.E. Koonin, *Computational Physics* (Benjamin/Cummings, 1985)
9.16 E. Gillman and H.R. Fiebig, Comput. Phys. **2** **(1)** (1988) 62
9.17 W.J. Lentz, Appl. Opt. **15** (1975) 668
9.18 W.J. Wiscombe, Appl. Opt. **19** (1980) 1505
9.19 W.D. Ross, Appl. Opt. **11** (1972) 1919
9.20 C.G. van der Laan and N.M. Temme, *Calculation of special functions: the gamma function, the exponential integrals and error-like functions* (CWI tract 10 Mathematisch Centrum, 1984)
9.21 P. Wynn, Proc. K. Ned. Akad. Wet. Ser. **A65** (1962) 127
9.22 J. Wimp, *Computation with Recurrence Relations* (Pitman, London, 1984)
9.23 A.R. Barnett, Comp. Phys. Comm. **21** (1981) 297
9.24 W. Gautschi, SIAM Review **9** (1967) 24
9.25 T. Jaaskelainen and J. Ruuskanen, Appl. Opt. **20** (1981) 3289
9.26 W.J. Lentz, Comput. Phys. **4** (1990) 403
9.27 W. Gautschi and J. Slavik, Math. Comp. **32** (1978) 865
9.28 L.C. Welch, Comput. Phys. **2** **(5)** (1988) 65
9.29 D.M. O'Brien, J. Comp. Phys. **36** (1980) 128
9.30 C.E. Fröberg, Rev. Mod. Phys, **27** (1955) 399
9.31 A.J. Strecock and J.A. Gregory, Math. Comp. **26** (1972) 955
9.32 D.H. Feng and A.R. Barnett, Comp. Phys. Comm. **11** (1976) 401
9.33 W.J. Cody and W. Waite, *Software Manual for the Elementary Functions* (Prentice Hall Series in Computational Mathematics, 1980)

10. Scattering Amplitudes for Electron–Atom Scattering

Klaus Bartschat

Department of Physics and Astronomy, Drake University,
Des Moines, Iowa 50311, USA

Abstract

Theoretical scattering amplitudes are introduced and related to scattering (**S**), transition (**T**), and reactance (**K**) matrices. They can be used to calculate various experimental observables, particularly differential cross sections. A computer program is presented to calculate these amplitudes for individual total spin channels in a non-relativistic angular momentum coupling scheme.

10.1 Introduction

Standard calculations for electron–atom scattering yield energy dependent scattering (**S**), transition (**T**), or reactance (**K**) matrices. (We use **K** as the symbol for the reactance matrix in order to avoid confusion with the **R** matrix defined in Chap. 7.) While they contain all the dynamical information about the collision process, the only observable that can be obtained directly from these matrices is the *total* collision cross section. Any (angle) *differential* observable, however, requires the calculation of scattering amplitudes which depend on the scattering angle of the projectile.

In order to define such amplitudes, it is necessary to choose a particular coordinate system and to write down the explicit form of the initial-state and final-state wavefunction, including the angular part of the projectile motion. This chapter is based on the very detailed analysis given in Sect. 2.3.1 of the book by Smith [10.1]. Instead of repeating all his steps, we will summarize the basic ideas in Sect. 10.2 and concentrate on some modifications that need to be introduced, depending on the asymptotic form of the wavefunction and the phase convention used in setting up the coupled angular momentum states. Section 10.3 deals with the problem of obtaining converged results from partial wave expansions, and some general symmetry properties of scattering amplitudes are discussed in 10.4. Finally, a computer program and some test runs to calculate these amplitudes are described in Sect. 10.5.

10.2 Definition of the Scattering Amplitudes

In this chapter, we are interested in transitions from an initial state denoted as $|\Gamma_0, L_0, M_{L_0}, S_0, M_{S_0}; \boldsymbol{k}_0 m_0\rangle$ to a final state $|\Gamma_1, L_1, M_{L_1}, S_1, M_{S_1}; \boldsymbol{k}_1 m_1\rangle$ where the projectile electrons (or other spin-$\frac{1}{2}$ particles) have initial (final) angular momentum $\boldsymbol{k}_0$ ($\boldsymbol{k}_1$) and spin component m_0 (m_1) with respect to a given quantization axis. Furthermore, the target states involved in the transition have total orbital and spin angular momenta L_0 (L_1) and S_0 (S_1) with components M_{L_0} (M_{L_1}) and M_{S_0} (M_{S_1}), respectively. Finally, Γ_0 and Γ_1 denote collectively all other quantum numbers that are used to construct the target states.

In all programs described in this book, it is assumed that the total spin S, the total angular momentum L, and the total parity $\Pi = (-1)^{\sum_i \ell_i}$ (the sum runs over the angular momenta ℓ_i of all individual electrons) of the combined target + projectile system are conserved during the collision. This LS approximation is very well fulfilled for electron scattering from light neutral targets such a hydrogen, helium or the first few alkali atoms in the periodic table. As outlined in detail by Smith [10.1], the scattering amplitudes for this case are constructed by making the following general assumptions:

1. When the projectile is far away from the target, the initial state can be written as

$$\Psi_{\text{in}} = \mathrm{e}^{ik_0 z} \chi(m_0)\, |L_0, M_{L_0}, S_0, M_{S_0}\rangle \,, \tag{10.1}$$

   i.e., a plane wave $\mathrm{e}^{ik_0 z}$ associated with a spin function $\chi(m_0)$ is multiplied by the initial target state $|\Gamma_0, L_0, M_{L_0}, S_0, M_{S_0}\rangle$.
2. The plane wave is expanded in terms of partial waves, and the spherical harmonics and the target states are coupled together to give properly anti-symmetrized functions with total orbital and spin angular momenta L and S for the combined system, respectively. Note that this expansion is simplified significantly by the fact that the incident plane wave is parallel to the quantization (z) axis.
3. Again far away from the target, the scattered part in the final-state wavefunction Ψ_{fin} is an outgoing spherical wave in the direction $\hat{\boldsymbol{k}}_1 \equiv (\theta, \phi)$. In the coupled representation, this means

$$\begin{aligned} \Psi_{\text{scatt}} &= \Psi_{\text{fin}} - \Psi_{\text{in}} \\ &= \frac{\mathrm{e}^{ik_1 r}}{r} \sum_{M_{L_1} M_{S_1} m_1} \chi(m_1)\, |L_1, M_{L_1}, S_1, M_{S_1}\rangle \\ &\quad \times f(\Gamma_1, L_1, M_{L_1}, S_1, M_{S_1}, \boldsymbol{k}_1, m_1; \Gamma_0, L_0, M_{L_0}, S_0, M_{0_1}, \boldsymbol{k}_0, m_0). \end{aligned} \tag{10.2}$$

With this normalization and our notation, Eq. (2.113) of Smith [10.1] corresponds to the following expression for the scattering amplitude:

$$
\begin{aligned}
&f(\Gamma_1, L_1, M_{L_1}, S_1, M_{S_1}, m_1; \Gamma_0, L_0, M_{L_0}, S_0, M_{S_0}, m_0; \theta, \phi) \\
&\quad = -\sqrt{\frac{\pi}{k_0 k_1}} \sum_{LM_LSM_S\ell_0,l_1,m_{\ell_1}} \mathrm{i}^{\ell_0-\ell_1+1}\sqrt{2\ell_0+1}\, Y_{\ell_1 m_{\ell_1}}(\theta,\phi) \\
&\qquad \times (L_0, M_{L_0}; \ell_0, 0|L, M_L)\ (S_0, M_{S_0}; \tfrac{1}{2}, m_0|S, M_S) \\
&\qquad \times (L_1, M_{L_1}; \ell_1, m_{\ell_1}|L, M_L)\ (S_1, M_{S_1}; \tfrac{1}{2}, m_1|S, M_S) \\
&\qquad \times \mathbf{T}^{LS\Pi}_{\Gamma_0,L_0,\ell_0;\Gamma_1,L_1,\ell_1}, \qquad (10.3)
\end{aligned}
$$

where $(j_1, m_1; j_2, m_2|j_3, m_3)$ is a standard Clebsch-Gordan coefficient and $Y_{\ell m}(\theta, \phi)$ is a spherical harmonic as defined in Edmonds [10.2].

The quantity $\mathbf{T}^{LS\Pi}_{\Gamma_0,L_0\ell_0;\Gamma_1,L_1,\ell_1}$ in (10.3) is the energy dependent transition (**T**) matrix element, and the superscripts indicate that the total angular momentum L, the total spin S, and the total parity Π of the combined target + projectile system are conserved during the collision. The **T** matrix is related to the scattering (**S**) and the reactance (**K**) matrices through

$$
\begin{aligned}
\mathbf{T} = \mathbf{S} - \mathbf{1} &= (\mathbf{1} + \mathrm{i}\mathbf{K})\,(\mathbf{1} - \mathrm{i}\mathbf{K})^{-1} - \mathbf{1} \\
&= -2\,\left(\mathbf{1} + \mathbf{K}^2\right)^{-1} (\mathbf{K}^2 - \mathrm{i}\mathbf{K}), \qquad (10.4)
\end{aligned}
$$

where $\mathbf{1}$ is the identity operator.

An important point to be made with respect to (10.3) concerns the spherical harmonics that are used to construct the coupled states. While Smith's derivation [10.1] (see his Eq. 2.8) assumes that these are the same kind as the $Y_{\ell_1 m_{\ell_1}}(\theta, \phi)$ in (10.3), some computer codes, such as the Belfast R-matrix programs [10.3,4] (see also (7.20)) construct these states by using the functions $\mathcal{Y}_{\ell m} \equiv \mathrm{i}^{\ell} Y_{\ell m}$ defined by Fano and Racah [10.5]. In this case, the factor $\mathrm{i}^{\ell_0-\ell_1}$ must be omitted in (10.3).

Similar equations can be derived for other angular momentum coupling schemes. The most important example is electron scattering from heavy targets where relativistic effects must be taken into account and only the total electronic (orbital + spin) angular momentum J of the system and the parity are conserved (see, for example, Bartschat and Scott [10.6]). Also, the necessary extensions for scattering from charged targets can be found in Smith [10.1] who applied the results of Lane and Thomas [10.7] to this case.

Note that the dependence on the spin quantum numbers is essentially limited to the Clebsch-Gordan coefficients in (10.3). Also, for the most important case of spherically symmetric interactions, the only azimuthal dependence on the angle ϕ enters through the factor $\mathrm{e}^{\mathrm{i}m\phi}$ in the spherical harmonic. One can therefore simplify this equation further by choosing the scattering plane, spanned by $\boldsymbol{k}_0$ and $\boldsymbol{k}_1$, as the xz plane. While this choice corresponds to $\phi \equiv 0$ in (10.3), it still contains all the physically relevant information.

Consequently, we rewrite (10.3) as

$$
\begin{aligned}
f(\Gamma_1, L_1, M_{L_1}, S_1, M_{S_1}, m_1; \Gamma_0, L_0, M_{L_0}, S_0, M_{S_0}, m_0; \theta, \phi) \\
&= \sum_{SM_S} (S_0, M_{S_0}; \tfrac{1}{2}, m_0 | S, M_S)\ (S_1, M_{S_1}; \tfrac{1}{2}, m_1 | S, M_S) \\
&\quad \times\ f^S(M_{L_1}, M_{L_0}; \theta)\,, \qquad (10.5)
\end{aligned}
$$

where

$$
\begin{aligned}
f^S(M_{L_1}, M_{L_0}; \theta) \\
&= -\tfrac{1}{2}\sqrt{\frac{1}{k_0 k_1}} \sum_{L,\ell_0,l_1} \mathrm{i}^{\ell_0-\ell_1+1} \sqrt{(2\ell_0+1)(2\ell_1+1)} \\
&\quad \times\ (L_0, M_{L_0}; \ell_0, 0 | L, M_L)\ (L_1, M_{L_1}; \ell_1, m_{\ell_1} | L, M_L) \\
&\quad \times\ (-1)^{m_{\ell_1}} \sqrt{\frac{(\ell - m_{\ell_1})!}{(\ell + m_{\ell_1})!}}\, P_{\ell_1 m_{\ell_1}}(\theta)\ \mathbf{T}^{LS\Pi}_{\Gamma_0, L_0, \ell_0; \Gamma_1, L_1, \ell_1}\,. \qquad (10.6)
\end{aligned}
$$

In (10.6), we have used the relationship between the spherical harmonics and the associated Legendre polynomials $P_{\ell_1 m_{\ell_1}}(\theta)$ in the form [10.8]

$$
Y_{\ell m}(\theta, \phi) = (-1)^m \sqrt{\frac{2\ell+1}{4\pi}\,\frac{(\ell-m)!}{(\ell+m)!}}\, P_{\ell m}(\theta)\, \mathrm{e}^{\mathrm{i}m\phi} \qquad (10.7)
$$

and set $\phi = 0$. Also, we have dropped the sums over the magnetic quantum numbers M_L and m_{ℓ_1}, since the selection rules for the Clebsch-Gordan coefficients imply that

$$
M_L = M_{L_0}\,, \qquad (10.8a)
$$

$$
m_{\ell_1} = M_L - M_{L_1} = M_{L_0} - M_{L_1}\,. \qquad (10.8b)
$$

The simplifications seen in (10.6,8) demonstrate the advantage of the "collision system" for explicit numerical calculations. In this coordinate frame, the incident beam axis determines the quantization (z) axis and the scattering takes place in the xz plane. Transformations to other coordinate systems, such as the "natural system" where the quantization axis coincides with the normal vector to the scattering plane and the x axis is defined by the incident beam direction, can be achieved by transforming the initial and final states through standard rotation matrices and by using the fact that the action of the **T** operator must be independent of the particular coordinate frame. For more details, see Sect. 10.4 and Chap. 11.

10.3 Convergence of Partial Wave Expansions

An important aspect in many practical calculations for the scattering amplitudes is the fact that the number of numerically available partial wave transitions or reactance matrices is limited. While the total cross sections may have converged to better than 1% with a few partial waves, this is often not the case for the scattering amplitudes, due to the interference terms between matrix elements for small and large angular momenta. This problem has been addressed by many authors, in particular for elastic scattering and for optically allowed transitions where long-range polarization potentials and slowly decreasing coupling potentials between various channels may cause a very slow convergence of the partial wave expansion. Some important solutions are discussed below.

10.3.1 Effective-Range Theory for Elastic Scattering

In elastic scattering from neutral atoms, the interaction potential corresponds to an induced dipole interaction when the projectile is far away from the target. Consequently, if α is the dipole polarizability of the target system, this adiabatic electron–atom potential behaves like $-\alpha/(2r^4)$ for large r. In close-coupling expansions with targets in a ground state without orbital angular momentum (i.e., $L_0 = 0$), this effect is represented by long-range coupling potentials between this S state and excited P states of opposite parity. However, since partial waves with large angular momenta will only "see" this induced dipole potential, it is possible to use "effective-range theory" (ERT) to calculate the corresponding phase shift. The special case mentioned above has been treated by O'Malley *et al* [10.9]. They found that

$$\tan\delta_{\ell_i} = K_{ii} = \frac{\pi\alpha k_i^2}{(2\ell_i+3)(2\ell_i+1)(2\ell_i-1)}\,, \tag{10.9}$$

where ℓ_i is the projectile orbital angular momentum for the diagonal reactance matrix element K_{ii} while k_i^2 is the channel energy in Rydberg. Although the expression is valid for $\ell_i \geq 1$ in the limit of very low energies, it is usually applied to cases where the angular momentum is large enough to ensure small phase shifts where $\delta_{\ell_i} \approx \tan\delta_{\ell_i}$ is also valid. More elaborate formulas, including additional effects such as a quadrupole polarizability, were derived by Ali and Fraser [10.10].

10.3.2 Unitarized Born Approximation for Excitation

Seaton [10.11] discussed various methods to solve the convergence problem for impact excitation, particularly for optically allowed transitions. One approach is the so-called "unitarized Born approximation" where the reactance matrix element between channels "i" and "j" is calculated as

$$K_{ij} = -\sqrt{k_i k_f}\int_0^\infty dr\, r^2 j_{\ell_i}(k_i r) V_{ij}(r) j_{\ell_j}(k_j r)\,. \tag{10.10}$$

Here k_i and k_j are the channel projectile momenta and $V_{ij}(r)$ is the long-range coupling potential matrix element between the channels. Furthermore,

j_{ℓ_i} and j_{ℓ_j} are standard spherical Bessel functions for the partial wave angular momenta ℓ_i and ℓ_j, respectively. They can be calculated with the program described in Chap. 9.

10.3.3 Exponential Fitting

It is important to note that the use of (10.9,10) up to very high partial wave angular momenta might give converged results that nevertheless exhibit unphysical oscillations in the angle differential parameters. This is due to the fact that switching between various approximations is often discontinuous, since not all the important physics is treated in the same way. Recently, Bartschat *et al* [10.12] introduced a semi-empirical fitting procedure for the **K** matrix elements that was found to be (i) sufficiently accurate, (ii) highly efficient, and (iii) very easy to implement.

Specifically, they replaced (10.10) in two-state and five-state close-coupling calculations for positron scattering from alkali atoms by assuming a dependence of the individual **K** matrix elements on the total angular momentum L of the form

$$K_{ij}(L) = a_{ij} \exp\{-b_{ij} L\}; \quad L \geq L_{\text{fit}}. \tag{10.11}$$

Extensive checks showed that (10.11) was indeed a very good approximation, i.e., least-squares fits of $\log |K_{ij}|$ for about ten L values around L_{fit} to a straight line and subsequent extrapolating for $L_{\text{switch}} \leq L \leq L_{\text{max}}$ produced converged results and smooth curves. The switching point L_{switch} was determined as the L value where the fitted and numerical results differed the least over the entire fitting range. The "effective" range formula (10.9) was still used for the elastic channel, but once more the switching point was determined by searching for the smallest discrepancy between the numerical and the analytic results.

The main advantage of the extrapolation method is the fact that a *smooth transition* between the two methods in the "hybrid approach" is essentially guaranteed. Since the overall effect of the high partial waves is basically limited to smoothing out remaining oscillations rather than changing the "wiggle averaged" results dramatically, such a smooth transition will avoid the above-mentioned unphysical oscillations associated with other hybrid methods. An explicit example will be shown in one of the test runs described in Sect. 10.5.2 below.

10.4 Symmetry Properties of Scattering Amplitudes

A important general aspect of scattering amplitudes is the fact that certain symmetry properties of the projectile–target interaction lead to conservation laws through the **T** operator. These, in turn, will cause interdependences between various scattering amplitudes or simply require some amplitudes to vanish. An important example can be seen in (10.3) where the Clebsch-Gordan coefficients reflect the conservation of the z component of the total projectile + target spin as

$$M_S = M_{S_0} + m_0 = M_{S_1} + m_1 \,. \tag{10.12}$$

Hence, amplitudes for transitions that do not fulfill (10.12) must vanish in the LS approximation, and the "sums" over M_S in (10.3,5) are actually restricted to only one term.

Another important example is the conservation of the total parity. This is an extremely good approximation for atomic collision physics. Consequently, computer programs in this field will generally include only projectile–target interactions that conserve the total parity, with the most important example being the Coulomb interaction.

For our case of interest, electron–atom scattering in a plane, the process must be invariant against reflection through a mirror that is placed in the scattering plane. This reflection operation can be constructed as the parity operation, followed by a 180° rotation around the normal axis (the y axis in our case) of the scattering plane. If we consider a general matrix element of the form $\langle j_1 m_1|\mathbf{T}|j_0 m_0\rangle$, reflection invariance implies that

$$\langle j_1 m_1|\mathbf{T}|j_0 m_0\rangle = \langle j_1 m_1|\mathbf{R}^\dagger \mathbf{T}\mathbf{R}|j_0 m_0\rangle \tag{10.13}$$

where

$$\mathbf{R} = \mathbf{D}_y(180^\circ) \circ \mathbf{P} \tag{10.14}$$

is the operator for the reflection, constructed as a product of the parity operator **P** and the rotation operator $\mathbf{D}_y(180^\circ)$.

Although the operation of the parity operator yields the same (eigen)state multiplied by the parity Π of the state, rotations are generally handled through the so-called "Euler angles", as defined, for example, in Edmonds [10.2]. Fortunately, a rotation around the y axis by 180° is straightforward. Using the general transformation

$$|j,m'\rangle_2 = \sum_m D(\gamma,\beta,\alpha)^j_{mm'}\;|j,m\rangle_1 \tag{10.15}$$

for states defined in coordinate systems (X_1, Y_1, Z_1) and (X_2, Y_2, Z_2) that evolve from each other through rotations by a set of Euler angles (γ, β, α), described by rotation matrix elements $D(\gamma,\beta,\alpha)^j_{mm'}$, and using the fact that the above rotation corresponds to $\alpha = \gamma = 0$ and $\beta = -\pi$, we can apply the second part of Eq. (4.2.1) in Edmonds [10.2] and obtain

$$\mathbf{R}|j_0 m_0\rangle = \mathbf{D}(0,-\pi,0) \circ \mathbf{P}|j_0 m_0\rangle = \Pi_0(-1)^{j_0-m_0}|j_0 - m_0\rangle \,. \tag{10.16}$$

Consequently, (10.13) becomes

$$\langle j_1 m_1|\mathbf{T}|j_0 m_0\rangle = \Pi_0 \Pi_1 (-1)^{j_0-m_0+j_1-m_1} \langle j_1 - m_1|\mathbf{T}|j_0 - m_0\rangle \,, \tag{10.17}$$

where Π_0 and Π_1 are the parities of the states. Finally, we note that applying the first part of Eq. (4.2.1) in Edmonds [10.2] would give the same result for scattering amplitudes, since $m_1 + m_0$ must be an integer number and, therefore, the signs for *both* m_0 and m_1 *simultaneously* in the exponent of (10.17) can be reversed.

The above symmetry relationship can be applied immediately to the amplitudes given in (10.3,6), yielding

$$\begin{aligned} f(M_{L_1}, M_{S_1}, m_1; M_{L_0}, M_{S_0}, m_0; \theta) \\ = \Pi_1 \Pi_0 (-1)^{L_1-L_0-\Delta_M+S_1-S_0} \\ \times f(-M_{L_1}, -M_{S_1}, -m_1; -M_{L_0}, -M_{S_0}, -m_0; \theta) \end{aligned} \tag{10.18a}$$

$$f(M_{L_1}, M_{L_0}; \theta) = \Pi_1 \Pi_0 (-1)^{L_1-L_0-\Delta_M} f(-M_{L_1}, -M_{L_0}; \theta) \tag{10.18b}$$

where $\Delta_M \equiv M_{L_1} - M_{L_0}$ and we have left out some of the quantum numbers for simplicity. Note that (10.18) have been simplified by using (10.12) together with the fact that $(-1)^n = (-1)^{-n}$ for integer values of n.

The same method can be applied to other cases, such as amplitudes in a relativistic coupling scheme (see, for example, Eq. (2.10) of Bartschat [10.13]) or for other coordinate systems such as the "natural frame" where the quantization axis is chosen to be perpendicular to the scattering frame. Without proof, we note that the equivalent of (10.17) in this frame is

$$\langle j_1 m_1|\mathbf{T}|j_0 m_0\rangle = \Pi_0 \Pi_1 (-1)^{m_1-m_0} \langle j_1 m_1|\mathbf{T}|j_0 m_0\rangle \,. \tag{10.19}$$

Instead of providing phase relationships ($\pm$) between amplitudes with a given set of magnetic quantum numbers and those with the sign-reversed quantum numbers in the collision frame, (10.19) shows that many amplitudes simply vanish in the natural frame (namely those where the exponent is an even or odd integer, depending on the product of the parities). This fact is one of the many advantages that can be used when formulating the general theory in the natural frame. Numerical calculations, on the other hand, are much simpler in the collision frame. If necessary, the resulting amplitudes can always be transformed into the natural frame. This will be further discussed in Chap. 11.

10.5 Computer Program

The computer program is written in standard FORTRAN77. It is an updated non-relativistic version of the computer code `SCATTAMPREL` [10.6] and calculates the amplitudes given in (10.6). The program consists of the main routine `SCATLS` and several subroutines for the following purposes: `ANALYS` analyzes the input **K** matrix file and re-writes it in a standardized format. (The input analysis should be changed by users to fit their individual needs.) Furthermore, `CLEGOR` calculates Clebsch-Gordan coefficients and `LEPOAS` associated Legendre polynomials; the factorials needed in these routines are set up by an initial call to `LFAK`. `MATMUP` and `MATINV` are matrix multiplication and inversion routines needed for the calculation of **T** matrices from input **K** matrices in `TMAT`. Both of these matrices can be printed by calls to `WRIMAT`. The scattering amplitudes are calculated in `FSCATT` where the summation over all the final-state symmetries and the channel angular momenta in (10.6) is performed. The contributions to the amplitudes from each individual final-state $LS\Pi$ symmetry is obtained in `FPART` for all scattering angles of interest. When all the amplitudes are known, they can be printed from subroutine `WRIMAT`. Finally, `SCATLS` calculates the differential cross section for each individual total spin channel as well as the result for scattering of unpolarized projectiles from unpolarized target atoms in the form

$$\sigma_{\rm u}(\theta) = \frac{k_1}{k_0}\,\frac{1}{\sum_S(2S+1)}\sum_S(2S{+}1)\sum_{M_{L1}M_{L_0}}\left|f^S(M_{L_1},M_{L_0};\theta)\right|^2 . \quad (10.20)$$

In addition to the computer code for calculating the scattering amplitudes, another program `FITMAT` is provided which can be used together with the output from the close-coupling program for positron scattering from alkali-like targets described in Chap. 6. This program reads the **K** matrix elements (called $R_{\nu\nu'}$ in Chap. 6), performs the fitting and extrapolation procedure described in Sect. 10.3.3 above, and writes out the matrix elements in the input format needed by `SCATLS`.

10.5.1 Description of the Input Data

We begin with the input data for the program `SCATLS`. Except for the first record (`FORMAT (1A80)`), the following control parameters are read in free format from unit 5:

1. `TITLE` Title for the run.

2. `ITAPE1,ITAPE2,ITAPE3`

`ITAPE1` Input file for the **K** matrices.
In order to simplify the calculation, each record of this file should contain one **K** matrix element and some additional information in the following form:

`IS,IL,NI,LI,NJ,LJ,AKMAT(IC,JC),ENERGY`

`IS`	$\Pi \cdot (2S+1)$ where $\Pi = \pm 1$ is the total parity.
`IL`	Total angular momentum L.
`NI`	Target state in channel "i".
`LI`	Continuum angular momentum in channel "i".
`NJ`	Target state in channel "j".
`LJ`	Continuum angular momentum in channel "j".
`AKMAT(I,J)`	Matrix element K_{ij}.
`ENERGY`	Collision energy (in Rydberg).

`ITAPE2` After the analysis of the original input file has been carried out, the (scratch) file assigned to `ITAPE2` contains the **K** matrix elements in standardized form for further use.

`ITAPE3` If `ITAPE3` $\neq 0$ and (at the moment) for an $^2\mathrm{S} \rightarrow {}^2\mathrm{P}^o$ transition only, the scattering amplitudes are written to a file assigned to this output unit – in a format that can be used directly as input for the observable program discussed in Chap. 11.

3. `IBUG1,IBUG2`

`IBUG1` 0: normal.
1: **K** matrices are printed.
2: **T** and **K** matrices are printed.

`IBUG2` If `IBUG2` $\neq 0$, the scattering amplitudes for every `IBUG2`-th angle are printed.

4. `NASTOT,IFIN,INIT,NMPTY,LPARMA,NANGLE,IFANO`

`NASTOT` Total number of states in the close-coupling expansion.

`IFIN` Final target state.

`INIT` Initial target state.

`NMPTY` Total number of LS symmetries for which **K** matrices are to be read.

`LPARMA` Highest ℓ value for which associated Legendre polynomials need to be calculated.

`NANGLE` Number of scattering angles.

`IFANO` If `IFANO` $= 1$, the Fano-Racah phase convention is to be used and the factor $\mathrm{i}^{\ell_0 - \ell_1}$ in (10.3,6) is omitted.

5. `LTAR(N), N=1,NASTOT`

L values of the target states.

6. `ISTAR(N), N=1,NASTOT`

$(2S+1)$ values of the target states.

7. `IPTAR(N), N=1,NASTOT`

Parities of the target states ($+1$ for "even", -1 for "odd").

8. `W(N), N=1,NASTOT`

Excitation energies of the target states in Rydberg. (The ground state corresponds to `W(1)` $= 0.0$.)

9. `ENERGY, ANGBEG,ANGDIF`

`ENERGY`	Total collision energy (with respect to the target ground state) in Rydberg.
`ANGBEG`	First scattering angle (in degrees).
`ANGDIF`	Interval for scattering angles (in degrees).

An example input file for a four-state close-coupling calculation for electron impact excitation of the $(3s)^2S \rightarrow (3p)^2P^o$ transition in sodium at an incident electron energy of 4.1 eV is listed below.

```
4-STATE CC  E-NA SCATTERING ... 3S --> 3P AT 4.1 EV  (LS-CALCULATION)
 1    2    3
 0    5
 4    2    1   40  240  181    1
 0    1    0    2
 2    2    2    2
 1   -1    1    1
  0.0000000     0.1546206     0.2345588     0.2658428
  0.3013600     0.0000000     1.0000000
```

The following input data are needed for the program `FITMAT`. Once again, free format is used for the control parameters read from unit 5.

1. `TITLE` Title for the run.

2. `ITAPE1,ITAPE2,NTARG,NDIFFC`

`ITAPE1`	Input unit for **K** matrix elements.
`ITAPE2`	Output unit for **K** matrix elements.
`NTARG`	Number of target states in the close-coupling expansion.
`NDIFFC`	Number of final-state symmetries for which the number of coupled channels is different from the maximum. (This is usually the case for the lowest few L values. For symmetry numbers beyond `NDIFFC`, the same number of coupled channels as for `NDIFFC` is assumed.

3. `LSTATE(N), N=1,NTARG`
   L values of the target states.

4. `LBEGIN,LENDMX,LEXTRP`

`LBEGIN`	Beginning of L range where the program tries to fit $\log\left(\lvert K_{ij}\rvert\right)$ according to (10.11).
`LENDMX`	End of L range for fitting.
`LEXTRP`	End of L range for extrapolation.

5. `LVAL,NCVAL`

`LVAL`	L value of final-state symmetry.
`NCVAL`	Number of coupled channels for final-state symmetry.
Note:	This record needs to be read `NDIFFC` times.

An example input file for a two-state close-coupling calculation for positron impact excitation of the $(3s)^2S \rightarrow (3p)^2P^o$ transition in sodium at an incident electron energy of 20 eV is listed below.

```
2-STATE CC  P-NA SCATTERING ... 3S --> 3P AT 20.0 EV  (LS-CALCULATION)
  1    2    2    2                  ITAPE1,ITAPE2,NTARG,NDIFFC
  0    1    0    2    1             LSTATE
 30   40  200                       LBEGIN,LENDMX,LEXTRP
  0    2                            LVAL,NCVAL
  1    3
```

10.5.2 Test Runs

The first test run deals with electron impact excitation of the $(3s)^2S \rightarrow (3p)^2P^o$ transition in sodium atoms. The scattering amplitudes for an incident electron energy of 4.1 eV were obtained in a four-state non-relativistic calculation, using a new version of the standard Belfast R-matrix codes [10.3,4]. The results for scattering angles of 0°, 5°, and 10° are shown in the condensed output file listed below.

```
ELECTRON SCATTERING FROM SODIUM ... 3S --> 3P AT 4.1 EV  (LS-CALCULATION)
***** FANO-RACAH PHASE CONVENTION *****
RESULTS FOR THE TRANSITION FROM STATE  1 TO STATE  2
TOTAL ENERGY:    0.301E+00 RYDBERG (  0.410E+01 EV )
INCOMING ELECTRON ENERGY:    0.301E+00 RYDBERG (  0.410E+01 EV )
OUTGOING ELECTRON ENERGY:    0.147E+00 RYDBERG (  0.200E+01 EV )
TOTAL SPIN CHANNEL 2S+1 =  1
 ANGLE   XML1   XMLO          REAL PART       IMAG. PART
   0.0    1.0    0.0        0.000000D+00     0.000000D+00
   0.0    0.0    0.0       -0.151660D+02    -0.862459D+01
   0.0   -1.0    0.0        0.000000D+00     0.000000D+00
   5.0    1.0    0.0       -0.290170D+01    -0.487013D+00
   5.0    0.0    0.0       -0.139883D+02    -0.846385D+01
   5.0   -1.0    0.0        0.290170D+01     0.487013D+00
  10.0    1.0    0.0       -0.478967D+01    -0.935529D+00
  10.0    0.0    0.0       -0.110710D+02    -0.801434D+01
  10.0   -1.0    0.0        0.478967D+01     0.935529D+00
TOTAL SPIN CHANNEL 2S+1 =  3
 ANGLE   XML1   XMLO          REAL PART       IMAG. PART
   0.0    1.0    0.0        0.000000D+00     0.000000D+00
   0.0    0.0    0.0       -0.123375D+02    -0.731732D+01
   0.0   -1.0    0.0        0.000000D+00     0.000000D+00
   5.0    1.0    0.0       -0.273057D+01    -0.483579D+00
```

5.0	0.0	0.0	-0.111689D+02	-0.713575D+01
5.0	-1.0	0.0	0.273057D+01	0.483579D+00
10.0	1.0	0.0	-0.445270D+01	-0.927288D+00
10.0	0.0	0.0	-0.827835D+01	-0.662506D+01
10.0	-1.0	0.0	0.445270D+01	0.927288D+00

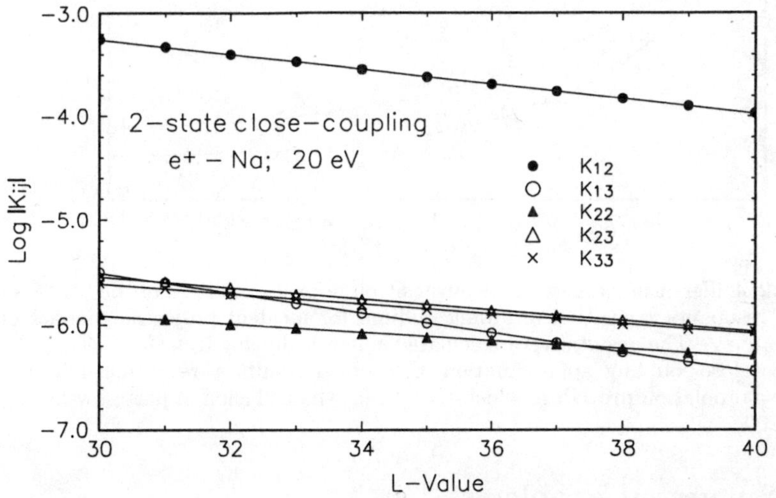

Fig. 10.1. Fits for **K** matrix elements obtained in the two-state close-coupling approximation for positron–sodium collisions at an incident energy of 20.0 eV.

The second test run is a two-state close-coupling calculation for positron impact excitation of the $(3s)^2S \rightarrow (3p)^2P^o$ transition in sodium at an incident electron energy of 20 eV. The results of the exponential fitting procedure described in Sect. 10.3.3 above for an L range between 30 and 40 are shown in Fig. 10.1. The corresponding differential cross sections, as obtained with and without the fitted **K** matrix elements for L values up to 200 are plotted in Fig. 10.2 for incident projectile energies of 20 and 40 eV.

10.6 Summary

In this chapter, we have presented a computer code to calculate scattering amplitudes in a non-relativistic LS coupling scheme. The general symmetry properties of these amplitudes were discussed, as well as approximations schemes to ensure convergence with respect to the number of partial waves included in the calculation.

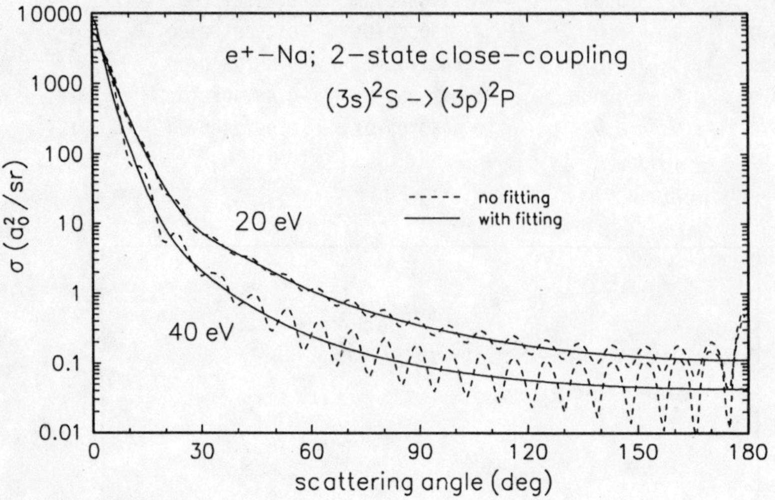

Fig. 10.2. Differential cross sections for positron impact excitation of the $(3s)^2S \rightarrow (3p)^2P^o$ resonance transition in atomic sodium for incident projectile energies of 20 and 40 eV. The results were calculated numerically for $0 \leq L \leq 40$ in the two-state close-coupling approximation. Converged results were obtained by the fitting/extrapolation procedure which allowed for the inclusion of partial waves up to $L = 200$.

10.7 Suggested Problems

1. Check the convergence of the results from the first test run with respect to the number of partial waves by truncating the number of final-state symmetries included in the summation.
2. Use the close-coupling program presented in Chap. 6 to calculate **K** matrices for other cases and calculate the scattering amplitudes. Compare your results with those of [10.12].
3. Extend the fitting program presented for the case of positron scattering to allow for the extrapolation of **K** matrices for individual spin channels in electron–atom collisions. Check your program with the numerical **K** matrices provided for the first test run.
4. Use the R-matrix program described in Chap. 7 and calculate scattering amplitudes for electron scattering from atomic hydrogen. Write a computer code that takes the **K** matrices printed in the program `PACK2` and produces an input file that can be used by the present program.
5. Replace the fitting/extrapolation routine currently implemented in the `DWBA1` program described in Chap. 4 by the method discussed in 10.3.3 above. Compare the accuracy of the two fitting algorithms and compare your results again with those of [10.12].

6. Prove (10.19) and generalize the result to more than one angular momentum in the initial and final states.
7. Write a program to transform scattering amplitudes from the "collision system" to the "natural frame". The corresponding Euler angles are $\alpha = -\pi/2$, $\beta = -\pi/2$, and $\gamma = 0$, respectively.
8. Use the time-reversal invariance of the **T** operator to derive a relationship between scattering amplitudes for inelastic excitation and "superelastic" de-excitation of a given transition. (The effect of time inversion on quantum states is discussed, for example, in the book by Taylor [10.14].) In general, such relationships will further reduce the number of independent scattering amplitudes for elastic scattering. For a detailed discussion of this problem for elastic electron scattering from spin-$\frac{1}{2}$ targets, see the paper by Burke and Mitchell [10.15].

Acknowledgments

This work has been supported, in part, by the National Science Foundation under grants PHY-9014103 and PHY-9318377.

References

10.1 K. Smith, *The Calculation of Atomic Collision Processes*, (New York: John Wiley, 1971)
10.2 A.R. Edmonds, *Angular Momentum in Quantum Mechanics*, (Princeton University Press, 1957)
10.3 K.A. Berrington, P.G. Burke, M. LeDourneuf, W.D. Robb, K.T. Taylor and Vo Ky Lan, Comp. Phys. Commun. **14** (1978) 367.
10.4 N.S. Scott and K.T. Taylor, Comp. Phys. Commun. **25** (1982) 347.
10.5 U. Fano and G. Racah, *Irreducible Tensorial Sets*, (New York: Academic Press, 1959)
10.6 K. Bartschat and N.S. Scott, Comp. Phys. Commun. **30** (1983) 369
10.7 A.M. Lane and R.G. Thomas, Rev. Mod. Phys. Commun. **30** (1958) 2911
10.8 H. Friedrich, *Theoretical Atomic Physics*, (Berlin: Springer, 1990)
10.9 T.F. O'Malley, L. Spruch and L. Rosenberg, J. Math. Phys. **2** (1961) 491
10.10 M.K. Ali and P.A. Fraser, J. Phys. B **10** (1977) 3091
10.11 M.J. Seaton, Proc. Phys. Soc. **77** (1961) 174
10.12 K. Bartschat, K.M. DeVries, R.P. McEachran and A.D. Stauffer, Hyperfine Interactions **89** (1994) 57
10.13 K. Bartschat, Phys. Rep. **180** (1989) 1
10.14 J.R. Taylor, *Scattering Theory* (New York: John Wiley, 1972)
10.15 P.G. Burke and J.F.B. Mitchell, J. Phys. B **7** (1974) 214

11. Density Matrices: Connection Between Theory and Experiment

Klaus Bartschat

Department of Physics and Astronomy, Drake University,
Des Moines, Iowa 50311, USA

Abstract

The basic concept of density matrices is introduced. It is demonstrated how reduced density matrices relate theoretical scattering amplitudes to experimental observables. A computer program is presented to calculate various observables from bilinear products of scattering amplitudes for impact excitation of (quasi) one-electron atoms.

11.1 Introduction

Besides the "classical" experiment of scattering physics, namely the measurement of cross sections, a wealth of other experiments are currently performed in modern atomic physics laboratories. While enormous progress is being made in terms of experimental techniques including, for example, the preparation of spin polarized projectile and target beams (for an introduction, see Kessler [11.1]) with narrow energy distributions and the detailed determination of the final state (by polarization analysis and particle–particle or particle–photon coincidences), the development of new theoretical methods has also been strongly encouraged. This is due, in part, to the construction of fast electronic computers which allow for increasingly complex programs to be run in shorter periods of time.

Through close interaction between experimental and theoretical work, a very detailed test of theoretical models is made possible. With regard to the variety of processes studied and the experimental and theoretical methods applied, the question arises whether it is possible to establish some connections in the description of the different phenomena. In identifying such links, more systematism can be achieved both in the general formulation and in the numerical calculation of the individual observables.

As outlined in the monograph by Blum [11.2] and more recent reviews by Andersen *et al* [11.3,4] and Bartschat [11.5], the density-matrix formalism is an ideal tool to achieve this goal. Although the density-matrix elements are calculated from theoretical scattering amplitudes, they can also be directly related to experimental observables, such as the "Stokes parameters" which describe the polarization of light emitted in optical transitions, or the "STU parameters" that determine the effect of an electron spin polarization on the outcome of a collision process.

We briefly review the definition of scattering amplitudes in Sect. 11.2 before introducing the density matrix and, more importantly, reduced density matrices for various experimental setups in 11.3. Section 11.4 deals with the parametrization of the density matrix in terms of irreducible tensor operators and the corresponding state multipoles, with special emphasis on their transformation properties and time evolution. Some frequently measured observables, such as the STU and Stokes parameters mentioned above, are introduced in Sect. 11.5. Since the special case of a ${}^2\mathrm{S} \to {}^2\mathrm{P}^\mathrm{o}$ transition in quasi one-electron targets can be studied directly with several of the computer codes presented in this book, it is discussed in more detail in Sect. 11.6, and a computer program to calculate various observables is presented in 11.7.

11.2 Scattering Amplitudes

The scattering amplitudes

$$f(M_1 m_1, M_0 m_0; \theta) \equiv \langle J_1 M_1; \boldsymbol{k}_1 m_1 \mid \mathbf{T} \mid J_0 M_0; \boldsymbol{k}_0 m_0 \rangle \tag{11.1}$$

describe the scattering of electrons (or, more general, spin-$\frac{1}{2}$ particles) with initial linear momentum $\boldsymbol{k}_0$ and spin component m_0 (with regard to a given quantization axis) from a target with total angular momentum J_0 and component M_0. The final state is characterized by the projectile momentum $\boldsymbol{k}_1$, the spin component m_1, and the target quantum numbers J_1 and M_1, respectively. The scattering angle θ is the angle between $\boldsymbol{k}_0$ and $\boldsymbol{k}_1$, and $\mathbf{T}$ is the transition operator, usually presented as the $\mathbf{T}$ matrix. Depending on the representation of the continuum wavefunctions, there may be a normalization factor in (11.1) which, for simplicity, is omitted in the general formulation presented here. It will be further discussed in the calculation of absolute differential cross sections from numerical scattering amplitudes below.

In all programs described in this book, it is assumed that the total spin S and the total angular momentum L of the combined target + projectile system is conserved during the collision. In this case, the scattering amplitudes depend in a purely algebraic way on the spin quantum numbers, and transitions between fine-structure levels are described by standard recoupling techniques. As a consequence, observables for such transitions are often related by simple factors. This will be discussed in detail for the special case of ${}^2\mathrm{S}_{1/2} \to {}^2\mathrm{P}^\mathrm{o}_{1/2,3/2}$ transitions in quasi one-electron systems in Sect. 11.6.

As shown by Bartschat and Madison [11.6], the scattering amplitudes (11.1) are then related to the amplitudes

$$f^S(M_{L_1}, M_{L_0}; \theta) \equiv \langle L_1 M_{L_1}; \boldsymbol{k}_1 \mid \mathbf{T}^S \mid L_0 M_{L_0}; \boldsymbol{k}_0 m_0 \rangle \tag{11.2}$$

defined in Chap. 10 as follows:

$$
\begin{aligned}
& f(M_1 m_1, M_0 m_0; \theta) \\
& \quad = \sum_{\substack{L_1, M_{L_1}, \\ L_0, M_{L_0}, \\ S, M_S}} (L_0, M_{L_0}; S_0, M_{S_0} | J_0, M_0)\ (S_0, M_{S_0}; \tfrac{1}{2}, m_0 | S, M_S) \\
& \qquad \times (L_1, M_{L_1}; S_1, M_{S_1} | J_1, M_1)\ (S_1, M_{S_1}; \tfrac{1}{2}, m_1 | S, M_S) \\
& \qquad \times f^S(M_{L_1}, M_{L_0}; \theta) ,
\end{aligned}
\tag{11.3}
$$

where S_0, S_1, M_{S_0}, M_{S_1}, L_0, L_1, M_{L_0}, and M_{L_1} are the spins as well as the angular momenta and the corresponding z components of the initial and final target states. Note that (11.3) expresses the conservation of the total spin S and its component

$$M_S = M_{S_1} + m_1 = M_{S_0} + m_0 \tag{11.4}$$

through the Clebsch-Gordan coefficients $(j_1, m_1; j_2, m_2 | j_3, m_3)$.

In practical applications, it is necessary to define the above scattering amplitudes with respect to a quantization axis for the angular momentum components. A standard choice for numerical calculations is the so-called "collision system" where the incident beam axis is the quantization (z) axis while the y axis is perpendicular to the scattering plane. On the other hand, many observables can be interpreted more easily in the "natural system" where the quantization axis coincides with the normal vector to the scattering plane and the x axis is defined by the incident beam direction. The transformation of the scattering amplitudes from one system to another can be achieved in a straightforward way by transforming the initial and final states through standard rotation matrices and using the fact that the action of the **T** operator must be independent of the particular coordinate system.

This transformation is performed as follows. Consider states $|j, m\rangle_1$ defined in a coordinate system (X_1, Y_1, Z_1). Suppose a second coordinate system (X_2, Y_2, Z_2) is obtained from (X_1, Y_1, Z_1) through a rotation by a set of three Euler angles (γ, β, α) as defined by Edmonds [11.7]. The corresponding states $|j, m'\rangle_2$ in system (X_2, Y_2, Z_2) are then obtained from the $|j, m\rangle_1$ according to

$$|j, m'\rangle_2 = \sum_m D(\gamma, \beta, \alpha)^j_{mm'}\ |j, m\rangle_1 , \tag{11.5}$$

where

$$D(\gamma, \beta, \alpha)^j_{mm'} = \mathrm{e}^{im\gamma}\ d(\beta)^j_{mm'}\ \mathrm{e}^{im'\alpha} \tag{11.6}$$

and $d(\beta)^j_{mm'}$ are rotation-matrix elements (for details, see [11.7]). Note that the angular momentum j is invariant against such rotations.

As an explicit example, consider amplitudes for an S → P transition without consideration of the electron spin. Omitting the arguments $\boldsymbol{k}_0$, $\boldsymbol{k}_1$, θ and S for simplicity, the corresponding amplitudes are

$$f(M) \equiv \langle 1, M|\mathbf{T}|0,0\rangle \,. \tag{11.7}$$

The natural coordinate system evolves from the collision system through rotation by the Euler angles $\alpha = -\pi/2$, $\beta = -\pi/2$ and $\gamma = 0$. Furthermore, only the P states have to be transformed since the spherically symmetric S states are invariant against such a rotation. Consequently, the amplitudes in the natural system (subscript "n") and the collision system (subscript "c") are related by

$$\begin{aligned}
\langle 1,1|\mathbf{T}|0,0\rangle_\mathrm{n} &= -\frac{\mathrm{i}}{2}\,\langle 1,1|\mathbf{T}|0,0\rangle_\mathrm{c} - \frac{1}{\sqrt{2}}\,\langle 1,0|\mathbf{T}|0,0\rangle_\mathrm{c} + \frac{\mathrm{i}}{2}\,\langle 1,-1|\mathbf{T}|0,0\rangle_\mathrm{c} \\
&= -\mathrm{i}\,\langle 1,1|\mathbf{T}|0,0\rangle_\mathrm{c} - \frac{1}{\sqrt{2}}\,\langle 1,0|\mathbf{T}|0,0\rangle_\mathrm{c}\,, \qquad (11.8\mathrm{a})\\
\langle 1,-1|\mathbf{T}|0,0\rangle_\mathrm{n} &= -\frac{\mathrm{i}}{2}\,\langle 1,1|\mathbf{T}|0,0\rangle_\mathrm{c} + \frac{1}{\sqrt{2}}\,\langle 1,0|\mathbf{T}|0,0\rangle_\mathrm{c} + \frac{\mathrm{i}}{2}\,\langle 1,-1|\mathbf{T}|0,0\rangle_\mathrm{c} \\
&= -\mathrm{i}\,\langle 1,1|\mathbf{T}|0,0\rangle_\mathrm{c} + \frac{1}{\sqrt{2}}\,\langle 1,0|\mathbf{T}|0,0\rangle_\mathrm{c}\,, \qquad (11.8\mathrm{b})\\
\langle 1,0|\mathbf{T}|0,0\rangle_\mathrm{n} &\equiv 0\,, \qquad (11.8\mathrm{c})
\end{aligned}$$

where we have used the reflection invariance of the interaction with respect to the scattering plane (see Chap. 10). Recall that this reflection invariance typically provides phase relationships (±) between amplitudes with a given set of magnetic quantum numbers and those with the sign-reversed quantum numbers in the collision frame, while approximately half of the amplitudes simply vanish in the natural frame. This fact is one of the many advantages that can be used when formulating the general theory in the natural frame. For more details, see [11.3,4].

11.3 Density Matrices

A thorough introduction to the theory of density matrices and their applications with emphasis in atomic physics can be found in the book by Blum [11.2], and extended applications to electron collisions have been given in [11.3–5]. The main advantage of the density-matrix formalism is its ability to deal with pure and mixed states in the same consistent manner. The preparation of the initial state as well as the details regarding the observation of the final state can be treated in a systematic way. In particular, averages over quantum numbers of unpolarized beams in the initial state and incoherent sums over non-observed quantum numbers in the final state can be accounted for by the "reduced density matrix". Furthermore, expansion of the density matrix in terms of "irreducible tensor operators" and the corresponding "state multipoles" (see Sect. 11.4) allows for the use of advanced angular momentum techniques.

Following Blum [11.2], the "complete" density operator after the collision process is given by

$$\boldsymbol{\rho}_{\text{out}} = \mathbf{T}\,\boldsymbol{\rho}_{\text{in}}\,\mathbf{T}^{\dagger}\,, \tag{11.9}$$

where $\boldsymbol{\rho}_{\text{in}}$ is the density operator before the collision and the dagger denotes the adjoint operator. The corresponding matrix elements are given by

$$\begin{aligned}(\boldsymbol{\rho}_{\text{out}})^{M_1' M_1}_{m_1' m_1;\theta} = &\sum_{m_0' m_0 M_0' M_0} \boldsymbol{\rho}_{m_0' m_0}\,\boldsymbol{\rho}_{M_0' M_0} \\ &\times\; f(M_1' m_1', M_0' m_0'; \theta)\, f^*(M_1 m_1, M_0 m_0; \theta)\,,\end{aligned} \tag{11.10}$$

where the $\boldsymbol{\rho}_{m_0' m_0}\boldsymbol{\rho}_{M_0' M_0}$ describe the preparation of the initial state and the star denotes the complex conjugate quantity. (We assume that the projectile and target beams are prepared independently, thereby allowing for this factorization.)

As pointed out above, "reduced" density matrices account for the fact that, in practical experiments, not all quantum mechanically allowed quantum numbers are determined simultaneously. For example, if only the scattered projectiles are observed, the corresponding elements of the reduced density matrix are obtained by summing over the atomic quantum numbers as follows:

$$(\boldsymbol{\rho}_{\text{out}})_{m_1' m_1;\theta} = \sum_{M_1'=M_1} (\boldsymbol{\rho}_{\text{out}})^{M_1' M_1}_{m_1' m_1;\theta}\,. \tag{11.11}$$

The differential cross section is then given by

$$\frac{\mathrm{d}\sigma(\theta)}{\mathrm{d}\Omega} = C \sum_{m_1'=m_1} (\boldsymbol{\rho}_{\text{out}})_{m_1' m_1;\theta}\,, \tag{11.12}$$

where C is a constant that depends on the normalization of the continuum waves in a numerical calculation. These reduced density-matrix elements contain information about the projectile spin. The information can be extracted through a measurement of the generalized STU parameters [11.5] which will be discussed in Sect. 11.5.1 below.

On the other hand, if only the atoms are observed (for example, by analyzing the light emitted in optical transitions), the elements

$$(\boldsymbol{\rho}_{\text{out}})^{M_1' M_1} = \int \mathrm{d}^3\boldsymbol{k}_1 \sum_{m_1'=m_1} (\boldsymbol{\rho}_{\text{out}})^{M_1' M_1}_{m_1' m_1;\theta} \tag{11.13}$$

determine the "integrated Stokes parameters", i.e., the polarization of the emitted light independent of the electron scattering angle. These parameters contain information about the angular momentum distribution in the excited target ensemble.

Finally, for electron–photon coincidence experiments without projectile spin analysis in the final state, the elements

$$(\rho_{\rm out})_\theta^{M_1' M_1} = \sum_{m_1'=m_1} (\rho_{\rm out})_{m_1' m_1;\theta}^{M_1' M_1} \tag{11.14}$$

simultaneously contain information about the projectiles and the target. This information can be extracted by measuring the angle-differential Stokes parameters (see Sect. 11.5.2).

The density-matrix formalism outlined above is very useful for obtaining a qualitative description of the geometrical and sometimes also of the dynamical symmetries of the collision process [11.5].

11.4 Irreducible Tensor Operators and State Multipoles

The general density-matrix theory can be formulated in a very elegant fashion by decomposing the density operator in terms of irreducible components whose matrix elements become the so-called "state multipoles". In such a formulation, full advantage can be taken of the most sophisticated techniques developed in angular momentum algebra. A thorough introduction to this method can be found, for example, in the book by Blum [11.2].

11.4.1 Basic Definitions

The density operator for an ensemble of particles in quantum states labeled as $|J, M\rangle$ where J and M are the total angular momentum and its magnetic component, respectively, can be written as

$$\rho = \sum_{M'M} \rho^J_{M'M} |JM'\rangle\langle JM| \,, \tag{11.15}$$

where

$$\rho^J_{M'M} = \langle JM'|\rho|JM\rangle \tag{11.16}$$

are the matrix elements. For simplicity, we have assumed that J is well defined, but coherences ($M' \neq M$) between different magnetic sublevels are possible.

Alternatively, one may write [11.2]

$$\rho = \sum_{KQ} \langle T(J)^{\dagger}_{KQ} \rangle \, \mathbf{T}(J)_{KQ} \,, \tag{11.17}$$

where the irreducible tensor operators (not to be confused with the **T** operator) are defined as

$$\mathbf{T}(J)_{KQ} = \sum_{M'M} (-1)^{J-M} \, (J, M'; J, -M|K, Q) \, |JM'\rangle\langle JM| \tag{11.18}$$

and the "state multipoles" or "statistical tensors" are given by

$$\langle T(J)^{\dagger}_{KQ} \rangle = \sum_{M'M} (-1)^{J-M} \, (J, M'; J, -M|K, Q) \, \langle JM'|\rho|JM\rangle. \tag{11.19}$$

Note that the selection rules for the Clebsch-Gordan coefficients imply

$$0 \le K \le 2J \,, \tag{11.20a}$$

$$M' - M = Q \,. \tag{11.20b}$$

Equation (11.19) can be inverted through the orthogonality condition of the Clebsch-Gordan coefficients to give

$$\langle J'M'|\rho|JM\rangle = \sum_{KQ} (-1)^{J'-M} \, (J, M'; J, -M|K, Q) \, \langle T(J)^{\dagger}_{KQ} \rangle. \tag{11.21}$$

The transformation properties of the tensor operators and the state multipoles defined in two different coordinate systems are given by expressions similar to (11.5) as

$$\mathbf{T}(J)_{KQ} = \sum_{q} \mathbf{T}(J)_{Kq} \, D(\gamma, \beta, \alpha)^{K}_{qQ} \tag{11.22a}$$

i.e., the rank K of the tensor operator is invariant, and

$$\langle T(J)^{\dagger}_{KQ} \rangle = \sum_{q} \langle T(J)^{\dagger}_{Kq} \rangle \; D(\gamma, \beta, \alpha)^{K*}_{qQ} \,. \tag{11.22b}$$

The irreducible tensor operators fulfill the orthogonality condition

$$\mathrm{tr} \, \{\mathbf{T}(J)_{KQ} \, \mathbf{T}(J)^{\dagger}_{K'Q'}\} = \delta_{K'K} \, \delta_{Q'Q}. \tag{11.23}$$

With

$$\mathbf{T}(J)_{00} = \frac{1}{\sqrt{2J+1}} \, \mathbf{1} \tag{11.24}$$

being proportional to the unit operator **1**, it follows that all tensor operators have vanishing trace, except for the monopole $\mathbf{T}(J)_{00}$.

The hermiticity condition for the density matrix implies

$$\langle T(J)^{\dagger}_{KQ} \rangle^{*} = (-1)^{Q} \langle T(J)^{\dagger}_{K-Q} \rangle. \tag{11.25}$$

Hence, the state multipoles $\langle T(J)^\dagger_{K0}\rangle$ are real numbers. Furthermore, the transformation property (11.22b) of the state multipoles imposes restrictions on non-vanishing state multipoles to describe systems with given symmetry properties. In detail, one finds the following.

(a) For spherically symmetric systems:

$$\langle T(J)^\dagger_{KQ}\rangle = \langle T(J)^\dagger_{Kq}\rangle_{\text{rot}} \tag{11.26}$$

for *all* sets of Euler angles. This implies that only the monopole term $\langle T(J)^\dagger_{00}\rangle$ can be different from zero.

(b) For axially symmetric systems:

$$\langle T(J)^\dagger_{KQ}\rangle = \langle T(J)^\dagger_{Kq}\rangle_{\text{rot}} \tag{11.27}$$

for *all* Euler angles ϕ that describe a rotation around the z axis. Since this angle enters through a factor $\exp\{-iQ\phi\}$ into the general transformation formula (11.22b), it follows that only state multipoles with $Q = 0$ can be different from zero in such a situation.

(c) For systems with properties invariant under reflection in the xz plane:

$$\langle T(J)^\dagger_{KQ}\rangle = (-1)^K \langle T(J)^\dagger_{KQ}\rangle^* \,. \tag{11.28}$$

In this case, state multipoles with even rank K are real numbers, while those with odd rank are purely imaginary.

The above results can be applied immediately to the description of atomic collisions in the "collision frame". For example, impact excitation of unpolarized targets by unpolarized projectiles without observation of the scattered projectiles is symmetric both with regard to rotation around the incident beam axis and with regard to reflection in any plane containing this axis. Consequently, the state multipoles $\langle T(J)^\dagger_{00}\rangle$, $\langle T(J)^\dagger_{20}\rangle$, $\langle T(J)^\dagger_{40}\rangle, \ldots$, fully characterize the atomic ensemble of interest. From (11.22b), similar relationships can be derived for state multipoles defined with regard to other coordinate frames, such as the "natural system".

As can be seen from the above discussion, the description of systems that do not exhibit spherical symmetry requires the knowledge of state multipoles with rank $K \neq 0$. Frequently, the multipoles with $K = 1$ and $K = 2$ are determined through the angular correlation and the polarization of radiation emitted from an ensemble of collisionally excited targets. Note, for example, that the state multipoles with $K = 1$ are proportional to the spherical components of the angular momentum expectation value [11.2] and, therefore, give rise to a non-vanishing circular light polarization (see also Sect. 11.5.2). This corresponds to a sense of rotation or an "orientation" in the ensemble which is therefore called "oriented". On the other hand, non-vanishing multipoles with rank $K \geq 2$ describe the "alignment" of the system.

11.4.2 Coupled Systems

Tensor operators and state multipoles for coupled systems are constructed as direct products of the operators for the individual systems. For example, the combined density operator for two subsystems in basis states $|L, M_L\rangle$ and $|S, M_S\rangle$ is constructed as [11.2]

$$\rho = \sum_{KQkq} \langle T(L)^\dagger_{KQ} \times T(S)^\dagger_{kq}\rangle \, [\mathbf{T}(L)_{KQ} \times \mathbf{T}(S)_{kq}]. \tag{11.29}$$

If the two systems are uncorrelated, the state multipoles factorize as

$$\langle T(L)^\dagger_{KQ} \times T(S)^\dagger_{kq}\rangle = \langle T(L)^\dagger_{KQ}\rangle \, \langle T(S)^\dagger_{kq}\rangle. \tag{11.30}$$

Furthermore, coupled operators can be defined as

$$\mathbf{T}(J', J)_{K'Q'} = \sum_{KQkq} \hat{K}\hat{k}\hat{J}\hat{J}' \, (KQ, kq|K'Q') \times \begin{Bmatrix} K & k & K' \\ L & S & J' \\ L & S & J \end{Bmatrix} \mathbf{T}(L)_{KQ} \times \mathbf{T}(S)_{kq}\,, \tag{11.31}$$

where $\hat{x} \equiv \sqrt{2x+1}$ and $\begin{Bmatrix} j_1 & j_2 & j_3 \\ j_4 & j_5 & j_6 \\ j_7 & j_8 & j_9 \end{Bmatrix}$ is a 9j-symbol.

Note that the coupling of fixed L and S values leads, in general, to several possible J values, in accordance with the rules for angular momentum coupling.

11.4.3 Time Evolution of State Multipoles: Quantum Beats

The time evolution of the density operator is determined by the Liouville equation [11.2]

$$\rho(t) = \mathbf{U}(t)\, \rho(0)\, \mathbf{U}^\dagger(t)\,, \tag{11.32}$$

where $\mathbf{U}(t)$ is the time evolution operator that relates wavefunctions at times $t_0 = 0$ and t according to

$$|\Psi(\boldsymbol{r}, t)\rangle = \mathbf{U}(t)\, |\Psi(\boldsymbol{r}, 0)\rangle\,. \tag{11.33}$$

From the general expansion

$$\rho(t) = \sum_{kq} \langle T(j;t)^\dagger_{kq}\rangle \mathbf{T}(j)_{kq} \tag{11.34}$$

and (11.19) for time $t_0 = 0$, it follows that

$$\langle T(j;t)^{\dagger}_{kq}\rangle = \sum_{JKQ} \langle T(J)^{\dagger}_{KQ}\rangle \, G(J,j;t)^{Qq}_{Kk}\,, \tag{11.35}$$

where the "perturbation coefficients" are defined as

$$G(J,j;t)^{Qq}_{Kk} = \mathrm{tr}\,[\mathbf{U}(t)\mathbf{T}(J)_{KQ}\mathbf{U}(t)^{\dagger}\mathbf{T}(j)^{\dagger}_{kq}]. \tag{11.36}$$

Hence, these coefficients relate the state multipoles at time t to those at $t_0 = 0$.

An important application of the perturbation coefficients is the coherent excitation of several quantum states which subsequently decay by optical transitions. Such an excitation may be performed, for example, in beam–foil experiments or electron–atom collisions where the energy width of the electron beam is too large to resolve the fine structure (or hyperfine structure) of the target states.

Suppose, for instance, that explicitly relativistic effects, such as the spin-orbit interaction between the projectile and the target, can be neglected *during* the collision process between an incident electron and a target atom. In this case, the orbital angular momentum (L) system of the collisionally excited target states may be oriented and/or aligned, depending on the scattering angle of the projectile. On the other hand, the spin (S) system remains unaffected; in particular, it remains unpolarized if both the target and the projectile beams are unpolarized before the collision.

During the lifetime of the excited target states, however, the spin-orbit interaction *within* the target produces an exchange of orientation between the L and the S systems, which results in a net loss of orientation in the L system. This effect can be observed directly through the intensity and the polarization of the light emitted from the excited target ensemble. The perturbation coefficients for the fine-structure interaction are found to be [11.2]

$$G(L;t)_K = \frac{\exp\{-\gamma t\}}{2S+1} \sum_{J'J} (2J'+1)(2J+1) \begin{Bmatrix} L & J' & S \\ J & L & K \end{Bmatrix}^2 \cos(\omega_{J'J} t)\,, \tag{11.37}$$

where $\begin{Bmatrix} j_1 & j_2 & j_3 \\ j_4 & j_5 & j_6 \end{Bmatrix}$ is a standard 6j-symbol and $\omega_{J'J} = \omega_{J'} - \omega_J$ corresponds to the (angular) frequency difference between the various multiplet states with total electronic angular momenta J' and J, respectively. Also, γ is the natural width of the spectral line; for simplicity, the same lifetime has been assumed in (11.37) for all states of the multiplet.

Note that the perturbation coefficients are independent of the multipole component Q in this case, and that there is no mixing between different multipole ranks K. Similar results can be derived for the hyperfine interaction and also to account for the combined effects of fine structure and hyperfine structure [11.2]. The cosine terms represent correlations between the signal from different fine-structure states, and they lead to oscillations in the light intensity as well as the measured Stokes parameters in a time-resolved experiment.

11.4.4 Time Integration over Quantum Beats

If the excitation and decay times are not resolved in a given experimental setup, the perturbation coefficients need to be integrated over time. As a result, the quantum beats will disappear, but a net effect may still be visible through a depolarization of the emitted radiation. For the case of atomic fine-structure interaction discussed above, one finds [11.2]

$$\bar{G}(L)_K = \int_0^\infty \mathrm{d}t\, G(L;t)_K = \frac{1}{2S+1} \sum_{J'J} (2J'+1)(2J+1) \times \begin{Bmatrix} L & J' & S \\ J & L & K \end{Bmatrix}^2 \frac{\gamma}{\gamma^2 + \omega_{J'J}^2} . \tag{11.38}$$

Note that the amount of depolarization depends on the relationship between the fine-structure splitting and the natural line width. For $|\omega_{J'J}| \gg \gamma$ (if $J' \neq J$), the terms with $J' = J$ will dominate and cause the maximum depolarization; for the opposite case $|\omega_{J'J}| \ll \gamma$, the sum rule of the 6j-symbols can be applied and no depolarization will be observed.

11.5 Observables

In this section, we will discuss in some detail the parametrization of the reduced density matrices that describe the spin polarization of the projectiles (11.11) and the angular momentum distribution of the excited target states (11.14) after the collision process. The former can be determined through a measurement of the "generalized *STU* parameters" [11.5] and the "generalized Stokes parameters" [11.8], respectively. All these functions depend on the scattering angle θ which will, however, be omitted for clarity of notation.

11.5.1 Generalized *STU* Parameters

We begin with the case of spin polarized electron scattering from unpolarized targets. Using the hermiticity of the density matrix and assuming parity conservation in the projectile–target interaction, it was shown in [11.5] that the reduced spin density matrix with the elements given by (11.11) can be fully characterized in terms of the eight independent parameters

$$\sigma_{\mathrm{u}} \equiv \frac{1}{2} \sum_{m_1 m_0} \langle m_1 m_0; m_1 m_0 \rangle = \frac{1}{2(2J_0+1)} \sum_{M_1 M_0 m_1 m_0} |f(M_1 m_1, M_0 m_0)|^2, \tag{11.39a}$$

$$\begin{aligned} S_{\mathrm{P}} &\equiv -\frac{2}{\sigma_{\mathrm{u}}} \mathrm{Im}\{\langle \tfrac{1}{2}\tfrac{1}{2}; -\tfrac{1}{2}\tfrac{1}{2} \rangle\} \\ &= \frac{-2}{\sigma_{\mathrm{u}}(2J_0+1)} \mathrm{Im} \left\{ \sum_{M_1 M_0} f(M_1 \tfrac{1}{2}, M_0 \tfrac{1}{2}) f^*(M_1 - \tfrac{1}{2}, M_0 \tfrac{1}{2}) \right\}, \end{aligned} \tag{11.39b}$$

$$\begin{aligned} S_{\mathrm{A}} &\equiv -\frac{2}{\sigma_{\mathrm{u}}} \mathrm{Im}\, \{\langle \tfrac{1}{2} - \tfrac{1}{2}; \tfrac{1}{2}\tfrac{1}{2} \rangle\} \\ &= \frac{-2}{\sigma_{\mathrm{u}}(2J_0+1)} \mathrm{Im} \left\{ \sum_{M_1 M_0} f(M_1 \tfrac{1}{2}, M_0 - \tfrac{1}{2}) f^*(M_1 \tfrac{1}{2}, M_0 \tfrac{1}{2}) \right\}, \end{aligned} \tag{11.39c}$$

$$T_x \equiv \frac{1}{\sigma_{\mathrm{u}}} \left[\langle -\tfrac{1}{2} - \tfrac{1}{2}; \tfrac{1}{2}\tfrac{1}{2} \rangle + \langle -\tfrac{1}{2}\tfrac{1}{2}; \tfrac{1}{2} - \tfrac{1}{2} \rangle \right], \tag{11.39d}$$

$$T_y \equiv \frac{1}{\sigma_{\mathrm{u}}} \left[\langle -\tfrac{1}{2} - \tfrac{1}{2}; \tfrac{1}{2}\tfrac{1}{2} \rangle - \langle -\tfrac{1}{2}\tfrac{1}{2}; \tfrac{1}{2} - \tfrac{1}{2} \rangle \right], \tag{11.39e}$$

$$T_z \equiv \frac{1}{\sigma_{\mathrm{u}}} \left[\langle \tfrac{1}{2}\tfrac{1}{2}; \tfrac{1}{2}\tfrac{1}{2} \rangle - \langle \tfrac{1}{2} - \tfrac{1}{2}; \tfrac{1}{2} - \tfrac{1}{2} \rangle \right], \tag{11.39f}$$

$$U_{xz} \equiv \frac{1}{\sigma_{\mathrm{u}}} \left[\langle -\tfrac{1}{2}\tfrac{1}{2}; \tfrac{1}{2}\tfrac{1}{2} \rangle - \langle -\tfrac{1}{2} - \tfrac{1}{2}; \tfrac{1}{2} - \tfrac{1}{2} \rangle \right] = \frac{2}{\sigma_{\mathrm{u}}} \mathrm{Re}\left\{ \langle \tfrac{1}{2}\tfrac{1}{2}; -\tfrac{1}{2}\tfrac{1}{2} \rangle \right\}, \tag{11.39g}$$

$$U_{zx} \equiv -\frac{1}{\sigma_{\mathrm{u}}} \left[\langle \tfrac{1}{2} - \tfrac{1}{2}; \tfrac{1}{2}\tfrac{1}{2} \rangle + \langle \tfrac{1}{2}\tfrac{1}{2}; \tfrac{1}{2} - \tfrac{1}{2} \rangle \right] = -\frac{2}{\sigma_{\mathrm{u}}} \mathrm{Re}\left\{ \langle \tfrac{1}{2} - \tfrac{1}{2}; \tfrac{1}{2}\tfrac{1}{2} \rangle \right\}, \tag{11.39h}$$

where $\mathrm{Re}\{Z\}$ and $\mathrm{Im}\{Z\}$ denote the real and imaginary parts of the complex quantity Z. Note that all the T parameters are real which follows from the hermiticity of the density matrix. Consequently, the most general form of Equation (3.78) of Kessler [11.1] is given by

$$\mathbf{P}' = \frac{(S_{\mathrm{P}} + T_y P_y)\hat{y} + (T_x P_x + U_{xz} P_z)\hat{x} + (T_z P_z - U_{zx} P_x)\hat{z}}{1 + S_{\mathrm{A}} P_y}. \tag{11.40}$$

The physical meaning of (11.40) and the generalized STU parameters is illustrated in Fig. 11.1. It should be noted that the differential cross section σ_{u} can be determined in a single scattering experiment, whereas S_{P} and S_{A} require "double", and T_x, T_y, T_z, U_{xz}, and U_{zx} even "triple", scattering experiments, corresponding to the production (1), change (2) and analysis (3) of the projectile polarization. Following the replacement of the first scattering process by a GaAs source for spin-polarized electrons [11.9], such triple scattering experiments have become experimentally feasible even at low energies and for physically interesting cases with small cross sections (see, for

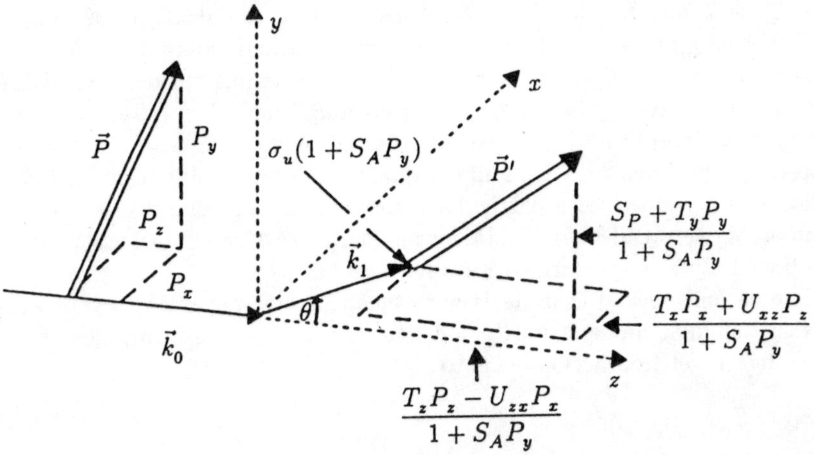

Fig. 11.1. Illustration of the generalized *STU* parameters: σ_u is the differential cross section for the scattering of unpolarized projectiles from unpolarized targets, the polarization function S_P gives the polarization of an initially unpolarized projectile beam after the collision, and the asymmetry function S_A determines a left–right asymmetry in the differential cross section for scattering of a spin-polarized beam. Furthermore, the contraction parameters (T_x, T_y, T_z) describe the change of an initial polarization component along the three cartesian axes while the parameters U_{xz} and U_{zx} determine the rotation of a polarization component in the scattering plane.

example, Berger and Kessler [11.10]). We also note that the measurement of the differential cross section σ_u — although possible with only one scattering process — is, in itself, very difficult, since it is the only *absolute* parameter that needs to be determined.

We finish this section by discussing two of the most important special cases, namely (i) the scattering from targets with zero angular momentum in both the initial and final states (i.e., $J_0 = J_1 = 0$) and (ii) elastic scattering from targets with arbitrary angular momentum $J = J_0 = J_1$. In the first case, it was shown in [11.5] that

$$S_P = \Pi_1\, \Pi_0\, S_A\,, \tag{11.41a}$$
$$T_y = \Pi_1\, \Pi_0\,, \tag{11.41b}$$
$$T_x = \Pi_1\, \Pi_0\, T_z\,, \tag{11.41c}$$
$$U_{xz} = \Pi_1\, \Pi_0\, U_{zx}\,, \tag{11.41d}$$

where Π_0 and Π_1 denote the parities of the atomic states involved in the transition. Hence, the number of independent *STU* parameters is reduced to three. Note that T_y can only become $+1$ or -1 depending on the product of the initial-state and final-state parities. If this product is $+1$ (as, for example, in elastic scattering), (11.41) yield the well-known *STU* parameters of Kessler [11.1] by setting $S_P = S_A \equiv S$ (the so-called "Sherman function"),

$T_x = T_z \equiv T$ and $U_{xz} = U_{zx} \equiv U$. Together with the absolute differential cross section σ_u, these are the four independent parameters that can be measured in a "complete" experiment to determine the magnitudes of and the relative phase between the "direct scattering amplitude" $f \equiv f(0\frac{1}{2}, 0\frac{1}{2}; \theta)$ and the "spin-flip amplitude" $g \equiv f(0 - \frac{1}{2}, 0\frac{1}{2}; \theta)$. Such "complete" or "perfect scattering experiments", where all the quantum mechanically available information is determined for a particular transition and a given set of kinematic parameters, were called for by Bederson many years ago [11.11–13], but to date have only been performed for a few special cases.

The second special case deals with *elastic* scattering from targets with arbitrary angular momentum $J_0 = J_1$. Here, time reversal invariance of the projectile–target interaction leads to

$$S_P = S_A \tag{11.42a}$$

and

$$U_{xz} - U_{zx} = (T_z - T_x) \tan\theta \tag{11.42b}$$

(for details, see [11.5]). Hence, the number of independent parameters is reduced to five relative generalized STU parameters and one absolute differential cross section σ_u. Although (11.42b) also applies to elastic scattering from targets with zero angular momentum, it does not yield any new information since, in this case, $T_x \equiv T_z$ and $U_{xz} \equiv U_{zx}$ for *all* scattering angles.

It can be seen from the discussion above that experiments with open-shell targets and non-vanishing angular momenta become more complicated already for elastic scattering and even more so for inelastic scattering. If, for example, relativistic effects are taken into account in the general description, six independent scattering amplitudes are needed to describe both elastic scattering of spin-$\frac{1}{2}$ particles from spin-$\frac{1}{2}$ targets (Burke and Mitchell [11.14]) as well as inelastic scattering involving target states with $J_0 = 0$ and $J_1 = 1$ or vice versa [11.15]. The determination of the six magnitudes and the five relative phases of these amplitudes requires a total of eleven independent observables. No "perfect" experiment can therefore be performed with spin-polarized projectiles and their observation alone. Nevertheless, the determination of the complete set of generalized STU parameters in "triple scattering" experiments provides one of the most detailed tests of theoretical models for electron–atom collisions.

11.5.2 Generalized Stokes Parameters

The state multipole description is also widely used for the parametrization of the Stokes parameters that describe the polarization of light emitted in optical decays of excited atomic ensembles. The general case of excitation by spin polarized projectiles was first treated by Bartschat *et al* [11.15], and their work was recently extended by Andersen and Bartschat [11.8]. The

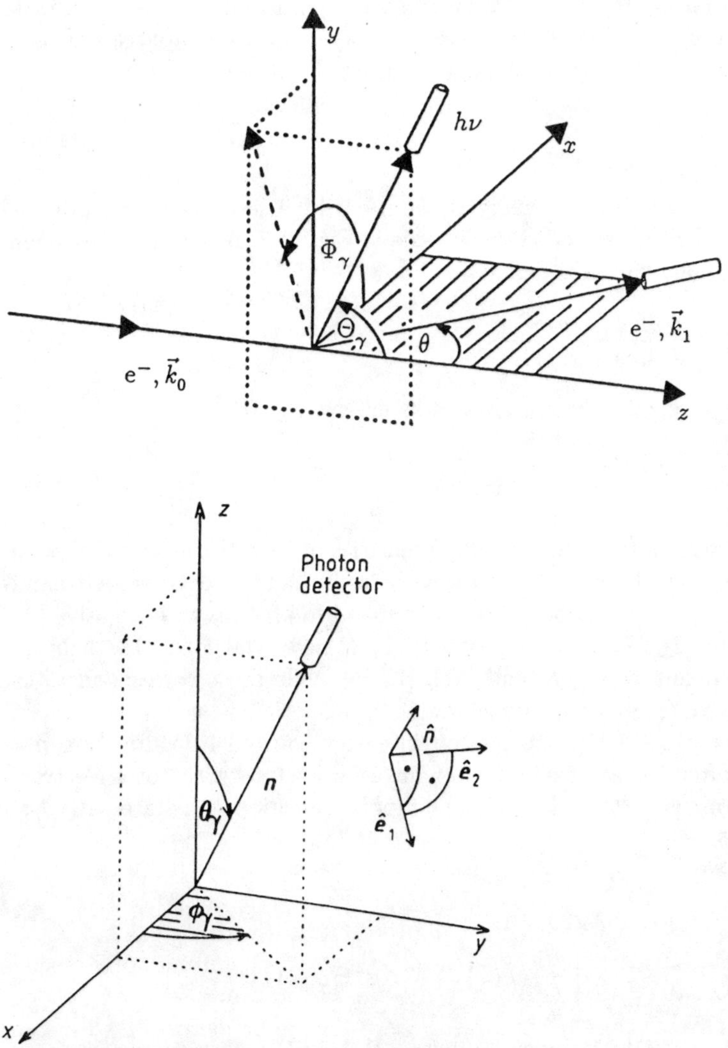

Fig. 11.2. Geometry of electron–photon coincidence experiments in the collision frame (top) and definition of the Stokes parameters (bottom). Photons are observed in a direction $\hat{\boldsymbol{n}}$ with polar angles $(\Theta_\gamma, \Phi_\gamma)$. The three unit vectors $(\hat{\boldsymbol{n}}, \hat{\mathbf{e}}_1, \hat{\mathbf{e}}_2)$ define the helicity system of the photons; $\hat{\mathbf{e}}_1 = (\Theta_\gamma + 90°, \Phi_\gamma)$ lies in the plane spanned by $\hat{\boldsymbol{n}}$ and $\hat{\boldsymbol{z}}$ and is perpendicular to $\hat{\boldsymbol{n}}$ while $\hat{\mathbf{e}}_2 = (\Theta_\gamma, \Phi_\gamma + 90°)$ is perpendicular to both $\hat{\boldsymbol{n}}$ and $\hat{\mathbf{e}}_1$. In addition to the circular polarization P_3, the linear polarizations P_1 and P_2 are defined with respect to axes in the plane spanned by $\hat{\mathbf{e}}_1$ and $\hat{\mathbf{e}}_2$ (see text). Counting in the direction of $\hat{\mathbf{e}}_1 \hat{=} 0^0$, the axes are located at $(0°, 90°)$ for P_1 and at $(45°, 135°)$ for P_2, respectively.

basic setup for electron–photon coincidence experiments and the coordinate system for the definition of the Stokes parameters are illustrated in Fig. 11.2.

To begin with, we define the linear light polarization

$$P_1 = \eta_3 \equiv \frac{I(0^o) - I(90^o)}{I(0^o) + I(90^o)}, \tag{11.43a}$$

where $I(\beta)$ denotes the intensity transmitted by a linear polarizer oriented at an angle β with respect to the $\hat{e}_1$ axis of Fig. 11.2. A second independent linear-light polarization is defined by

$$P_2 = \eta_1 \equiv \frac{I(45^o) - I(135^o)}{I(45^o) + I(135^o)}. \tag{11.43b}$$

Finally, the circular-light polarization is determined by

$$P_3 = -\eta_2 \equiv \frac{I(\mathrm{RHC}) - I(\mathrm{LHC})}{I(\mathrm{RHC}) + I(\mathrm{LHC})} = -\frac{I_+ - I_-}{I_+ + I_-}, \tag{11.43c}$$

where $I(\mathrm{RHC})$ and $I(\mathrm{LHC})$ are the intensities of light transmitted by polarization filters which only admit light with "right-hand" (RHC) or "left-hand" (LHC) circular polarization. These correspond to photons with negative (I_-) and positive (I_+) helicity, respectively. In (11.43), the P are the light polarizations defined by Born and Wolf [11.16] while the η correspond to the notation used, for example, by Blum [11.2].

As shown in [11.15], the photon intensity emitted in a direction $\hat{n} = (\Theta_\gamma, \Phi_\gamma)$, after impact excitation of an atomic state with total electronic angular momentum J and an electric dipole transition to a state with J_f, is given by

$$\begin{aligned} I(\Theta_\gamma, \Phi_\gamma) = C \Bigg(& \frac{2\,(-1)^{J-J_\mathrm{f}}}{3\sqrt{2J+1}} \langle T(J)^\dagger_{00}\rangle \\ & - \begin{Bmatrix} 1 & 1 & 2 \\ J & J & J_\mathrm{f} \end{Bmatrix} \Bigg[\sqrt{\tfrac{1}{6}} \langle T(J)^\dagger_{20}\rangle \, (3\cos^2\Theta_\gamma - 1) \\ & + \mathrm{Re}\{\langle T(J)^\dagger_{22}\rangle\} \, \sin^2\Theta_\gamma \cos 2\Phi_\gamma - \mathrm{Re}\{\langle T(J)^\dagger_{21}\rangle\} \, \sin 2\Theta_\gamma \cos\Phi_\gamma \\ & - \mathrm{Im}\{\langle T(J)^\dagger_{22}\rangle\} \, \sin^2\Theta_\gamma \sin 2\Phi_\gamma + \mathrm{Im}\{\langle T(J)^\dagger_{21}\rangle\} \, \sin 2\Theta_\gamma \sin\Phi_\gamma \Bigg] \Bigg), \end{aligned} \tag{11.44}$$

where

$$C = \frac{e^2\omega^4}{2\pi c^3} \, |\langle J_\mathrm{f} \| \mathbf{r} \| J \rangle|^2 \; (-1)^{J-J_\mathrm{f}} \tag{11.45}$$

is a constant containing the frequency ω of the transition as well as the reduced radial dipole matrix element originating from the Wigner-Eckart theorem. (For details of the derivation, see [11.2] and [11.15].)

Similarly,

$$I\,P_3(\Theta_\gamma,\Phi_\gamma) = -C \begin{Bmatrix} 1 & 1 & 1 \\ J & J & J_\mathrm{f} \end{Bmatrix} \Big[\langle T(J)^\dagger_{10}\rangle \,\cos\Theta_\gamma$$

$$+\,\mathrm{Im}\{\langle T(J)^\dagger_{11}\rangle\}\,2\sin\Theta_\gamma\sin\Phi_\gamma - \mathrm{Re}\{\langle T(J)^\dagger_{11}\rangle\}\,2\sin\Theta_\gamma\cos\Phi_\gamma\Big]\,, \qquad (11.46)$$

so that the circular-light polarization can be calculated as

$$P_3(\Theta_\gamma,\Phi_\gamma) = I\,P_3(\Theta_\gamma,\Phi_\gamma)/I(\Theta_\gamma,\Phi_\gamma)\,. \qquad (11.47)$$

Finally, the two linear light polarizations can be obtained from

$$I\,P_1(\Theta_\gamma,\Phi_\gamma) = C \begin{Bmatrix} 1 & 1 & 2 \\ J & J & J_\mathrm{f} \end{Bmatrix} \Big[\sqrt{\tfrac{3}{2}}\langle T(J)^\dagger_{20}\rangle\,\sin^2\Theta_\gamma$$

$$+\,\mathrm{Re}\{\langle T(J)^\dagger_{22}\rangle\}\,(1+\cos^2\Theta_\gamma)\cos 2\Phi_\gamma + \mathrm{Re}\{\langle T(J)^\dagger_{21}\rangle\}\,\sin 2\Theta_\gamma\cos\Phi_\gamma$$

$$-\,\mathrm{Im}\{\langle T(J)^\dagger_{22}\rangle\}\,(1+\cos^2\Theta_\gamma)\sin 2\Phi_\gamma - \mathrm{Im}\{\langle T(J)^\dagger_{21}\rangle\}\,\sin 2\Theta_\gamma\sin\Phi_\gamma\Big]$$

$$(11.48)$$

and

$$I\,P_2(\Theta_\gamma,\Phi_\gamma) = C \begin{Bmatrix} 1 & 1 & 2 \\ J & J & J_\mathrm{f} \end{Bmatrix}$$

$$\times \Big[\mathrm{Re}\{\langle T(J)^\dagger_{22}\rangle\}\,2\cos\Theta_\gamma\sin 2\Phi_\gamma + \mathrm{Re}\{\langle T(J)^\dagger_{21}\rangle\}\,2\sin\Theta_\gamma\sin\Phi_\gamma$$

$$+\,\mathrm{Im}\{\langle T(J)^\dagger_{22}\rangle\}\,2\cos\Theta_\gamma\cos 2\Phi_\gamma + \mathrm{Im}\{\langle T(J)^\dagger_{21}\rangle\}\,\sin 2\Theta_\gamma\cos\Phi_\gamma\Big]\,.$$

$$(11.49)$$

Note that each state multipole gives rise to a characteristic angular dependence in the formulas for the *absolute* Stokes parameters, i.e., the light polarizations multiplied by the intensity. Hence, (11.47) is not trivial, since the numerator and denominator have to be calculated separately. Furthermore, perturbation coefficients (see Sects. 11.4.3,4) may need to be applied to deal, for example, with depolarization effects due to internal or external fields.

As pointed out before, some of the state multipoles may vanish, depending on the experimental arrangement. A detailed analysis of the information contained in the state multipoles for spin-polarized incident electrons has been given by Andersen and Bartschat [11.8]. They defined the "generalized" Stokes parameters for specific values of the projectile spin polarization. With a photon detector placed in the $\hat{\boldsymbol{n}}$ direction, four light intensities are measured for orthogonal positions of the light-polarization analyzers and electron-beam polarizations $\pm\boldsymbol{P}$. These intensities are then combined in a Stokes *matrix*. The elements $\left(\mathbf{Q}^{\hat{\boldsymbol{n}}}_{1j}\right)_{\boldsymbol{P}}$, $j=\{1,2,3\}$, in the first row are defined as follows [11.8]:

$$I_u^{\hat{n}}(Q_{11}^{\hat{n}})_{\boldsymbol{P}} \equiv I_{\boldsymbol{P}}^{\hat{n}}(0^\circ) + I_{-\boldsymbol{P}}^{\hat{n}}(0^\circ) - I_{\boldsymbol{P}}^{\hat{n}}(90^\circ) - I_{-\boldsymbol{P}}^{\hat{n}}(90^\circ)\,, \tag{11.50a}$$

$$I_u^{\hat{n}}\left(Q_{12}^{\hat{n}}\right)_{\boldsymbol{P}} \equiv I_{\boldsymbol{P}}^{\hat{n}}(0^\circ) - I_{-\boldsymbol{P}}^{\hat{n}}(0^\circ) - I_{\boldsymbol{P}}^{\hat{n}}(90^\circ) + I_{-\boldsymbol{P}}^{\hat{n}}(90^\circ)\,, \tag{11.50b}$$

$$I_u^{\hat{n}}\left(Q_{13}^{\hat{n}}\right)_{\boldsymbol{P}} \equiv I_{\boldsymbol{P}}^{\hat{n}}(0^\circ) - I_{-\boldsymbol{P}}^{\hat{n}}(0^\circ) + I_{\boldsymbol{P}}^{\hat{n}}(90^\circ) - I_{-\boldsymbol{P}}^{\hat{n}}(90^\circ)\,, \tag{11.50c}$$

where

$$I_{\mathrm{u}}^{\hat{n}} \equiv I_{\boldsymbol{P}}^{\hat{n}}(0^\circ) + I_{-\boldsymbol{P}}^{\hat{n}}(0^\circ) + I_{\boldsymbol{P}}^{\hat{n}}(90^\circ) + I_{-\boldsymbol{P}}^{\hat{n}}(90^\circ) \tag{11.51}$$

is proportional to the light intensity measured with unpolarized electrons, independent of the light-analyzer setting. Similarly, one defines the elements $\left(Q_{2j}^{\hat{n}}\right)_{\boldsymbol{P}}$ and $\left(Q_{3j}^{\hat{n}}\right)_{\boldsymbol{P}}$, $j = \{1,2,3\}$, of the second and third rows by replacing $(0^\circ, 90^\circ)$ by $(45^\circ, 135^\circ)$ and, for circular polarization analysis, by (RHC,LHC) or (σ^-, σ^+), respectively.

The first column of the generalized Stokes parameter matrix, i.e., the parameters $\left(\mathbf{Q}_{i1}^{\hat{n}}\right)_{\boldsymbol{P}}$, $i = \{1,2,3\}$, is the standard Stokes vector (P_1, P_2, P_3) for unpolarized incident electrons. Similarly, the parameters in the third column, $\left(\mathbf{Q}_{i3}^{\hat{n}}\right)_{\boldsymbol{P}}$, $i = \{1,2,3\}$, yield an "optical asymmetry" which compares light intensities measured with spin-up and spin-down electrons, independent of the light-analyzer setting. Finally, the elements of the second column, $\left(\mathbf{Q}_{i2}^{\hat{n}}\right)_{\boldsymbol{P}}$, $i = \{1,2,3\}$, correspond to observables whose measurement requires both spin-polarized incident projectiles and photon-polarization analysis.

As an example, Andersen and Bartschat [11.8] re-analyzed the experiment performed by Sohn and Hanne [11.17] for electron impact excitation of the $(6s6p)^3P_1^o$ state of mercury and showed how the state multipoles of the excited atomic ensemble can be determined by a measurement of the generalized Stokes parameters. In this case, such measurements almost correspond to a "perfect scattering experiment", thereby demonstrating that this goal is now within reach even for fairly complex excitation processes.

11.6 ^{2}S → ^{2}P^o Transitions

11.6.1 General Description of an Excited ^{2}P^o State

The targets of interest for this section, typically light alkali atoms or hydrogen, have an electron spin of $S_0 = S_1 = \frac{1}{2}$ in the initial and final states. Recently, this problem has been analyzed in detail by Andersen and Bartschat [11.18]. The discussion below closely follows their formulation.

If electron exchange is the only important spin-dependent process, there are four independent scattering amplitudes, two each for the triplet (t) and singlet (s) channels of the total spin in the combined projectile + target system. Using the simplified notation $f_{M_L}^{\mathrm{s,t}} \equiv \langle 1, M_L | \mathbf{T}^{S=0,1} | 0, 0 \rangle_{\mathrm{n}}$ for scattering amplitudes in the *natural* frame, the amplitudes of interest are:

$$f^{\mathrm{t}}_{+1} = \alpha_+ \mathrm{e}^{\mathrm{i}\phi_+} , \tag{11.52a}$$
$$f^{\mathrm{t}}_{-1} = \alpha_- \mathrm{e}^{\mathrm{i}\phi_-} , \tag{11.52b}$$
$$f^{\mathrm{s}}_{+1} = \beta_+ \mathrm{e}^{\mathrm{i}\psi_+} , \tag{11.52c}$$
$$f^{\mathrm{s}}_{-1} = \beta_- \mathrm{e}^{\mathrm{i}\psi_-} , \tag{11.52d}$$

where $\alpha_\pm$, $\beta_\pm$, $\phi_\pm$, and $\psi_\pm$ are real numbers. Neglecting an overall phase, we thus need seven independent parameters for each scattering angle θ to characterize the amplitudes completely. One is traditionally chosen as the differential cross section σ_{u} corresponding to scattering of unpolarized beams. In addition, six dimensionless parameters may be defined, three to characterize the relative lengths of the four complex vectors (see Fig. 11.3), and three to define their relative phase angles. We use the following definitions:

$$\delta^{\mathrm{t}} = \phi_+ - \phi_- , \tag{11.53a}$$
$$\delta^{\mathrm{s}} = \psi_+ - \psi_- , \tag{11.53b}$$
$$\Delta^+ = \phi_+ - \psi_+ , \tag{11.53c}$$
$$\Delta^- = \phi_- - \psi_- . \tag{11.53d}$$

In this case, the following density matrices can be introduced for the individual total spin channels [11.19]:

$$\rho^{\mathrm{t}} = \sigma^{\mathrm{t}}\,\frac{1}{2}\begin{pmatrix} 1+L^{\mathrm{t}}_\perp & 0 & -P^{\mathrm{t}}_\ell\,\mathrm{e}^{-2i\gamma^{\mathrm{t}}} \\ 0 & 0 & 0 \\ -P^{\mathrm{t}}_\ell\,\mathrm{e}^{2i\gamma^{\mathrm{t}}} & 0 & 1-L^{\mathrm{t}}_\perp \end{pmatrix} \tag{11.54a}$$

and

$$\rho^{\mathrm{s}} = \sigma^{\mathrm{s}}\,\frac{1}{2}\begin{pmatrix} 1+L^{\mathrm{s}}_\perp & 0 & -P^{\mathrm{s}}_\ell\,\mathrm{e}^{-2i\gamma^{\mathrm{s}}} \\ 0 & 0 & 0 \\ -P^{\mathrm{s}}_\ell\,\mathrm{e}^{2i\gamma^{\mathrm{s}}} & 0 & 1-L^{\mathrm{s}}_\perp \end{pmatrix} , \tag{11.54b}$$

where

$$\sigma^{\mathrm{t}} = \alpha_+^2 + \alpha_-^2 , \tag{11.55a}$$
$$\sigma^{\mathrm{s}} = \beta_+^2 + \beta_-^2 , \tag{11.55b}$$
$$L^{\mathrm{t}}_\perp = \frac{1}{\sigma^{\mathrm{t}}}\,(\alpha_+^2 - \alpha_-^2) = -P^{\mathrm{t}}_3 , \tag{11.55c}$$
$$L^{\mathrm{s}}_\perp = \frac{1}{\sigma^{\mathrm{s}}}\,(\beta_+^2 - \beta_-^2) = -P^{\mathrm{s}}_3 , \tag{11.55d}$$
$$\gamma^{\mathrm{t,s}} = -\tfrac{1}{2}\,(\delta^{\mathrm{t,s}} \pm \pi) , \tag{11.55e}$$
$$P^{\mathrm{t}}_\ell\,\mathrm{e}^{2\mathrm{i}\gamma^{\mathrm{t}}} = P^{\mathrm{t}}_1 + \mathrm{i}\,P^{\mathrm{t}}_2 = -\frac{2\alpha_+\alpha_-}{\sigma^{\mathrm{t}}}\,\mathrm{e}^{-\mathrm{i}\delta^{\mathrm{t}}} , \tag{11.55f}$$
$$P^{\mathrm{s}}_\ell\,\mathrm{e}^{2\mathrm{i}\gamma^{\mathrm{s}}} = P^{\mathrm{s}}_1 + \mathrm{i}\,P^{\mathrm{s}}_2 = -\frac{2\beta_+\beta_-}{\sigma^{\mathrm{s}}}\,\mathrm{e}^{-\mathrm{i}\delta^{\mathrm{s}}} . \tag{11.55g}$$

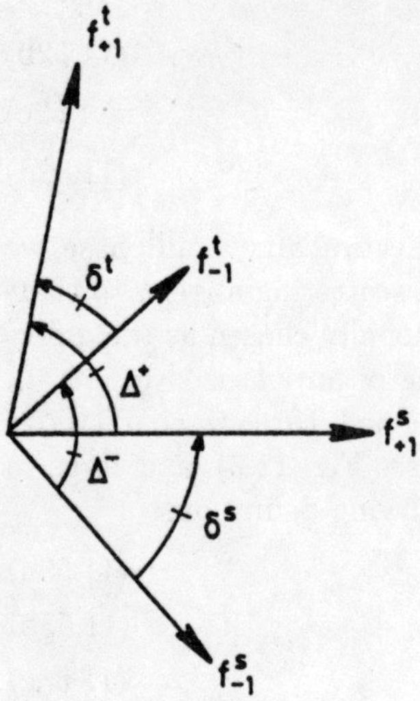

Fig. 11.3. Schematic diagram of triplet (t) and singlet (s) scattering amplitudes in the natural frame for $^2\mathrm{S} \to {}^2\mathrm{P}^{\mathrm{o}}$ transitions by electron impact. Note that $\Delta^+ + \delta^s = \Delta^- + \delta^t$.

In (11.55), P_1, P_2, and P_3 are the light polarizations measured with a photon detector perpendicular to the scattering plane, i.e., $\theta_\gamma = \phi_\gamma = 90°$ in the collision frame or $\theta_\gamma = 0°$ in the natural frame.

In the case where unpolarized beams are used, the total density matrix becomes the weighted sum of the two matrices $\boldsymbol{\rho}^{\mathrm{s}}$ and $\boldsymbol{\rho}^{\mathrm{t}}$, i.e.,

$$\begin{aligned}
\boldsymbol{\rho}_{\mathrm{u}} &= \sigma_{\mathrm{u}}\,\frac{1}{2}\begin{pmatrix} 1+L_\perp & 0 & -P_\ell\,\mathrm{e}^{-2\mathrm{i}\gamma} \\ 0 & 0 & 0 \\ -P_\ell\,\mathrm{e}^{2\mathrm{i}\gamma} & 0 & 1-L_\perp \end{pmatrix} \\
&= 3w^{\mathrm{t}}\,\sigma_{\mathrm{u}}\,\frac{1}{2}\begin{pmatrix} 1+L^{\mathrm{t}}_\perp & 0 & -P^{\mathrm{t}}_\ell\,\mathrm{e}^{-2\mathrm{i}\gamma^{\mathrm{t}}} \\ 0 & 0 & 0 \\ -P^{\mathrm{t}}_\ell\,\mathrm{e}^{2\mathrm{i}\gamma^{\mathrm{t}}} & 0 & 1-L^{\mathrm{t}}_\perp \end{pmatrix} \\
&\quad + w^{\mathrm{s}}\,\sigma_{\mathrm{u}}\,\frac{1}{2}\begin{pmatrix} 1+L^{\mathrm{s}}_\perp & 0 & -P^{\mathrm{s}}_\ell\,\mathrm{e}^{-2\mathrm{i}i\gamma^{\mathrm{s}}} \\ 0 & 0 & 0 \\ -P^{\mathrm{s}}_\ell\,\mathrm{e}^{2\mathrm{i}\gamma^{\mathrm{s}}} & 0 & 1-L^{\mathrm{s}}_\perp \end{pmatrix},
\end{aligned} \tag{11.56}$$

where

$$w^{\mathrm{t}} = \frac{\sigma^{\mathrm{t}}}{\sigma^{\mathrm{s}} + 3\sigma^{\mathrm{t}}} = \frac{\sigma^{\mathrm{t}}}{4\sigma_{\mathrm{u}}}, \tag{11.57a}$$

$$w^{\mathrm{s}} = \frac{\sigma^{\mathrm{s}}}{\sigma^{\mathrm{s}} + 3\sigma^{\mathrm{t}}} = \frac{\sigma^{\mathrm{s}}}{4\sigma_{\mathrm{u}}} = 1 - 3w^{\mathrm{t}}, \tag{11.57b}$$

$$\sigma_{\mathrm{u}} = (3w^{\mathrm{t}} + w^{\mathrm{s}})\,\sigma_{\mathrm{u}} = \tfrac{3}{4}\,\sigma^{\mathrm{t}} + \tfrac{1}{4}\,\sigma^{\mathrm{s}}. \tag{11.57c}$$

The parameters w^{t} and w^{s} are related to the parameter $r = \sigma^{\mathrm{t}}/\sigma^{\mathrm{s}}$ used by Hertel *et al* [11.19] through

$$r = \frac{w^{\mathrm{t}}}{1-3w^{\mathrm{t}}} = \frac{1-w^{\mathrm{s}}}{3w^{\mathrm{s}}} . \tag{11.58}$$

At this point we have thus introduced a total of six independent parameters, namely $\sigma_{\mathrm{u}}, w^{\mathrm{t}}, L_{\perp}^{\mathrm{t}}$, $L_{\perp}^{\mathrm{s}}, \gamma^{\mathrm{t}}$, and γ^{s}, leaving one parameter, a relative phase, still to be chosen. Inspection of Fig. 11.3 suggests, for example, the angle Δ^{+}. As can be seen from the figure and also from (11.53), the fourth angle, Δ^{-}, is then fixed through the relation

$$\Delta^{+} - \Delta^{-} = \delta^{\mathrm{t}} - \delta^{\mathrm{s}} = 2\left(\gamma^{\mathrm{s}} - \gamma^{\mathrm{t}}\right) . \tag{11.59a}$$

Consequently, the parameters

$$\sigma_{\mathrm{u}}; w^{\mathrm{t}}, L_{\perp}^{\mathrm{t}}, L_{\perp}^{\mathrm{s}}; \gamma^{\mathrm{t}}, \gamma^{\mathrm{s}}, \Delta^{+} \tag{11.59b}$$

allow for a complete description of the scattering process. Although other sets of parameters may serve the same purpose, an important advantage of the set (11.59) is the fact that these parameters can be interpreted in simple physical pictures. This is illustrated in Fig. 11.4 which shows that the γ correspond to the alignment angles of the atomic charge cloud while the $L_{\perp}$ correspond to the angular momentum transfers perpendicular to the scattering plane. Also, these parameters are accessible in "partial" (i.e., non-complete) experiments (for details, see [11.18,19]) and they represent a natural generalization of the parameters that have been used extensively in the description of unpolarized beam experiments [11.3].

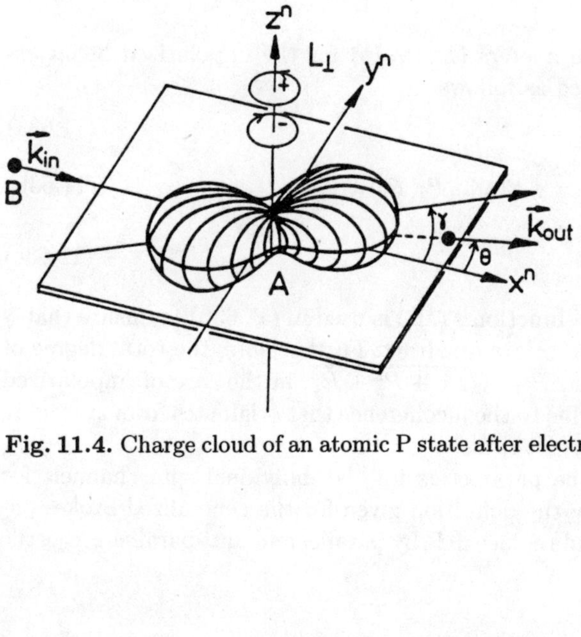

Fig. 11.4. Charge cloud of an atomic P state after electron impact excitation [11.3].

11.6.2 Relationship to Experimental Observables

The density-matrix parameters can be determined in electron–photon coincidence experiments or, alternatively, in superelastic collision processes between electrons and laser-excited atoms. The basic information is equivalent for both experimental setups; for details see [11.3,4,15,18,19]. Note, however, that in most practical cases (such as hydrogen or sodium targets) depolarization effects due to the atomic fine structure and hyperfine structure must be taken into account. For the special case of fine-structure depolarization of radiation, emitted from an excited $^2\mathrm{P^o}$ state in a direction perpendicular to the scattering plane, it is possible to define "reduced" light polarizations which are related to the measured light polarizations by simple factors. One finds [11.3]

$$\bar{P}_{1,2} = \frac{7}{3} P_{1,2} \,, \tag{11.60a}$$

$$\bar{P}_3 = P_3 \,, \tag{11.60b}$$

where the bar denotes the reduced polarizations that would be measured without fine-structure depolarization.

For the individual spin channels, the reduced Stokes vectors are unit vectors, i.e.,

$$|\bar{\boldsymbol{P}}^{\mathrm{t}}| = |\bar{\boldsymbol{P}}^{\mathrm{s}}| = 1 \,. \tag{11.61}$$

On the other hand, the reduced Stokes vector $\bar{\boldsymbol{P}}$ for the unpolarized beam experiment is given as the weighted sum of the singlet and triplet Stokes vectors

$$\bar{\boldsymbol{P}} = 3\, w^{\mathrm{t}} \bar{\boldsymbol{P}}^{\mathrm{t}} + w^{\mathrm{s}} \bar{\boldsymbol{P}}^{\mathrm{s}} \tag{11.62}$$

from which the set of parameters $(L_\perp, \gamma, \bar{P}_\ell)$ for the unpolarized beam experiment may be evaluated as follows:

$$L_\perp = 3\, w^{\mathrm{t}}\, L^{\mathrm{t}}_\perp + w^{\mathrm{s}}\, L^{\mathrm{s}}_\perp = -\bar{P}_3 \,, \tag{11.63a}$$

$$\gamma = \tfrac{1}{2} \arctan\left(\frac{P_2}{P_1}\right) = \tfrac{1}{2}\, \mathtt{ATAN2}(P_2, P_1) \,, \tag{11.63b}$$

$$\bar{P}_\ell = \sqrt{\bar{P}_1^2 + \bar{P}_2^2} \,. \tag{11.63c}$$

Note that the FORTRAN function **ATAN2** is used in (11.63b) to ensure that γ is obtained from the appropriate quadrant. Furthermore, the total degree of reduced light polarization, $\bar{P} = \sqrt{\bar{P}_1^2 + \bar{P}_2^2 + \bar{P}_3^2}$, in the case of unpolarized beams is less than unity, due to the incoherence that originates from averaging over the projectile and target spins.

In order to extract the parameters for the individual spin channels for this case, one can modify the definition given for the generalized Stokes parameters in (11.50,51) and replace $\pm\boldsymbol{P}$ by parallel and anti-parallel projectile and target spin-polarization vectors, respectively. Depending on the actual experimental setup, it may also be necessary to replace the total electronic angular momenta J and J_f in (11.44–49) by the corresponding orbital angular momenta L and L_f. For a detailed analysis of the equivalent "time-reversed" superelastic scattering experiment, which has actually been used to determine some of the spin resolved parameters, see the paper by Hertel *et al.* [11.19].

We conclude this section by pointing out that the generalized STU parameters discussed in Sect. 11.5.1 can also be expressed in terms of the amplitudes (11.52) and the complete parameter set (11.59). The reason is the so-called "fine-structure effect" discussed in detail by Hanne [11.20]. Mathematically, it is contained in the transformation given in (11.3), which shows that excitation amplitudes to different fine-structure levels are not independent of each other. This approximation is expected to be fulfilled very well for alkali-like targets, where electron exchange is the most important spin-dependent effect during the excitation process. (One must also assume that the energy splitting of the fine-structure levels is small compared to the initial and final energies of the projectile.) In this case, the generalized STU parameters get vastly simplified, since the seven asymmetry, polarization, contraction, and rotation parameters for each fine-structure level reduce to the following set of only four independent parameters (the superscripts denote the J value of the excited target state) [11.18]:

$$S_{\mathrm{P}} \equiv S_{\mathrm{P}}^{1/2} = -2\, S_{\mathrm{P}}^{3/2}\,, \tag{11.64a}$$

$$S_{\mathrm{A}} \equiv S_{\mathrm{A}}^{1/2} = -2\, S_{\mathrm{A}}^{3/2}\,, \tag{11.64b}$$

$$T \equiv T_x^{1/2} = T_y^{1/2} = T_z^{1/2} = T_x^{3/2} = T_y^{3/2} = T_z^{3/2}\,, \tag{11.64c}$$

$$U \equiv U_{zx}^{1/2} = U_{xz}^{1/2} = -2\, U_{zx}^{3/2} = -2\, U_{xz}^{3/2}\,. \tag{11.64d}$$

The final results in terms of the amplitudes (11.52) and the parameter set (11.59) are [11.4]:

$$S_{\mathrm{P}} = -\frac{1}{4\sigma_{\mathrm{u}}}\left[\alpha_+^2 - \alpha_-^2 + \beta_+^2 - \beta_-^2\right] = -w^{\mathrm{t}}\, L_\perp^{\mathrm{t}} + w^{\mathrm{s}}\, L_\perp^{\mathrm{s}}\,, \tag{11.65a}$$

$$\begin{aligned} S_{\mathrm{A}} &= \frac{1}{4\sigma_{\mathrm{u}}}\left[\mathrm{Re}\,\{2\alpha_+\beta_+ \mathrm{e}^{i(\phi_+-\psi_+)} - 2\alpha_-\beta_- \mathrm{e}^{i(\phi_--\psi_-)}\} - 2(\alpha_+^2 - \alpha_-^2)\right] \\ &= \frac{1}{2\sigma_{\mathrm{u}}}\left[\alpha_+\beta_+ \cos\Delta^+ - \alpha_-\beta_- \cos\Delta^-\right] - 2\,w^{\mathrm{t}}\, L_\perp^{\mathrm{t}}\,, \end{aligned} \tag{11.65b}$$

$$\begin{aligned} T &= \frac{1}{4\sigma_{\mathrm{u}}}\left[\mathrm{Re}\,\{2\alpha_+\beta_+ \mathrm{e}^{i(\phi_+-\psi_+)} + 2\alpha_-\beta_- \mathrm{e}^{i(\phi_--\psi_-)}\} + 2(\alpha_+^2 + \alpha_-^2)\right] \\ &= \frac{1}{2\sigma_{\mathrm{u}}}\left[\alpha_+\beta_+ \cos\Delta^+ + \alpha_-\beta_- \cos\Delta^-\right] + 2\,w^{\mathrm{t}}\,, \end{aligned} \tag{11.65c}$$

$$\begin{aligned} U &= \frac{1}{4\sigma_{\mathrm{u}}}\left[\mathrm{Im}\,\{2\alpha_+\beta_+ \mathrm{e}^{i(\phi_+-\psi_+)} - 2\alpha_-\beta_- \mathrm{e}^{i(\phi_--\psi_-)}\}\right] \\ &= \frac{1}{2\sigma_{\mathrm{u}}}\left[\alpha_+\beta_+ \sin\Delta^+ - \alpha_-\beta_- \sin\Delta^-\right]. \end{aligned} \tag{11.65d}$$

The amount of information contained in the atomic density matrix (i.e., the Stokes parameters) and the reduced density matrix of the scattered electrons (i.e., the STU parameters) can be summarized as follows. From a generalized Stokes parameter analysis one can obtain information about the relative phase between the two f^{t}_{+1} and f^{t}_{-1} amplitudes and the relative phase between the two f^{s}_{+1} and f^{s}_{-1} amplitudes, as well as the relative sizes of all four amplitudes. However, none of the relative phases between any triplet and singlet amplitude can be determined. The STU parameters, on the other hand, determine the relative phase between the two f^{t}_{+1} and f^{s}_{+1} amplitudes and the relative phase between the two f^{t}_{-1} and f^{s}_{-1} amplitudes if the relative sizes of all four amplitudes are known from Stokes parameter measurements. In this case, however, none of the relative phases between $M_L = +1$ and $M_L = -1$ amplitudes can be obtained. Consequently, a minimum set of both generalized Stokes and STU parameters needs to be determined in a "perfect scattering experiment", but additional measurements offer many opportunities for valuable consistency checks.

11.7 Computer Program

The computer program is written in standard FORTRAN77. It is a simplified version of the more general code `OBSALK` [11.21]. It consists of the main routine `ALKALI` and the subroutines `CLEGOR`, `SIXJ`, and `LFAK`. The main program reads some basic control data from unit `IREAD` (set to 5 at the moment), as well as singlet and triplet scattering amplitudes calculated in the collision frame.

The calculation of state multipoles and the Stokes parameters requires Clebsch-Gordan coefficients and 6j-symbols which are obtained by subroutines `CLEGOR` and `SIXJ`; the factorials needed in these routines are set up by an initial call to `LFAK`. Since the variable names (P_1, P_2, P_3) are reserved for photon observation perpendicular to the scattering plane, the light polarizations for unpolarized beams and arbitrary direction of observation are called `Q11`, `Q21`, and `Q31` (corresponding to the first column of the generalized Stokes matrix) in the program.

The input scattering amplitudes are transformed into the natural coordinate system, and the parameter set (11.59) is calculated. The generalized STU parameters are then obtained directly from (11.65). Alternatively, they could be calculated after generating the amplitudes (11.3) from the non-relativistic input, followed by the expressions defined in (11.39).

11.7.1 Description of the Input Data

The following control parameters are read in free format from unit 5:

1. `ITIN,ITOUT,NANGLE,NBLANK`

`ITIN` Input unit for scattering amplitudes.

Complex scattering amplitudes for the triplet (variable name AMPTRP) and singlet (variable name AMPSIN) channels are read from unit ITIN for magnetic quantum numbers $M_L = 0, +1$ in the format F5.1,1P,4(1X,2E11.3). The first variable is the scattering angle, followed by the real and imaginary parts of f_0^{t}, f_{+1}^{t}, f_0^{s}, and f_{+1}^{s}, respectively. Users may want to define their own way of reading these amplitudes.

ITOUT Output unit for the various parameters.

NANGLE Number of scattering angles for which amplitudes are available.

NBLANK Number of blank lines at the top of the input file assigned to unit ITIN that describe the model used for the calculation of the scattering amplitudes.

2. EIN,ELOSS,NORM

EIN Energy of incident electron.

ELOSS Energy loss in the excitation.
Since only the ratio of EIN and ELOSS is used to normalize the cross section, these parameters can be read in arbitrary units, as long as the same unit is used for both.

NORM Control parameter for the normalization factor in the differential cross section; see (11.12).
1: $C = 1$
2: $C = k_{\mathrm{f}}/k_{\mathrm{i}}$ where $k_{\mathrm{i}} \propto \sqrt{\mathtt{EIN}}$ and $k_{\mathrm{f}} \propto \sqrt{\mathtt{EIN} - \mathtt{ELOSS}}$. This is the convention used in the output of the program SCTAMP described in Chap. 10.
3: $C = 4\pi^4\, k_{\mathrm{f}}/k_{\mathrm{i}}$. This is the convention used in the output of the program DWBA1 described in Chap. 4.

3. THEGAM,PHIGAM,G1,G2

THEGAM Photon detector angle θ_γ (in degrees), defined in the collision frame.

PHIGAM Photon detector angle ϕ_γ (in degrees), defined in the collision frame.

G1 Perturbation coefficient for the state multipoles with rank $K = 1$.

G2 Perturbation coefficient for the state multipoles with rank $K = 2$.

An example input file for electron impact excitation of the $(3s)^2S \rightarrow (3p)^2P^o$ transition in sodium at an incident electron energy of 10 eV is listed below. The photon detector is placed perpendicular to the scattering plane, and the perturbation coefficients are set to unity to yield the reduced light polarizations.

```
      11         12        181         10
  10.000      2.100          1
   1.000      1.000       90.0       90.0
```

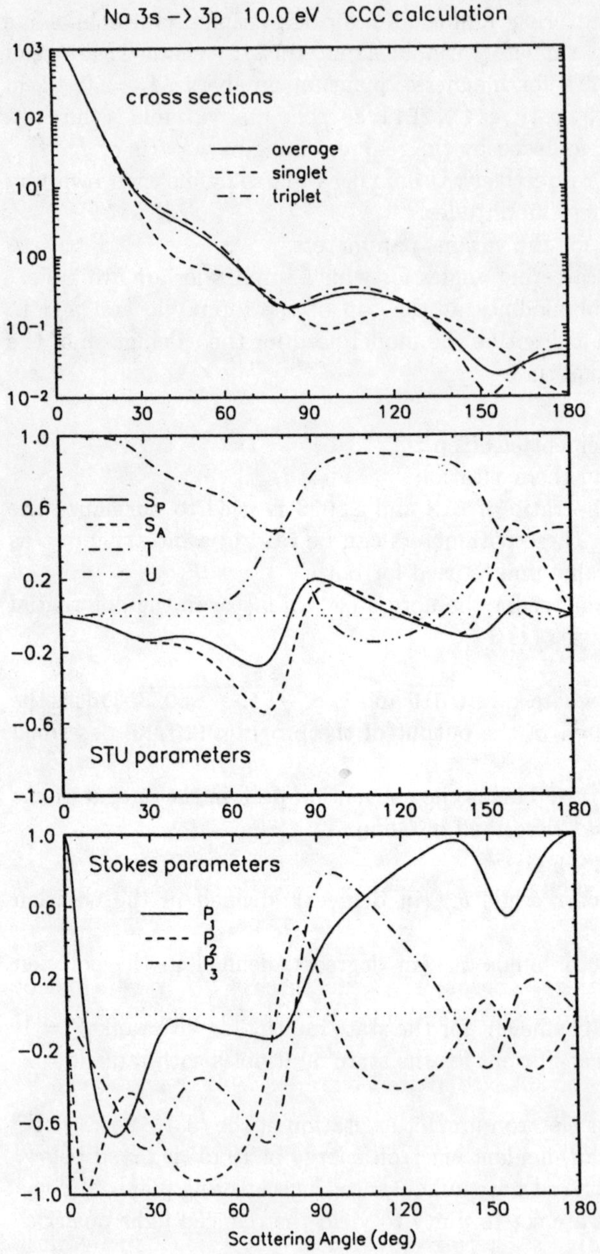

Fig. 11.5. Differential cross sections (top), generalized *STU* parameters (center), and reduced light polarizations (bottom) for electron impact excitation of the $(3s)^2S \rightarrow (3p)^2P^o$ transition in sodium at an incident electron energy of 10 eV, calculated from scattering amplitudes obtained in the CCC approach of Bray [11.22]. The photon detector is positioned perpendicular to the scattering plane.

11.7.2 Test Run

The test run deals with electron impact excitation of the $(3s)^2S \to (3p)^2P^o$ transition in sodium atoms. The scattering amplitudes for an incident electron energy of 10.0 eV were obtained by Bray [11.22] with the "convergent close-coupling (CCC) method". Details of this numerical approach are described in Chap. 8. Example results for differential cross sections, the generalized *STU* parameters, and the reduced light polarizations for light emission in the direction perpendicular to the scattering plane are shown in Fig. 11.5.

11.8 Summary

In this chapter, we have summarized the basic ideas behind the (reduced) density-matrix approach that can be used to connect theoretical scattering amplitudes to experimentally measured observables. Special emphasis was given to the case of $^2S \to {}^2P^o$ transitions, and a computer program to calculate various observables was presented.

11.9 Suggested Problems

1. Use the `DWBA1` code of Chap. 4 to generate scattering amplitudes for electron scattering from hydrogen, sodium, and other alkali targets at various collision energies and calculate the observables with this program. If possible, compare with experimental data.
2. Use the program of Chap. 10 to obtain scattering amplitudes from the output of the close-coupling and R-matrix codes described in Chaps. 6 and 7 and calculate the observables. If possible, compare with experimental data.
3. Write a new subroutine to obtain the generalized *STU* parameters by first calculating the amplitudes (11.3) and then using (11.39). Verify your results by comparing with those obtained directly from (11.65).
4. Extend the program to calculate angle-integrated Stokes parameters. Be careful to only integrate bilinear products of scattering amplitudes which do not vanish according to selection rules for integration over spherical harmonics. (For details of the theory, see [11.15].)
5. Extend the program to allow for more general atomic transitions, including the possibility of reading scattering amplitudes that were calculated in a relativistic coupling scheme. (For example, see [11.23,24].) For such a case, a more general routine for the *STU* parameters, such as the one developed for Problem 3 above, will be needed.
6. Write a program to calculate perturbation coefficients for fine-structure and hyperfine-structure depolarization effects, and also for the case where

both of those interactions affect the degree of polarization. Formulas for these cases can be found in [11.2].

Acknowledgments

The author would like to thank Nils Andersen for his contributions to the general description of the $^2S \rightarrow {}^2P^o$ transitions, Igor Bray for providing the input scattering amplitudes, and John Broad and Jean Gallagher for providing excellent working conditions at the JILA Data Center where some of the theory and an early version of the computer program were developed. This work has been supported, in part, by the National Science Foundation under grants PHY-9014103 and PHY-9318377.

References

11.1 J. Kessler, *Polarized Electrons*, 2nd edition (Springer, Berlin, 1985)
11.2 K. Blum, *Density Matrix Theory and Applications* (Plenum, New York, 1981)
11.3 N. Andersen, J.W. Gallagher and I.V. Hertel, Phys. Rep. **180** (1988) 1
11.4 N. Andersen, K. Bartschat, J.T. Broad and I.V. Hertel, Phys. Rep. (est. 1996), in preparation
11.5 K. Bartschat, Phys. Rep. **180** (1989) 1
11.6 K. Bartschat and D.H. Madison, J. Phys. B **21** (1988) 2621
11.7 A.R. Edmonds, *Angular Momentum in Quantum Mechanics* (Princeton University Press, Princeton, 1957)
11.8 N. Andersen and K. Bartschat, J. Phys. B **27** (1994) 3189
11.9 D.T. Pierce, R.J. Celotta, G.-C. Wang, W.N. Unertl, A. Galejs, and C.E. Kuyatt, Rev. Sci. Instrum. **51** (1980) 478
11.10 O. Berger and J. Kessler, J. Phys. B **19** (1986) 3539
11.11 B. Bederson, Comments At. Mol. Phys. **1** (1969) 41
11.12 B. Bederson, Comments At. Mol. Phys. **1** (1969) 65
11.13 B. Bederson, Comments At. Mol. Phys. **2** (1970) 160
11.14 P.G. Burke and J.F.B. Mitchell, J. Phys. B **7** (1974) 214
11.15 K. Bartschat, K. Blum, G.F. Hanne, and J. Kessler. J. Phys. B **14** (1981) 3761
11.16 M. Born and E. Wolf, *Principles of Optics*, 4th edition (Pergamon, New York, 1970)
11.17 M. Sohn and G.F. Hanne. J. Phys. B **25** (1992) 4627
11.18 N. Andersen and K. Bartschat, Comments At. Mol. Phys. **29** (1993) 157
11.19 I.V. Hertel, M.H. Kelley, and J.J. McClelland. Z. Phys. D **6** (1987) 163
11.20 G.F. Hanne. Phys. Rep. **95** (1983) 95
11.21 K. Bartschat and N. Andersen, Comp. Phys. Commun. **84** (1994) 335
11.22 I. Bray, Phys. Rev. A. **49** (1994) 1066
11.23 K. Bartschat and N.S. Scott, Comp. Phys. Commun. **30** 369 (1983)
11.24 K. Bartschat, Comp. Phys. Commun. **30** 383 (1983)

Subject Index

Springer-Verlag and the Environment

We at Springer-Verlag firmly believe that an international science publisher has a special obligation to the environment, and our corporate policies consistently reflect this conviction.

We also expect our business partners – paper mills, printers, packaging manufacturers, etc. – to commit themselves to using environmentally friendly materials and production processes.

The paper in this book is made from low- or no-chlorine pulp and is acid free, in conformance with international standards for paper permanency.